Engineering the Eternal City

Frontispiece Anonymous (formerly attributed to Stefano [Étienne] Duperac), panoramic view of Rome from the roof of the Cancelleria. Ca. 1567–1568. Vatican City, BAV, Disegni Ashby 131.

Engineering the Eternal City

Infrastructure, Topography, and the Culture of Knowledge in Late Sixteenth-Century Rome

Pamela O. Long

The University of Chicago Press
Chicago and London

The University of Chicago Press, Chicago 60637
The University of Chicago Press, Ltd., London

Published 2018
Printed in the United States of America

27 26 25 24 23 22 21 20 19 18   1 2 3 4 5

ISBN-13: 978-0-226-54379-6 (cloth)
ISBN-13: 978-0-226-59128-5 (paper)
ISBN-13: 978-0-226-59131-5 (e-book)
DOI: https://doi.org/10.7208/chicago/9780226591315.001.0001

Published with support of the Susan E. Abrams Fund.

Cover illustration: Detail from Giovanni Guerra and Natale Bonifacio, *Disegno del modo di condurre l'Obelisco detto volgarmente la Guglia*. Rome, 1586 (fig. 8.8). Collection of Vincent J. Buonanno.

Library of Congress Cataloging-in-Publication Data
Names: Long, Pamela O., author.
Title: Engineering the Eternal City : infrastructure, topography, and the culture of knowledge in late sixteenth-century Rome / Pamela O. Long.
Description: Chicago : The University of Chicago Press, 2018. | Includes bibliographical references and index.
Identifiers: LCCN 2018019731 | ISBN 9780226543796 (cloth : alk. paper) | ISBN 9780226591285 (pbk. : alk. paper) | ISBN 9780226591315 (e-book)
Subjects: LCSH: Urban renewal—Italy—Rome—History—16th century. | Municipal engineering—Italy—Rome—History—16th century. | Civic improvement—Italy—Rome—History—16th century. | Rome (Italy)—History—16th century.
Classification: LCC DG812.4 .L59 2018 HT178.I82 R6 | DDC 307.3/4160945632031—dc23
LC record available at https://lccn.loc.gov/2018019731

♾ This paper meets the requirements of ANSI/NISO Z39.48-1992 (Permanence of Paper).

For

Bob Korn

Allison Rachel Korn

Marco Yunga Tacuri

Lucas Samay Yunga Korn

and

Tiago Asha Yunga Korn

Contents

Money, Weights, and Measures

MONEY

baiocco = a base metal coin of varying value; in accounts worth one-tenth of a giulio
giulio = equivalent to ten baiocchi; valued at 2.9 grams of silver
quattrino = a coin worth a small amount (about one-third of a baiocco)
scudo = a coin worth ten giulii, or one hundred baiocchi
scudo d'oro = a money of account valued in gold, not a coin like the scudo

WEIGHTS AND MEASURES FOR ROME IN THE LATE SIXTEENTH CENTURY

Length

1 *canna* = 10 Roman *palmi* or 7.33 feet or 2.234 meters
1 *palmo* = 8.796 inches or 0.2234 meters
1 *passo* = 4.887 feet or 1.4895 meters
1 *piede* = 1½ Roman *palmo* = 11.729 inches or 0.2979 meters

Weights

1 *libbra* = 12 *unciae* = 0.748 pounds or 0.3391 kilograms

Author's Note

Unless otherwise noted, all translations are mine (sometimes made in consultation with others, most often Chiara Bariviera). For primary source material, the original is given in the footnote. Where I have occasionally cited modern French, Italian, or German scholarship in the text, I have translated them into English but not given the original in the footnote because the originals are readily available. In many archival documents the folios have been numbered more than once, giving the same folio two or more numbers. Different scholars citing the same passage will therefore sometimes cite different folio numbers, but all give the date, which is identifying. I have used the most recent archival foliations where available. I have given the original language of translated citations in the footnotes. I have expanded abbreviations but have not otherwise changed or modernized the original. (For example, I have not added accent marks where there are none in the original document.) Proper names of individuals are sometimes given in Latin in the documents and sometimes in Italian with variations. I have given all Italian names in the standard (modern) Italian form.

Introduction

Rome

Portrait of the Late Sixteenth-Century City

Rome circa 1560 was a city of resplendent processions, ancient ruins, and crumbling infrastructure. It was also a city in the process of being reimagined visually by means of images, including cartographical images, and physically by means of construction and large-scale engineering projects. Between 1560 and 1590, the city bustled with activity—the building and renovation of churches, palaces, and walls; the repair and reconstruction of two great aqueducts; the creation of new fountains made possible by the greatly augmented water supply; the widening and paving of streets; the redesign of streets and piazze; the transport of obelisks from their ancient Roman resting places to new locations; and projects aimed at preventing the periodic catastrophic flooding of the Tiber River.[1]

Rome was a city headed by popes but also governed by the traditional city government known as the Capitoline government. As a city undergoing multiple transformations, it served as a magnet for learned humanists and upwardly striving clerics and for painters, sculptors, "architect/engineers," and other workers seeking employment in projects of urban engineering and building construction. It was also home to numerous elites—cardinals, ambassadors, and Roman noble families. Despite lip service to papal absolutism, the city was characterized by multiple centers of power and patronage.

The pope stood at the head of the Papal States (in central Italy), the city of Rome, and the Catholic Church. Papal governance was influenced by the Council of Trent (1542–1563), a church council assembled to reform a Catholic Church under attack by the Protestants. Pius IV ended the council in 1563 and promulgated its decrees in 1564. Included among them was the encouragement of the material enhancement of divine worship, including the use of paintings and other visual images in churches.[2]

In the post-Tridentine era, as the years after the Council of Trent were called, such reforms were accompanied in Rome by the vision that a reformed church should be represented by a renewed capital city. This city would be as magnificent as imperial Rome had been. The ideals of *renovatio imperii* and *renovatio Romae* lay behind the reconstruction and renewal of Rome. The popes should rule—as the Roman emperors had—a splendid and now also Christian city.[3]

Giovanni Botero (1540–1617), author of a treatise on cities published in 1588, insisted that princes were crucial to flourishing cities just as the popes were to Rome:

> Would not Rome, the capital of the world, be more like a desert than a city if the Supreme Pontiff did not reside there, and magnify the city with his splendid court and the ambassadors, prelates and princes who flock there? And if he did not populate it with the infinite number of people from every nation? [. . .] And if he did not beautify it with magnificent buildings, aqueducts, fountains and streets?[4]

Botero was correct, but the pope was far from the sole agent in the many infrastructure projects of late sixteenth-century Rome.

In this book I investigate major projects of engineering and urban redesign. I treat both the physical city and visual representations of the city. The chronological range, with exceptions mostly having to do with cartography, focuses on the period between the great flood of 1557 and the death of the great urbanizing pope Sixtus V in 1590.[5]

My focus on this well-defined period facilitates an approach that emphasizes the *processes* of urban construction and engineering, including decision making and financing. I include failed and incomplete projects as well as successfully completed ones. This approach eschews the frequently encountered teleological methodology in the history of technology, which jumps from one ingenious invention to another or focuses only on successfully completed projects. My study also includes the *maintenance* of infrastructure. It links engineering and construction to broader political and cultural aspects of the city of Rome, including a flourishing print culture, the proliferation of maps and other urban images, and the culture of knowledge.[6]

In some ways Roman urban development in the late sixteenth century reflected that of other cities in Italy and elsewhere on the continent. This included rising populations and multiple projects of construction and renovation influenced by the rediscovery of the ancient world and its urban vocabulary (including classical forms of architecture). Rome's cityscape encompassed a growing concept of elite family honor associated with great palaces often sited on dramatic urban spaces.[7] These developments tied power, authority, and family prestige to conspicuous consumption, grand physical structures, open piazze framed by palatial facades, and straight, wide streets that were paved.

While Rome was being transformed physically, cartographical and other kinds of visual images of the city flooded the urban marketplace. Mapmaking and printing came to be closely tied to the study of antiquities, surveying, engineering, construction, and urban redesign. Often accompanying these activities were intense discussions about how to proceed, ardent competition for contracts, and acrimonious arguments concerning such topics as the location of the ancient Roman Forum. Humanists, antiquarians, painters, designers, engineers, architects, engravers, woodblock cutters, and printers created maps, views, and plans of the ancient city and its contemporary counterpart as well as images of streets and squares, statues, columns, and buildings. These images were presented to potential patrons or sold to residents of the city or to eager visitors including pilgrims and antiquarians.[8]

Engineering and urban reconstruction, the study of Roman antiquities and Roman topography, and the creation of city plans and maps functioned as deeply interrelated activities and practices. The ways in which engineering projects were undertaken encouraged interaction on many levels between practically trained men on the one hand and learned, university-educated men on the other. It was a time of intense engineering and construction activity that occurred well before the modern professionalization of engineering, architecture, and archaeology. These activities were not clearly separated one from another. Although the word *engineer* (*ingegnere*) was part of sixteenth-century Italian vocabulary, individuals engaged in what we would call engineering were usually called architects (*architetti*). Such individuals came from diverse backgrounds and typically engaged in a far wider range of activities than do engineers and architects today.

Historians of architecture in particular have posited fifteenth- and sixteenth-century Italy, including Rome, as the locus for the origin of the modern profession of architecture. They point to the separation of building design from building construction; the acquisition of a particular skill set, including the skills of drawing, model making, surveying, and other mathematical proficiencies; and rising social status. The same period incubated the concept of the engineer, associated particularly with military matters, especially triangular bastion fortification.[9]

Accompanying these developments, writings proliferated about architecture and military topics, especially fortification. These included Vitruvian commentaries and other tracts that depicted the architect as learned in a number of disciplines, knowledgeable about mathematics, and separate from artisanal workers while also showing that architecture and engineering were topics suitable for learned written discourse.[10]

This view of the origins of architectural and engineering professionalism is valid as long as it is understood that modern professionalism in any developed sense did not then exist in these disciplines. Although skill was certainly valued, no apparatus of required learning or licensing existed for either architecture or engineering (the latter often referred to as architecture). The backgrounds of men given charge of building and engineering projects were highly variable. The ability to win contracts depended as much on patronage as on any specific training. Men called *architetti* were also given other descriptive titles and engaged in activities far more diverse than the usual activities of modern architects. Historians of architecture have tended to anachronistically project their modern profession onto early modern counterparts, thereby ignoring the various tasks and occupations undertaken by these sixteenth-century men that were outside of the purview of modern architecture centered on design.[11] The fluidity of disciplinary identification and lack of educational and licensing requirements engendered broad participation in energetic debates about whether and how sixteenth-century Roman infrastructure projects were to be carried out.

Competition among individuals and groups for major urban contracts was the norm. Romans often debated, decided on, and carried out engineering and construction projects in contentious and competitive ways. Similarly, enthusiasts from a variety of backgrounds studied Roman topography, ruins, and antiquities with intensity, often cooperatively, but at times within the context of bitter argument. Such modes of investigation and construction contributed to a culture of knowledge in which engineering practice gained legitimacy as a topic of interest in both practical and learned cultures.

The mid- to late sixteenth century was characterized by a growing number of what I have called "trading zones"—arenas in which substantive communication occurred between university-educated people and those trained in artisanal workshops or in other practical/technical venues. An openness developed in discussions concerning problems of infrastructure and engineering. People from different backgrounds offered opinions, suggested alternatives, conversed and argued with one another, and produced advisory writings, drawings, and maps. What passed for "expertise" could vary from one situation to another and was far more diverse than has been the case since the development of modern professionalism and its requisite educational and licensing requirements. Further, certain individuals trained as practitioners experienced rising status and a multiplicity of roles that rendered less sharp — occasionally even eliminated — the distinction between them and learned individuals. At the same time, learned humanists and elite patrons increasingly appreciated and sometimes acquired practical skills.[12]

Such interactions between practical/artisanal cultures and learned cultures were important for the development of the new empirically based sciences of the late sixteenth and seventeenth centuries. My suggestion here counters the traditional view of Rome as a backwater for the sciences, a reputation that resulted from the trial of Galileo in 1633 — about forty years after the period of my study ends.[13]

Rome: City of Popes

To understand late sixteenth-century Rome, it is necessary to grasp, at least in broad outline, the structures and workings of urban power and the complexities of patronage and governance. The most powerful person in Rome was the pope. The pope's governance of the city was carried out primarily by the part of the papal bureaucracy known as the Camera Apostolica. Yet the power of the popes, no matter how much they may have wished it, was not absolute. Moreover, the popes were individuals with very different backgrounds and abilities whose interests in the city of Rome as a physical site varied widely.

More than any other European city, Rome possessed multiple centers of power and patronage. It was a unique characteristic of the city that the central figure in the patronage network, the pope, usually belonged to a family entirely different from the families of both his immediate predecessor and his successor. At each new papal succession, radical changes in patronage could occur — including new courtiers, new officers, new favorites, new cardinals, and new patronage networks — as an entirely new family reaped the huge benefits that accrued to the papal crown. It was not uncommon for a new pope to be the actual enemy of his immediate predecessor. At the same time, a complex group of rituals accompanied the death of one pope and the crowning of his successor, providing a stable structure of transition from one papacy to the next.[14]

Late sixteenth-century popes also shared a common history in which their predecessors in varying degrees had reformed the physical city. Post-Tridentine activities of urban reconstruction emerged from the prior accomplishments of fifteenth- and early sixteenth-century Renaissance Rome beginning in 1420, when the papacy moved back to Rome from Avignon under Otto Colonna, Pope Martin V (ruled 1417–1431).[15]

Efforts to reform and renovate the city came to an abrupt halt in 1527 because of the brutal sack of Rome and the assault on its people by an unpaid army of the Hapsburg emperor Charles V, an army led by Charles III, Duke of Bourbon (1490–1527), who was killed in the initial attack. The leaderless soldiers looted, burned, raped, castrated, and murdered in a rampage that lasted for ten long months. The catastrophic Sack was followed in 1530 by a disastrous Tiber River flood.[16] Slowly the city recovered, helped initially by the dedicated efforts of a pope from a Roman family, Alessandro Farnese, Paul III (ruled 1534–1549). However, Pope Paul was distracted from urban concerns by the great resource-draining reception into the city of the Holy Roman emperor, Charles V, in 1536 (planned as a kind of reconciliation after the Sack) as well as by the time-consuming efforts to reform the Catholic Church, resulting in the opening of the Council of Trent in 1542.[17]

Some of Paul III's successors were far less interested in the physical city than he had been. Julius III (ruled 1550–1555) concentrated on building his great palace, the Villa Giulia, on the Pincian Hill, and he redirected scarce city water from the barely functioning Acqua Vergine to its nymphaeum—an elaborate, multistoried fountain and pool structure designed following ancient examples.[18] Nevertheless, Julius III also saw to the repair of one of the city's important bridges, the Ponte Santa Maria.

Julius III's successor, the stern and unyielding Paul IV Carafa (ruled 1555–1559) focused on enforcing strict notions of piety, enlarging his family fortunes, and fighting the Spanish. A fanatical inquisitor, he first targeted the Jews, issuing the bull *Cum nimis absurdum* in 1555, requiring that the Jews be confined to a ghetto on the left bank of the Tiber River. (A papal bull is an official papal decree that is issued by the pope and sealed with a bulla or lead seal.) During Paul IV's harsh rule, the city was dealt a terrible blow by another flood—the great Tiber River flood of 1557.[19]

Four Individuals, Four Popes

This book focuses on a period (1557–1590) dominated by four popes. Each was strikingly different from (if not hostile to) his predecessor, and each shaped the city in his own particular way. Each played an important role in the renewal of urban infrastructure, the redesign of streets, hydraulic engineering projects, and urban construction and reconstruction. Knowledge of their backgrounds, personalities, and ideals is necessary to a contextual urban history of Rome.

Pius IV Medici (Ruled 1559–1565)

At Paul IV Carafa's death two years after the destructive flood of 1557, a compromise candidate was finally elected after the longest conclave ever held to that date (fig. 0.1). The urbane Giovanni Angelo Medici (1499–1565) became Pope Pius IV. Pius IV Medici immediately reversed some of the harsh policies of his predecessor, and further, prosecuted two Carafa nephews for acts of corruption, theft, abuse of power, and murder. Ultimately, he ordered the execution of one Cardinal Carlo Carafa (1519–1561) by strangulation in the Castel Sant'Angelo; the execution of his brother, Duke Giovanni of Paliano (d. 1561), by beheading in the nearby prison, the Tor di Nona; and the confiscation of their property. Another family member, Cardinal Alfonso Carafa (1540–1565) was absolved after paying a huge fine.[20]

Figure 0.1 Pius IV Papa Mediolanensis. Copperplate engraving in Onofrio Panvinio, *XXVII Pontificum Maximorum elogia et imagines: Accuratissime ad vivum aeneis typeis delineatae* (Rome: Antonio Lafreri, 1568). Folger Shakespeare Library Call no. BX950.A2 1568 Cage. Used by permission of the Folger Shakespeare Library.

Pius IV was the son of a notary, born into a midlevel family from Marignano (modern Melegnano) sixteen kilometers southeast of Milan. His family was unrelated to the princely Florentine Medici, although they happily adopted him as he rose in the church hierarchy. He studied jurisprudence at Bologna, acquired doctorates in civil and canon law, and came to Rome in 1527. His career was aided by his protector, Cardinal Alessandro Farnese, who had become Pope Paul III in 1534. The Farnese pope appointed him governor of several cities within the Papal States—including Città di Castello in 1535, Fano in 1539, and Parma in 1540. He was a highly competent manager who advanced only slowly up the curial ladder—having fathered a son and two daughters, his personal life left something to be desired from the point of view of the church. Still, Paul III, shortly before his death in 1549, made him a cardinal.[21]

Pius IV's extensive experience as an administrator of cities undoubtedly contributed to his interest in renovating the city of Rome itself. Upon election, he forced the long-running Council of Trent to conclusion and turned his attention to urban matters.[22] He surrounded himself with humanists and men of letters. Throughout his career Pius IV showed himself to be a thoroughly practical realist, often taking a middle position and successfully mediating conflicts, a style that was antithetical to that of his immediate predecessor, the hardheaded Paul IV Carafa. Pius IV was an affable man whose life remained a focus of gossip among Roman wags. He believed in exercise, and early in his papacy he was often to be found unattended either walking or on horseback on the streets of Rome. The ambassador to Rome from Venice, Girolamo Soranzo, reported that "the pope is by nature inclined to a private and free life [...] and in all his actions he shows rather sweetness than gravity, leaving himself to be seen by all, at all hours, and going by horse and by foot through the city with very little company."[23]

Pius IV was a great urbanizing pope, and he initiated many construction and engineering projects—each of which came with its own price tag. As the humanist Onofrio Panvinio (1529–1568) noted, "he built many public buildings in the whole city and through ecclesiastic authority."[24]

Pius's urban projects created a great need for money. Funding sources included hefty taxes on food and wine in the city itself, a burden deeply resented by the Roman people, among whom the pope became increasingly unpopular. In July 1562 pamphlets and broadsheets appeared around the city denouncing him as a tyrant who deserved death. In August a bullet was fired from the street into a palace where he had been shortly before. He temporarily refrained from walking through the city and increased his bodyguard. An assassination plot in 1564 was thwarted, and the conspirators were publicly executed. This frightening episode brought about an increase in papal troops and a halt to Pius IV's urban mobility.[25]

Pius V (Ruled 1565–1572)

Pius IV's successor, Michele Ghislieri (1504–1572), radically differed from Giovanni Angelo Medici both in his personality and his interests (fig. 0.2). His election on January 7, 1566, was facilitated by the Medici pope's pious nephew, Carlo Borromeo (1538–1584), after Ghislieri had acceded to Carlo's requirement that he take the same name—Pius—as his predecessor. Pius V was born into a poor family in the Piedmont region of Italy. Initially named Antonio Ghislieri, he was a shepherd until he entered the Dominican order at the age of fourteen and took the name Michele. He studied at the University of Bologna, lectured at Pavia for sixteen years, and then began a zealous career as an inquisitor. Paul IV Carafa had made him a cardinal and later, in 1551, the inquisitor general. He favored Paul IV's inquisitorial policies. At his election as pope, he immediately reversed (to the extent possible) Pius IV's strictures against the Carafa family.[26]

Figure 0.2 Pius V Papa Alexandrinus. Copperplate engraving in Onofrio Panvinio, *XXVII Pontificum Maximorum elogia et imagines: Accuratissime ad vivum aeneis typeis delineatae* (Rome: Antonio Lafreri, 1568). Folger Shakespeare Library call no. BX950.A2 1568 Cage. Used by permission of the Folger Shakespeare Library.

Although he oversaw important urban projects, Pius V took it as his main task to reform the church and put the decrees of the Council of Trent into effect. He persecuted heretics harshly. He established the Congregation of the Index to review and censor books. He, like his predecessors, battled the Turks, and he received the news of the victory of the combined naval force over the Turkish fleet on October 7, 1571 (the Battle of Lepanto), a few months before he died. Although he was ascetic, zealous, and uninterested in antiquities, he also presided over the repair of the ancient aqueduct, the Acqua Vergine, which would bring an abundant and continuous flow of much-needed fresh spring water into the city. He was beatified in 1672 and made a saint in 1712.[27]

Gregory XIII (Ruled 1572–1585)

Pius V was succeeded by Ugo Boncompagni (1502–1585) who became Gregory XIII (fig. 0.3). Ugo was the fourth son of a Bolognese merchant in a family that only recently had joined the aristocracy through marriage. He graduated from the University of Bologna in law and became a professor of law for eight years. He arrived in Rome in 1539 and eventually was ordained a priest and then, in 1558, made a bishop. He had previously (and deliberately) fathered a son, Giacomo Boncompagni (1548–1612), to insure continuation of the family line. In Rome he was highly regarded as a lawyer, and Paul IV sent him on diplomatic missions. Later, under Pius IV he attended the final years of the Council of Trent as an expert in canon law. Pius IV then made him cardinal and sent him as a legate to Spain. He became a confidant of the Spanish monarch, Philip II (ruled 1556–1598), and, in a brief conclave, he was elected pope largely because of Spanish influence. Although not as harsh as his immediate predecessor (Pius V), Gregory was dedicated to putting the decrees of the Council of Trent into practice.[28]

Concerned with the education of the clergy, Gregory created and supported many colleges in Rome — the English, Greek, Maronite, Armenian, and Hungarian colleges — and he also secured the future of the German College by richly endowing it. The new colleges ensured a steady stream of educated priests to various locales in Europe and also initiated numerous Roman building programs to house them. Gregory also supported the Jesuits and their Roman building programs that created their church, the Gesù, and the building for their college, the Collegio Romano, both massive buildings that transformed the center of the city.[29]

Much as Gregory encouraged enormous building projects in Rome, he also cut off the benefices of one of the cardinals, Cardinal Montalto (Felice Peretti, 1521–1590), in order to curb what he considered the cardinal's overly ostentatious building program involving Montalto, the enormous palace and garden complex that Peretti was building on the Esquiline Hill. The palace was adjacent to the basilica

Figure 0.3 Gregory XIII surrounded by his works. Alphonsus Ciaconius, *Vitae et res gestae pontificorum romanum tomus quartus* (Rome: Philippus et Antonius De Rubeis, 1677), cols. 39–40. Vatican City, BAV, Barberini U. IV.6. © 2018 Biblioteca Apostolica Vaticana, by permission of the Biblioteca Apostolica Vaticana with all rights reserved.

of Santa Maria Maggiore. Peretti responded to Gregory XIII's censure by withdrawing from the papal court and associating only with a group of (well-placed) friends who duly elected him pope in 1585. Thereafter, Felice Peretti, now pope Sixtus V did everything possible to besmirch his predecessor's reputation.[30]

Sixtus V (Ruled 1585–1590)

Sixtus V was a farmworker's son born in the town of Grottammare in the Marches (fig. 0.4). Schooled by the Franciscans, he joined the order at the age of twelve in the nearby town of Montalto. After a brilliant student career he was ordained in 1547, and the next year he earned a doctor of theology at the University of Fermo (in the town of Fermo, also in the Marches). He came to Rome in 1552 and gave Lenten sermons that earned him fame. Paul IV made him inquisitor of Venice, from which he was recalled because of his severity only to be reappointed by Pius IV in 1560. The next pope, Pius V, appointed him vicar general of the Franciscans and in 1570 named him a cardinal. As pope from 1585, he ruthlessly suppressed banditry, executing numerous brigands—displaying their heads on the Ponte Sant'Angelo—and punishing the nobles who protected them. He was an ardent urban builder—he remodeled the city streets, completed the construction of the new aqueduct begun by Gregory XIII (the Acqua Felice), rebuilt the Lateran Palace, completed the dome of St. Peter's, constructed a new Vatican Library, drained the Pontine marshes to the south of the city, and had four immense obelisks moved to new sites to mark important piazze, basilicas, and streets.[31]

◆ ◆

It is notable that the four popes from Pius IV through Sixtus V shared relatively middle or low-level backgrounds—only the family of Gregory XIII had recently joined the aristocracy and at that through marriage.

The popes served as heads of the Christian church and of the city of Rome. Yet how they actually governed the city is the focus of much scholarly debate. Recent Roman scholarship has responded variously to the still influential concept of the "papal prince" formulated by Alberto Caracciolo, Mario Caravale, Paolo Prodi, and Wolfgang Reinhard. Much of their scholarship was carried out in the late 1970s and early 1980s. These scholars differed in their respective approaches but in general stressed the transformation of papal rule to a new territorial monarchy in the fifteenth and sixteenth centuries. Prodi in particular saw the papacy from the mid-fifteenth century not as a further development of medieval universalism but as developing toward a territorial state—a prototype of the modern bureaucratic state in which ecclesiastical matters were subordinated to the interests of the secular government.[32]

Figure 0.4 Ambrogio Brambilla, portrait of Sixtus V. Published by Nicolaus van Aelst, 1589. Etching on paper, 20½ × 14 in. (517 × 358 mm). Rijksmuseum, Amsterdam.

More recent research has modified the Prodi thesis, emphasizing on the one hand the ongoing interest and effectiveness of the papacy in traditional ecclesiastical, pastoral, and sacred matters and on the other the multifaceted complexity of the governance of Rome and the limits and incompleteness of papal power and authority in the city itself and in the papal territories. Cases in point include the continuing unruliness of the barons and ongoing banditry. They also include the continuation and importance of the "communal" government—the term often used for the government of Rome, also called the Capitoline government. They include the extensive power and authority of Capitoline officials and the importance of the Capitoline judicial system, which created overlapping jurisdictions with the papal bureaucracy and the papal judicial system.[33] This book supports this revised picture. The popes were powerful, but their power was far from absolute. In certain instances the popes were remarkably unsuccessful in achieving their urban aims.

Structures of Government

Adding to the complexity of Roman patronage and governance was the circumstance that Rome was governed by the two separate and overlapping bureaucracies. The first was the pope and the papal bureaucracy. The second was the traditional government of the city of Rome, called the Capitoline government, that had ruled the city since the twelfth century.

The part of the papal bureaucracy (the papal curia) that governed Rome, the Camera Apostolica, was administered by a cardinal called the cardinal camerlengo. The pope also appointed the governor of Rome (*governatore di Roma*) who was in charge of the city's most important judicial tribunal. In addition, he called the cardinals together in consistories—regular meetings of all the cardinals—to deal with matters of papal governance. The cardinals themselves were involved in a complex web of factions and alliances influenced by Roman politics and international interests, especially those emanating from Florence, France, and Spain.[34]

From the fifteenth century, the popes began to create special committees called congregations—committees headed by cardinals—to deal with specific issues. From the mid-sixteenth century the increased number of these congregations signaled the decline of the College of Cardinals as a governing body. Some congregations were long standing, and some were of brief duration. In 1568, Pius V created an especially long-standing example, the Congregation on Streets, Bridges, and Fountains (*Congregatio super viis pontibus et fontibus*), which dealt with a wide range of urban concerns. Sixtus V terminated this congregation in 1588, when on January 22, 1588, he issued the bull *Immensa aeterni Dei*, creating fifteen permanent congregations, which remain as structures of papal governance to this day.[35]

The Capitoline government was established in 1143. It was headed by three conservators (elected every three months) and a prior. These four officials governed with an elected Roman council, which met both in a smaller, "private" or secret assembly (*consulium secretum* or *ordinatum*) and in a larger, public assembly (*consulium publicum*). The Capitoline government oversaw its own tribunal. In addition to the conservators and the prior, government officials included the *marescialli*, or marshals, one of which was assigned to the conservators, one to

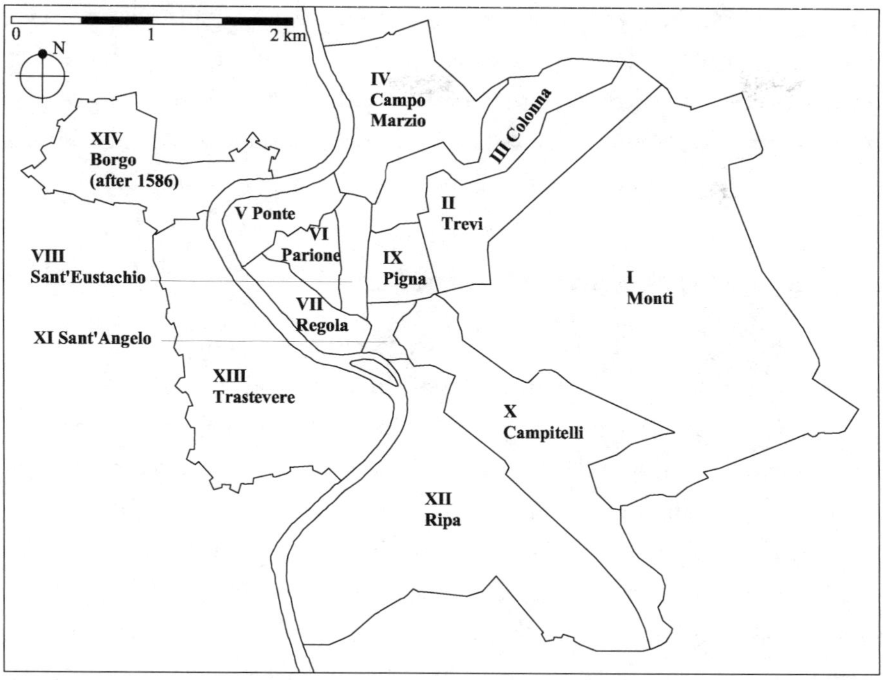

Figure 0.5 The fourteen *rioni* of Rome after the creation of the Borgo as Rioni XIV in 1586. Courtesy of Chiara Bariviera with kind permission.

the senator (by the late sixteenth century a largely symbolic officer), and one to the masters of the streets. The *marescialli* were in charge of the security of the government and also helped the *caporioni* (heads of the thirteen districts, or *rioni,* into which Rome was divided).[36]

Numerous courts and tribunals under the purview of both papal and Capitoline governments, often with overlapping jurisdictions, characterized the legal processes of the city. In the late sixteenth century, the papacy exerted predominant power over the Capitoline government, but the authority and range of action exerted by the commune was never negligible. Duplicated bureaucratic and legal functions were the norm. At times the papacy and the Capitoline Council cooperated—often facilitated by the heavy hand of the pope—and at other times they were at loggerheads.[37]

Of the city's thirteen *rioni,* twelve were on the left bank of the Tiber (facing downstream), and the thirteenth (Trastevere) was on the right (fig. 0.5). The Borgo—the area surrounding the Vatican—was considered to be outside of Rome itself until Sixtus V made it the fourteenth *rione* in 1586. The *caporione* was in charge of public order within his *rione* and commanded twenty or thirty men under him, called the *conestabili,* who functioned as civic police. The prior was the *caporione,* usually (although not always) from the largest *rione*—Monti on the Esquiline Hill. He helped govern the city along with the three conservators. Officers of the city included two masters of the streets—Roman patricians who were elected for a year and often reelected. It should be noted that in sixteenth-century Roman documents, the frequent references to "the Roman People" (*Popolo Romano* or *Populus Romanus*) refer to the elite citizens who participated in the Capitoline government, not to the entire population.[38]

Beyond the Structures of Government

Rome was a growing city that encompassed far more than the long arm of the papal bureaucracy and the shorter arm of the Capitoline Council. As the city gradually recovered from the 1527 Sack, the population reached about one hundred thousand by the year 1600. But the population fluctuated wildly as a result of events such as jubilees. For example, in the jubilee of 1575, an estimated four hundred thousand pilgrims flooded into the city.[39] The city was inhabited by a large number of cardinals, noble Roman families, ambassadors, elite residents, and visitors from all over Europe. This diverse group ensured multiple, complex patronage networks beyond the papacy itself. Roman elites included the "civic nobility," or *municipali*, who controlled the Roman communal government and held most communal offices, and the baronial or feudal nobility led by the Colonna and Orsini families.[40]

The city also accommodated numerous religious orders—some traditional, such as the Dominicans and the Franciscans, and others new, such as the Theatines and the Jesuits, as well as the lay brotherhood of the Oratorians (created by Filippo Neri, 1515–1595). Religious groups included numerous confraternities—associations often aligned with guilds or other organizations and usually dedicated to religious and charitable functions. In the late sixteenth century, many of these religious orders embarked on ambitious building programs.[41]

Further, cardinals, each of whom was chosen by a particular pope and thereby became the "creature" of that pope, often responded to their elections by initiating building programs to create palatial accommodations suitable for their new position. The cardinals were also assigned titular churches from which they accrued benefits and which often included an attached residence. Cardinals habitually repaired, renovated, and expanded those residences as their palaces, and they also initiated building improvements and ornamentation of the church itself. The Roman building boom of the late sixteenth century was fueled as much by cardinals as by the popes.[42]

The elite classes also included powerful ambassadors from other cities and regions as well as papal and curial elites, including cardinalate families such as the Farnese, formerly based in Orvieto. Many of them undertook palatial construction projects as a way of extending their family's status and power. They often had extensive connections to families outside of Rome throughout Italy and beyond, connections cemented by bonds of marriage. Within the city itself, civic elites undertook extensive projects of purchasing, selling, and building within the *rione* as a way of extending their power and authority. Women from elite families often became important patrons of the arts and architecture. Elite men and women participated in the papal court, but they also functioned as patrons on their own, thereby contributing to the complex, multicentered Roman context of power and patronage.[43]

Rome's expanding elites, with their urban and palatial aspirations, and the resulting Roman building boom ensured that farther down the social scale, thousands of skilled artisans and practitioners of all kinds, as well as unskilled workers, traveled to Rome to take advantage of opportunities for contracts and other kinds of employment. Painters, sculptors, stonecutters, lime workers, carpenters, architects, engineers, engravers, and printers arrived in droves, often developing patronage relationships and alliances, formal and informal, often with families from their own regions.[44] These skilled practitioners were both geographically and occupationally mobile. Sculptors and painters became architects, carpenters became woodcutters, printers became mapmakers or designers of gardens.

Learned humanists also journeyed to Rome to study ancient ruins and antiquities and to seek positions in the papal curia or as members of the "families" of cardinals or other elite patrons.[45] And, over the decades, an increasing number of pilgrims traveled to the city, as did increasing numbers of impoverished beggars and vagabonds.[46] Further, as a city populated by a disproportionate number of males, it supported numerous well-off courtesans as well as less well-off and even destitute prostitutes. In a city of about one hundred thousand in 1600, there were no more than seventy women to every hundred men.[47] Finally, Rome's substantial Jewish population was both essential to the life of the city and regularly harassed, especially after Paul IV's creation of the ghetto in 1555.[48]

A fluidity of artisanal identity was especially marked in the city of Rome. It was matched by the mobility of the Roman elite classes, whose power was tied to complex relationships with the artisanal classes living in the *rioni*.[49] Skilled artisans as well as unskilled laborers found employment at the construction sites of palaces, churches, and cathedrals and at the great infrastructure projects such as aqueduct repair projects and in hundreds of projects involving the ornamentation of palaces and churches with painting, stucco, gilding, and statuary. Many came as young men, not fully trained, and learned on the job. Rome did possess traditional guilds for many of the skilled trades, which were themselves changing, as the emergence of the Academy of St. Luke for painters, sculptors, and architects at the end of the century exemplifies. In this new academy, founded in 1593, some traditional guild functions were retained, but lectures and other more academic pursuits were added. Although some of the guilds required that newcomers be licensed, many newly arrived artisans neglected to apply for a license, being able to find work beyond guild control within the extensive and complex patronage networks of the city. Further, many artisans developed varied skills and changed occupations as they took advantage of opportunities for work and patronage.[50]

◆ ◆

Two central events constitute the essential prehistory of the topics discussed in depth in this book. The first is the Sack of Rome in 1527, and the second is the great Tiber River flood of 1530. These events ended the prosperous decades of Renaissance Rome and in a sense forced a new beginning, particularly with regard to the urban landscape. The chapters that follow proceed topically rather than in strict chronological order. The topics include flooding, aqueduct construction and repair, streets, sewers, topography, and cartography.

To begin, we turn to the terrible day of Monday, September 15, 1557.

1

Troubled Waters

The Tiber River

On September 15, 1557, after days of torrential rain, the raging waters of the Tiber River broke its banks and created a devastating flood that inflicted catastrophic damage and death throughout Rome. An *avviso,* one of the anonymous news bulletins sent out from Rome to various rulers and diplomats, reported that peace had just been concluded (in a war between Spain and the harsh ruling pope, Paul IV Carafa) when the river, "wishing to celebrate at night having grown proud, amused itself through the whole city." It rose "continuously all day and into the following night, running through the piazze and streets," coming within a palm (about 8½ in.) of the marker indicating the water level of the great flood of 1530. Gradually "it restrained its fury" and returned to its riverbed, leaving Rome "full of mud and filthy."[1]

More than a thousand people drowned. So did many animals. The Roman-born poet and playwright Angelo Oldradi (b. 1525) reported these grim facts and noted that the people of Rome had no chance to rejoice over the end of the war because the flood "put everything into confusion and fear." Half the bridge of Santa Maria (now Ponte Rotto) was destroyed "together with that beautiful little chapel of Julius III that was there in the middle constructed with such great art and cost." Farther up river, large blocks of marble had embanked on another Tiber River bridge,

Figure 1.1 Flood markers on facade of Santa Maria sopra Minerva, Rome, with 5 ft. 5 in. person standing in front. Detail of marker for flood of September 15, 1557. Photograph May 2017 © Bob Korn with kind permission.

the Ponte Sant'Angelo. On Tiber Island—the small island in the Tiber River at the center of Rome connected to the shores by two bridges—half the church and the entire monastery of St. Bartholomew had been swept away. The flood scattered grain, legumes, vines, and olives, and it ruined houses. Walking through Rome, Oldradi lamented, one could see houses propped up, palaces and shops flooded and deserted, the paving on the streets removed, and indeed, "a miserable spectacle with everything confused, poorly organized, and evocative of pity."[2]

The flood of 1557 came to 18.90 meters above sea level, an estimate that can be made based on flood markers (fig. 1.1) that the Romans embedded in walls around the city. Oldradi reported that around ten Tiber river mills had been damaged or destroyed. Rome depended on river mills powered by the river's fast-flowing current for grinding its grain. Their destruction caused extraordinary hardship. Three weeks after the flood, an *avviso* related that because of the loss of mills (fig. 1.2) and grain supplies, the situation of the population had become desperate and that scarcity of bread had been aggravated by the bakers engaging in price gouging.[3]

The flood added a further dismal episode to the history of Tiber River flooding that extended back to antiquity.[4] The severity of the flooding resulted from the fact that most of the population of Rome lived in the Tiber's flood plain. The river originates far north of Rome at Monte Funaiolo, east of Florence, in the Apennines, and it meanders south for 406 kilometers (about 252 mi.) until it empties into the Tyrrhenian Sea at Ostia, about 20 kilometers (12½ mi.) west of Rome. Its drainage basin is the largest in Italy. Other rivers drain into it, including the Aniene (in the sixteenth century called the Teverone), just north of Rome.[5]

The damaged Ponte Santa Maria, which lost two of its arches, had connected the urban quarter of Trastevere and the Ripa Grande (the main river port of the city) to the urban center on the left bank.[6] The river could still be traversed by rafts and boats and by the *traghetti,* small ferryboats drawn back and forth from one bank to the other by ropes. But without the crucial Ponte Santa Maria, the Tiber Island bridges immediately up river (the Ponte Cestio and Ponte Fabricio, some-

Figure 1.2 Hieronymous Cock, "Pontis nunc 'Quatuor Capitum' olim Fabricii prospectis," in Hieronymous Cock, *Operum antiquorum Romanorum libellus,* 1550. Etching. View of Ponte Fabricio at Tiber Island with two mills. Rijksmuseum, Amsterdam.

Figure 1.3 Pieter Breugel the Elder, *View of the Ripa Grande in Rome*. Ca. 1552. Pen and ink on paper. Collection of the Duke of Devonshire, Chatsworth House, UK © Devonshire Collection, Chatsworth. Reproduced by permission of Bridgeman Images.

times called Quattro Capi) became clogged and overburdened with carters hauling grain and other goods from the Ripa Grande (fig. 1.3).[7]

The Waters of Rome

The main source of Rome's water through the 1560s was the Tiber River. Water was also available from wells, cisterns, and springs within the city; two barely functioning aqueducts (the Acqua Vergine that ended at the Trevi Fountain, and another, the Acqua Damasena, which served the Vatican and two public fountains in front of St. Peter's); the Acqua Marrana Mariana, an aqueduct introduced by Pope Calixtus II in 1122 to supply the area of the Lateran; and a natural stream, the Acqua Crabra (often confused with the Marrana).[8]

Individuals fetched water themselves or purchased it from water carriers called *acquaroli* (fig. 1.4). *Acquaroli* drew water from sources such as the Trevi Fountain and at specific locations on the river—such as the Tor di Nona at the left bank by Ponte Sant'Angelo and outside Piazza del Popolo to the north of the city.

The *acquaroli* might also sell spring water, such as that from the medicinal spring newly discovered in May 1567 two miles outside of the city near the church of San Sebastiano. An *avviso* of May 30, 1567, reported that many sick and infirm people gathered at the spring and that "the *acquaroli* who take it sell it very well." After drawing the water—especially if from the Tiber River—the *acquaroli* stored it in decanting barrels, letting sediment settle to the bottom, then placed the barrels on the backs of their donkeys and walked the streets to sell it.[9]

Sewage and refuse of all kinds drained into the river or was dumped there. We would be aghast at the thought of drinking Tiber River water precisely because of its pollution and undoubted pathogens, but sixteenth-century Romans had different understandings of disease and were often more concerned about the effects of bad air than bad water, especially if the water was free from visible filth.[10]

Nevertheless, physicians in Rome carried on a heated debate on the potability of Tiber River water that continued for decades. Some, such as Alessandro Traiano Petroni (d. 1585), physician to Paul IV Carafa and later to Gregory XIII, extolled the good qualities of Tiber water provided it was properly cleaned. Others, such as the physician Giovanni Battista Modio, denounced Tiber River water as unhealthy and undrinkable and assailed any physician who defended it. Modio was a follower of Filippo Neri and the Oratorians, a community of lay brothers and priests bound together by the mission of charity. He lived in the house of Cardinal Montepulciano Giovanni Ricci (1498–1574) in Via Giulia, that is, the Palazzo Ricci (now Sachetti) for at least five years until his death in 1560.[11]

Cardinal Montepulciano may well have agreed with Modio—he would be instrumental in getting one of Rome's ancient aqueducts, the Acqua Vergine, repaired, providing abundant spring water to central Rome. (The availability of fresh spring water would have seemed urgent to anyone who deemed Tiber River water

Figure 1.4 Detail of a Roman *acquarolo* with donkey carrying water barrels. From *Ritratto di quelli che vanno vendendo e lavorando per Roma con la nova aggiunta di tutti quelli che nelle altre mancavano sino al presente anno* (Rome: Andrea Vaccario, 1612). Folger Shakespeare Library call no. ART 231749 (size L). By permission of the Folger Shakespeare Library.

undrinkable.) The debate about whether Tiber River water was healthy to drink permeated other discussions and debates about water in general and the problem of Tiber River flooding in particular. These issues were of vital importance in the Roman context, but interest in water as a topic could extend geographically well beyond the city itself and chronologically to the waters of the ancient world.

The individual who perhaps best exemplified this wide-ranging interest was Pirro Ligorio (ca. 1513–1583). Ligorio was an intensely argumentative and combative architect and investigator of antiquities (see fig. 3.2). Despite his irascible personality and voluminous writings, little is known about his early background. Born into a noble family in Naples, he arrived in Rome in 1534 at around age twenty. His first known employment in Rome (or indeed anywhere) was painting decorative house facades with historical or mythical scenes, suggesting that his training had been primarily of a practical nature. Over the years, he became known for his numerous investigations of Roman antiquities and ruins, and eventually (in 1549) he was appointed court archaeologist for Cardinal Ippolito II d'Este (1509–1572).[12] Ligorio's famous garden at the Villa d'Este in Tivoli demonstrates both his obsession with water and his ability to manipulate it.[13]

Ligorio's manuscript writings comprise hundreds of pages on Roman antiquities and related topics, including rivers, springs, lakes, and aqueducts. They include an alphabetically arranged encyclopedia on waters written in the 1550s, which listed more than a thousand entries on rivers, springs, and lakes in the ancient and modern worlds. The work is filled with references to ancient sources. Water had particular symbolic significance for Ligorio, as his elaborate hydraulic water systems and fountains at the Villa d'Este in Tivoli demonstrate. Ligorio defended Tiber water as good for drinking, but he did not complete his alphabetical tract on water. He never reached the letter T, thereby failing to provide an extensive discussion of the Tiber under "Tevere" or, for that matter, anywhere else.[14]

Floods and Flood Prevention

Romans of varying occupations and backgrounds took it on themselves to write on the topic of water, including ways to prevent the flooding of the Tiber River. By far the most prolific was the physician Andrea Bacci (1524–1600). Other contributors to riverine discussions included a military engineer (Antonio Trevisi, d. 1566), a physician (Paolo Clarante, fl. 1560s–1570s) who worked as a supervisor to an engineering project under Pius V, the Roman magistrate and legal expert Luca Peto (1512–1581), and a priest (Lorenzo Parigioli, fl. in Rome 1560s–1580s).[15]

The varied solutions of these men to flooding are foreshadowed in the anonymous Ms. 153 in the Biblioteca Angelica (BA) in Rome, probably by the previously mentioned Oldradi. It consists of handwritten notes and compiled materials, including a copy of the 1531 tract on the great Roman flood of 1530 by Ludovico Gómez (i.e., Luis Gómez, ca. 1484–ca. 1542, auditor of the Vatican Palace under Clement VII). Manuscript 153 contains a catalog of Roman flood markers and their inscriptions. It recorded various opinions on how to remedy Tiber River flooding: dredge the river, restore the office of curator of the Tiber and the sewers (who would clean the water of sand and other impediments), enlarge the bed of the river, make more drains like the Cloaca Maxima that ran from the Roman Forum (as we understand its location today) to the Tiber, remove unnecessary bridges and superfluous arches of bridges, and create new channels to provide runoff for river water during floods.[16]

In the arena of actual practice, more than a year after the 1557 disaster, Pope Paul IV called on the Capitoline Council to study the issue of Tiber River flooding. On March 8, 1559, at a public meeting of the council, we learn that the pope had asked the *Popolo* to elect "some gentlemen" who, together with the papacy, would consider methods of flood prevention. The council duly chose four councilors; their names are not mentioned, and we do not know what their deliberations or conclusions were.[17] But it is notable that the pope's response to an urban problem involving physical topography and infrastructure was to call a broadly based council meeting where a wide-ranging discussion could take place.

The city was still in great disrepair when the hated Paul IV Carafa died a few months later in August 1559. His death sparked an urban riot. The people of Rome pulled down his statue that stood on the Capitoline Hill. An executioner cut off the statue's nose, ears, and right arm, cut off its tiara, and put the yellow hat of the Jews on its head. Then the executioner threw the head out of the window of the Palace of the Conservators. Children dragged it through the city for several days until one of the conservators paid them to throw it into the river. (The statue turned up again in the twentieth century; the nose and beard were repaired with plaster, and it resides in the collection of the Museum of Castel Sant'Angelo.) Crowds also smashed Carafa coats of arms wherever they appeared in the city and burned down the palace of the inquisition, releasing its prisoners.[18]

Despite the threat of continuing disorder, it took almost four months for the conclave to elect Pius IV. The new pope's many projects of urban reconstruction included his attempts to ameliorate the problem of flooding. The problem attracted ambitious men, some seeking patronage and urban contracts. They advocated plans, proffered advice, and wrote tracts and proposals.

Andrea Bacci and the Tiber River

One of the most prolific of such proposers was Andrea Bacci (fig. 1.5). Bacci, born in the town of Sant'Elpidio a Mare in the Marches, arrived in Rome as a young

Figure 1.5 Portrait of Andrea Bacci (1524–1600). On title page of *De naturali vinorum historia de vinis Italiae et de conviviis antiquorum libri septem* (Rome: Nicholai Mutis, 1597). Engraving.

man. His father, Antonio Bacci (dates unknown) was an architect, which was perhaps significant given the son's interests in hydraulic engineering. Bacci studied medicine in Siena and arrived in Rome in 1551. In 1567 he was made chair of medicinal simples (herbal and botanical medicine) at the Roman university, La Sapienza. He served as the personal physician to Felice Peretti (who was also from the Marches), both before and after Peretti became Sixtus V. Bacci investigated numerous topics in medicine and natural philosophy, and his primary focus was water in the form of springs, baths, and rivers. In the debate over the potability of the Tiber River, he was firmly on the side of those who advocated drinking it.[19]

In 1558, a year after the devastating flood, Bacci wrote a tract in which he defended the river and suggested remedies for flooding. His 1558 *Del Tevere,* the first of three successive (and very different) versions, was divided into two parts: the first treated water in general, including that of the Tiber; the second focused on flooding. Bacci insisted that "the water of the Tiber with regard to goodness has few equals in the world." He provided explanations for flooding, reported the remedies of the ancients for Tiber flooding, and suggested solutions for the present day.[20]

In his discussion of the nature of water and the causes of flooding, Bacci combined philosophical interests with practical proposals. In the second part of his treatise, he began by outlining differing opinions about the causes of flooding in general. They included the sea, winds, sand at the mouth of the river, the burgeoning of water "through rain or through nature," all of the above intermittently, and the influence of the stars. Different floods could have different causes, the most important being excessive rainfall.[21] Bacci's approach took him to fundamental issues of natural philosophy, including the origins of water itself. He disagreed with Aristotle's view that water was made inside the earth, arguing that all water on the earth was from the sea, its natural place.[22]

Turning to the terrible flood of 1557, Bacci provided a detailed account of the weather in Italy and Europe in the months before the catastrophe. The spring season had been serene with north winds. The summer was very dry. Then came heavy air and a dark and foggy autumn, infecting all of Italy from Sicily to the Alps. It was almost like a plague. It irritated the head, obscured the senses, and dried the mouth and chest. It gave fevers that never left and a violent cough, "not so much deadly as dreadful." In mid-September the skies darkened, and great rains began. Rivers all around Italy became extraordinarily full and overflowed. In Rome the flood came gradually. Because of the low level of the riverbed and the resistance of bridges and other buildings, water first began to fill the drains and sewers and then the streets. Most of the city became navigable. Because the Rome of his day lacked the diligent care that had been provided by the ancients, Bacci feared that the city might remain "sunken in a marsh."[23]

Bacci's recommendations amounted to the reestablishment of ancient practices. The banks of the river needed to be fortified and augmented. The riverbed should be lowered to its ancient level—by removing debris and dredging, one presumes. Perhaps those put in charge of the massive dredging task could then serve as permanent caretakers of the river. The drains and sewers should be kept clean, as they had been in antiquity. An officer in charge of these operations, a curator of the Tiber and of the urban sewers and drains, should be appointed once again. Another consideration was the narrowness and low level of the arches of the bridges as an impediment. All these measures would eliminate floods; would make the river more navigable, cleaner, and less dreadful to drink; and would improve the pestilent air of Rome. When floods did come, the water would flow out through the clean drains and freely into the city; Rome would remain washed and clean from all filth. If in addition the foundations of the buildings were well constructed as well as the walls, they would not be harmed by floods.[24]

At least some of Bacci's recommendations may well have been taken seriously. On June 10, 1564, the Capitoline Council appointed a committee of three—Luca Peto, Paolo del Bufalo, and Gentile Albertoni—to consider one of his recommendations. They were charged with studying writings and the law pertaining to the office of custodian of the Tiber and afterward to report back to the council.[25] Their report does not appear to be extant.

Antonio Trevisi and the Tiber River

In contrast to the learned Bacci, Antonio Trevisi (d. 1566) was a military engineer who had gained his expertise primarily through practical experience. He first aired his proposal in a treatise on flood control published in 1566. Trevisi came from the southern Italian city of Lecce, where he worked on the construction of walls, palaces, and fortifications in the region of Apulia. He traveled to Rome in 1559 under the auspices of the governor of the States of the Church, Camillo Orsini (1491–1559), probably to work on flood control.[26]

Trevisi's literary efforts probably contributed to his success in attaining the patronage of Pius IV. He dedicated his treatise on flood control to Federico Borromeo (1535–1562), nephew of Pius IV and brother of the better-known Carlo Borromeo (1538–1584). Federico was soon to be made captain general of the church at the age of twenty-six (in 1561, a year before his premature death), suggesting his value as a potential patron.[27]

In the dedication, Trevisi emphasized the continuous and excessive damage to the city that flooding had caused. He claimed to have measured the relevant parts of the city and to have made a *modello* (model) of the work to be done. Reflecting the treatise of Vitruvius (*De architectura,* 1.1.2) written in the first century BCE, Trevisi wrote that neither skill without learning nor learning without skill could make any perfect artifice. Further, "I, unlearned, cannot dedicate a learned work to you." Nevertheless, he did dedicate to Federico "that which comes from the simple source of my rude and low intellect." He emphasized his reliance on Federico himself to supplement and correct the insufficiencies of his low work, perfect it, and defend it from malevolent persons.[28]

Notwithstanding this initial self-effacement, Trevisi began his treatise on flooding with natural philosophy, a discussion of "this machine that is called the sphere of the world," into which "that highest Architect that created it" included all things needed for "the governing of universal nature." He described the Aristotelian cosmos, divided into the incorruptible sphere and the sphere of the elements made up of fire, air, water, and earth. Following Bacci as well as Aristotle's *Meteorologia* (354b1–359b1), Trevisi discussed the different kinds of water inside the earth and on the surface, including seas, springs, rivers, rainwater, and well water, emphasizing the divine creation of the world in opposition to the Aristotelian view of its eternity.[29]

Concerning the origins of floods, Trevisi explained that after heavy rains the waters from above and those from the springs below become too abundant to be contained underground. He noted that, against contrary opinions, forty-two streams and rivers flowed into the Tiber before it reached Rome. These rivers constituted one reason for the Tiber's propensity to flood, especially given the impediments of bridges, mills, and other obstacles as it flowed through the city. Trevisi claimed to have measured the flood area, to have spent eighteen months discussing the flood, and to have measured the site and circumference of the city. The flood of 1557 had raised the water ten feet above the river bed; flooding had occurred for about forty miles along the river.[30]

Trevisi recorded the height of the floodwaters during floods of 864, 1374, 1422, 1495, 1530, and 1557. He detailed the terrible damage caused by the recent flood of 1557 to structures, including the Ponte Santa Maria, many churches and palaces, and a bastion of Castel Sant'Angelo. Further, "many and infinite persons drowned." He applauded the strenuous efforts of the masters of the streets, Giordano Boccabella and Camillo Pignanello, to clean the mud that was deposited on the streets, thereby preventing disease from coming to the city as a result of bad air. He estimated the cost of the cleanup at a million gold ducats and noted that from one flood to another, mud had never been cleaned from the cellars.[31]

In his final chapter, Trevisi turned from floods to another topic of engineering—how to raise sunken ships to the surface, and he concluded with a dialogue between a master and student on this subject. Using the dialogue form favored by the humanists, Trevisi thus treated a topic of long-standing interest to Renaissance learned culture, the raising of the two ancient ships resting on the bottom of Lake Nemi.[32] By concluding with this topic, Trevisi forcefully connected issues of hydraulic engineering to the learned antiquarian interests of Renaissance humanism.

Trevisi and Bufalini's Map of Rome

In addition to his treatise on flood prevention, Trevisi communicated his views through letters attached to the lower edge of a large (78¾ × 74¾ in./200 × 190 cm) wall map of Rome, the Bufalini map (fig. 1.6). The map is extant only in Trevisi's 1560 republication. Trevisi seemingly undertook the republication of this map in order to circulate his flood proposal more widely. Bufalini was a surveyor and military engineer, who had spent years surveying parts of the city to create his large wall map. The map was published in 1551, not long before Bufalini's death in 1552. It was the earliest printed, accurately measured wall map of Rome and one of the most influential Roman maps of the sixteenth century. Trevisi had a patron with close ties to the papal printer, Antonio Blado (1490–1567). Because Blado had printed the map in 1551 and had kept the woodblocks, Trevisi was able to republish it in 1560.[33]

Trevisi's edition of Bufalini's map exists in three copies: two in the Vatican Library and one in the British Library. The large wall map was published in twenty separate sections. Trevisi replaced two of Bufalini's sections on the bottom of two separate maps with four of his own letters on the topic of Tiber River flooding.

Figure 1.6 Leonardo Bufalini, *Roma*. London: British Library. Maps S. T. R. 175. Woodcut on 24 sheets, approx. 78¾ × 74¾ in. (200 × 190 cm). First published in 1551 (no copies extant); this is one of three extant copies published by Antonio Trevisi in 1560. Copyright © The British Library Board.

Each of the letters argues in a way persuasive to the particular addressee that Trevisi's flood prevention proposal—which consisted of digging a large diversionary canal for water runoff in flood conditions—should be accepted.[34]

In his first letter, Trevisi addressed the three conservators of Rome (fig. 1.7). He explained that he had shown a model of his idea to the previous lord of the conservators (i.e., Pope Paul IV), and had spoken to many people in authority on the subject, including one of the conservators of Rome, Orazio Naro (1506–1575). Trevisi urged that the cost of his proposed canal would be less than the enormous costs of cleaning up after Tiber River floods.[35] In the second letter, addressed to Pius IV's nephew, Cardinal Carlo Borromeo, Trevisi noted that he had presented his plan for flood prevention to Pius IV, who had instructed him to negotiate with the cardinal. He also had discussed the plan with the conservators of Rome, resolving objections one by one. Now with one word, the cardinal "can make perfect all my imperfection" and bring about the execution of the project.[36]

Two other letters, "To Readers" and "To Virtuous Architects," are found only in the copy of the map in the British Library. In his letter "To Readers," Trevisi described the area of Rome that had flooded and provided measurements for the area. He gave the details of his planned canal—it would start at a point below the Ponte Milvio north of the city, would run through the uninhabited area of Prati (the low-lying area adjacent to the Vatican), and end near the bastion of the Belvedere Palace. During floods, excess water would run out of the river via the ditch behind the Vatican and into a low-lying area called the *Vallis Infernis*.[37] Finally, in his letter "To Virtuous Architects," Trevisi emphasized the benefits of his plan: it would fortify the Borgo and prevent bad air in the area adjacent to the Borgo called Prati (because excess water would be drained).[38] Although Trevisi's specific proposal for digging the massive trench was not carried out, Pius IV's fortification projects did include efforts to ameliorate Tiber River flooding.

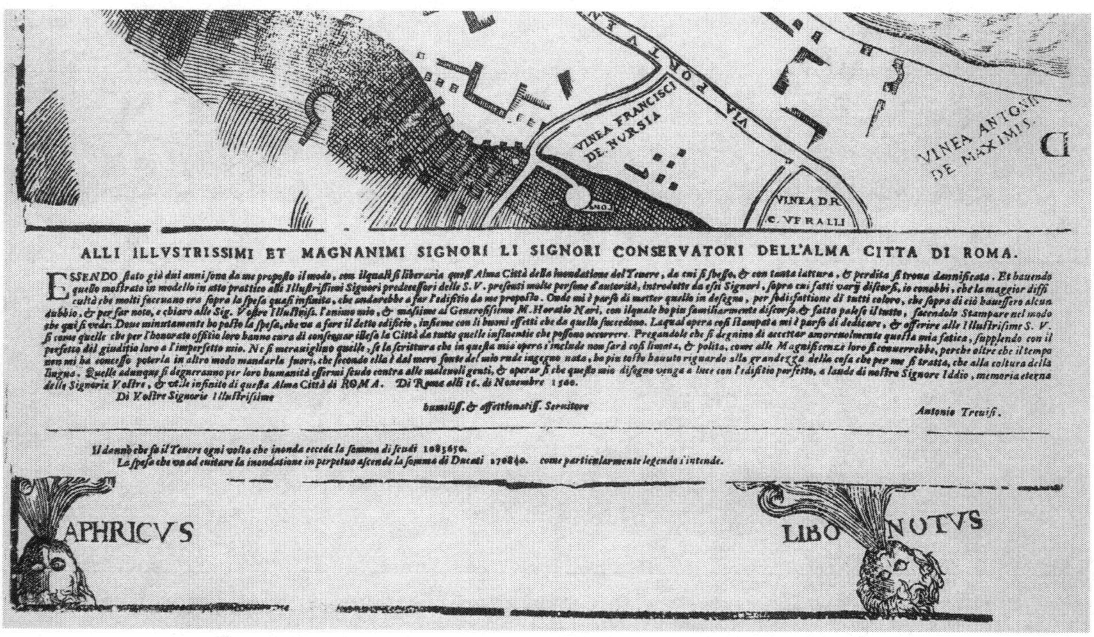

Figure 1.7 Detail of Bufalini, *Roma*: Letter to the conservators by Antonio Trevisi, from Francesco Ehrle, *Roma al tempo di Giulio III: La pianta di Roma di Leonardo Bufalini del 1551 riprodotta dall'esemplare esistente nella Biblioteca Vaticana* (Vatican City: Biblioteca Apostolica Vaticana, 1911).

Pius IV's Flood Control Measures

Fragmentary evidence suggests that Pius IV wanted to solve the problem of Tiber River flooding and that intense discussions and debates surrounded possible solutions. Pius IV seems to have accepted the idea of a canal running through the city that could be used for floodwater runoff. On June 27, 1561, minutes of a consistory (the council of cardinals presided over by the pope) reported that Pius IV had appointed a group of cardinals to be in charge of repairs and a solution to the flooding of the Tiber, including new canals (*novi alvei*). The next day an *avviso* reported that the pope wanted to make a branch (*ramo*) of the Tiber run through Prati and return to the river at Magliana (southeast of the city) "in order to put the Borgo in a peninsula and to ameliorate flooding." Two drawings in the Uffizi attributed to an obscure military engineer and traceable to 1560–1561 show the detailed plans for such a canal (fig. 1.8).[39]

Another piece of evidence is an anonymous handwritten account that discusses proposed remedies for flooding under Pius IV. The author stated that the pope very much wanted to devise a remedy for Tiber flooding—flooding that was so devastating that it was "just like the sack of the city." Discussions were held on possible remedies "with all the skillful Roman men who were experts" and also "with all those who arrived, who were many." All agreed that either of two major remedies would be effective: either enlarging the river bed or constructing a new channel. Each would be enormously expensive. Other ideas were to open up the closed arches of the Ponte Sant'Angelo, to make openings for a channel exit at the Castel Sant'Angelo, and finally, to prohibit garbage being thrown into the river upstream from the bridge.[40]

Pius IV's hydraulic plans were at least partially carried out. They were part of his project of expanding the geographic area of the Borgo (calling it after himself, the Borgo Pia), fortifying it, and continuing to fortify the Castel Sant'Angelo. From the time of Antonio da Sangallo the Younger in the 1530s, plans for fortifying the Castel Sant'Angelo often included a pentagonal outer ring of walls with five bastions and a moat around the whole—a plan finally settled on by 1555 (at the latest), when Pope Paul IV Carafa had such a structure built as an earthen construction, an edifice that was washed away by the flood of 1557. Pius IV attempted to refortify Castel Sant'Angelo and incorporate flood prevention into the new construction. He had a long wall built demarcating the newly expanded Borgo and alongside had a new trench or canal excavated, which helped drain excess water into the moats of the Castel Sant'Angelo.[41]

The extensive work of fortification and hydraulic engineering was supervised by Francesco Laparelli (1521–1570). Laparelli was from a noble family in Cortona, a town in Tuscany. At an early age he became interested in military and mechani-

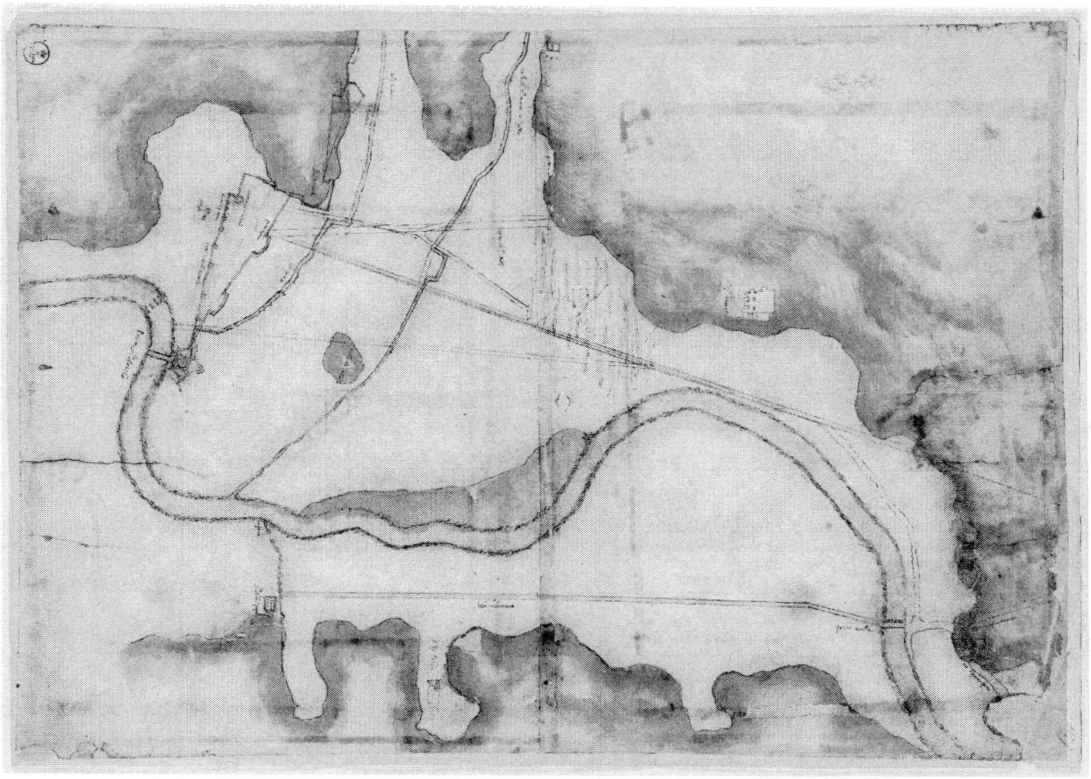

Figure 1.8 Depiction of canalization of Tiber River. Copy of drawing from 1560–1561. Ponte Molle crosses the river on the lower right, Castel Sant'Angelo on the left, proposed canal drawn above the river through the area of Prati and the Borgo. Ink on paper. Florence, Galleria degli Uffizi, Gabinetto dei Disegni e delle Stampe, Uffizi 288A. Attributed to Bartolomeo de Rocchi. Su concessione del Ministero per i Beni e le Attività Culturali e del Turismo.

cal things. Laparelli was profoundly influenced by his friend Gabrio Serbelloni (1508–1580), a nephew of Giovanni Angelo Medici and an expert in fortification. As soon as Giovanni Angelo became Pope Pius IV, he called his cousin Serbelloni to Rome and made him captain general of the pontifical guard. Serbelloni oversaw all papal fortifications and called on his friend Laparelli to assist. Between 1562 and 1565, Laparelli supervised construction of the wall of the new Borgo Pia, which was bordered by a moat, and the restoration and construction of the bastions and moats around the Castel Sant'Angelo.[42]

Some of the moat still exists today, and it is decorated with commemorative inscriptions crediting Pius IV for work that was not actually completed until the seventeenth century. The trenches were designed to improve the defensive capacity of the fort and also to allow runoff from the Tiber in times of flooding. They can be seen in a 1569 engraving of the Castel Sant'Angelo by Giovanni Antonio Dosio (1533–1609), where a moat around the northern side of the castle is visible (fig. 1.9), and in another etching by Stefano (Étienne) Duperac (Dupérac) (ca. 1525–

1604), where two low drainage arches are visible on the southwest side where the water from the moat drained back into the Tiber (fig. 1.10).[43] That the exit was constructed is confirmed by a letter of instruction by Serbelloni. The letter states that the moat of the *castello* should be cleaned out so that when there is a flood, water can flow in without impediment. The entrance to the moat must be kept narrow and the exit wider so that the waters can easily flow out.[44]

Pius IV's new trenches may have done some good. An intriguing *avviso* written on January 4, 1567, two years after the pope's death, described "a pleasing flood" that did much less damage than the floods of 1530 and 1557. In this "pleasing" 1567 flood, the water went to level areas of the city, and "the frightened people began to remove goods from dangerous places and to withdraw to the secure areas." Traders of stalls (*mercanti de banchi*) removed bales into attics and fortified barrels that were in cellars, "and all night the street ran with people who with the light of torches brought goods, horses, and people to the hills." After describing the damage, the *avviso* notes that "it is believed that the new ditches around the *castello* [Castel Sant'Angelo] made by Pius IV were very much of use against the impetus

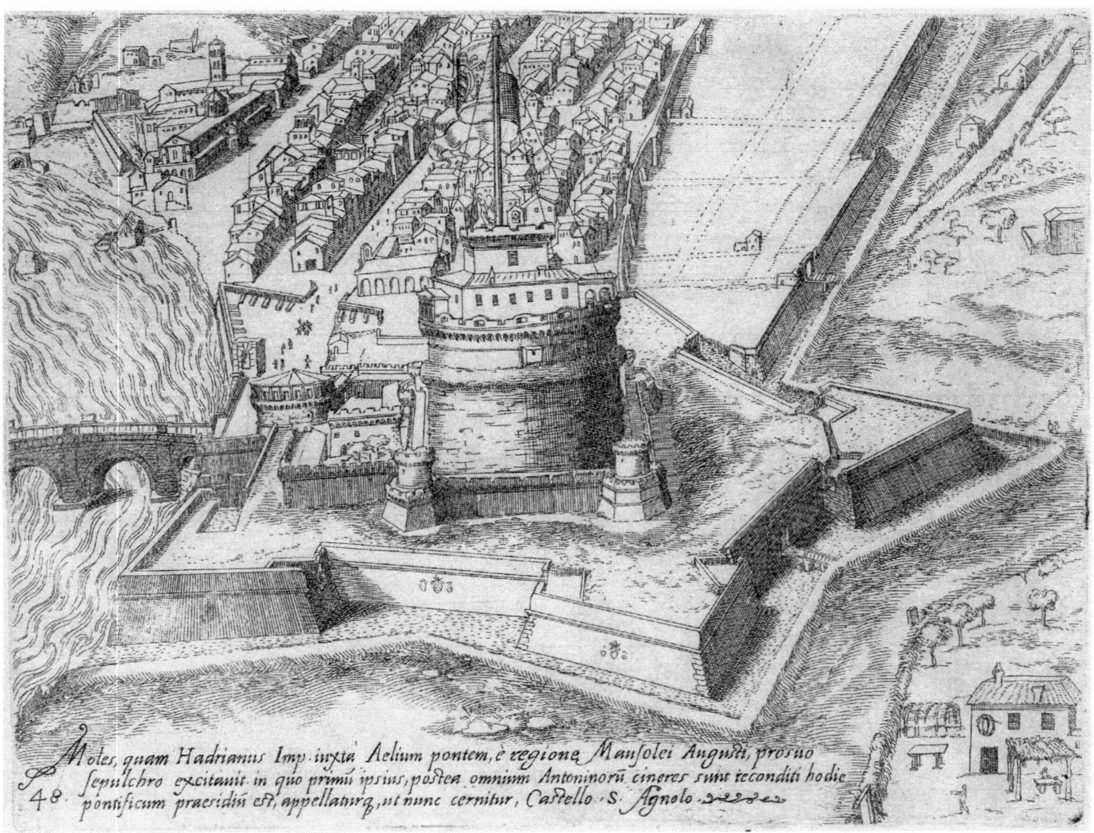

Figure 1.9 View of the Castel Sant'Angelo with moat. Giovanni Antonio Dosio, *Urbis Romae aedificiorum illustriumquae supersunt reliquiae* (Florence, 1569), 48. Etching. Collection of Vincent J. Buonanno.

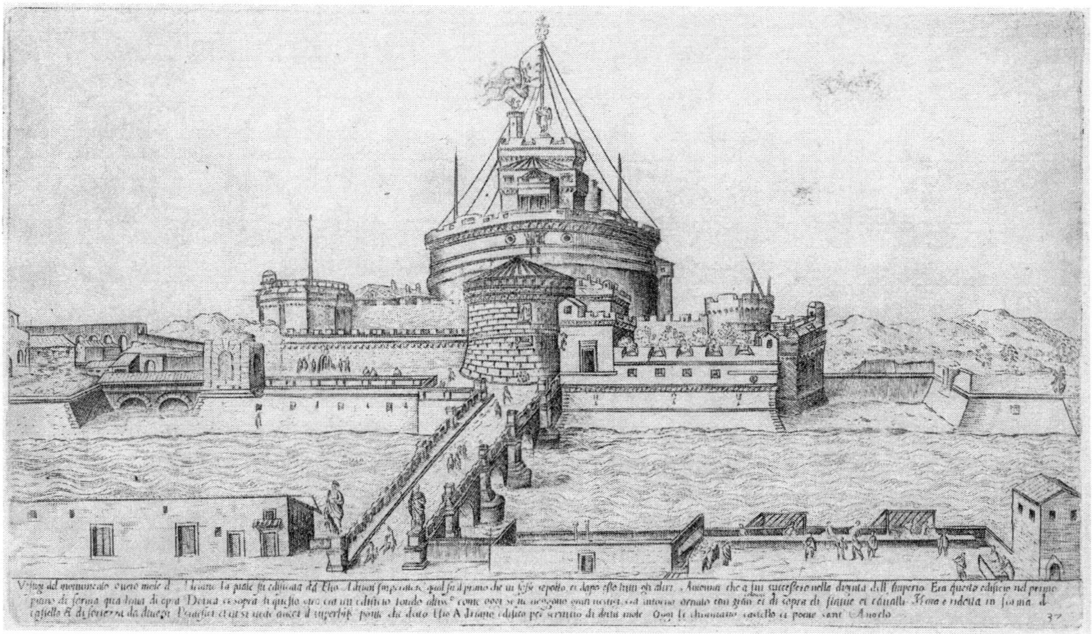

Figure 1.10 Stefano (Étienne) Duperac, *Castel Sant'Angelo*. From *I vestigi dell'antichità di Roma* (Rome: Lorenzo della Vaccheria, 1575), No. 37. Engraving/etching. Rijksmuseum, Amsterdam.

of the river."[45] Pius IV's trenches may have ameliorated damage, but the 1567 flood itself was undoubtedly far less serious than the one that had devastated the city ten years before. What is certain is that Tiber River floods continued to plague the city.

Luca Peto: A Roman Magistrate's Advice

Luca Peto was a Roman-born patrician and jurist, a hardworking magistrate active in the affairs of the commune most of his adult life. As a young man, between 1531 and 1537 he studied law at the University of Bologna, listening to the lectures of Ugo Boncompagni, who in 1572 would become Pope Gregory XIII. Peto served on the Capitoline Council sixteen times, as one of the three conservators for two three-month terms in 1571 and 1575, and as *caporione* (the officer who headed a *rione* or district) of Ripa for three months in 1549. From the 1540s he served on a committee to reform the chaotic statutes of Rome, which over the centuries had become redundant, confusing, and contradictory. It was a difficult task on which he worked for decades and that he finally completed single-handedly with the new law code published in 1580.[46]

Peto wrote tracts on practical, historical, and even literary subjects, including a short commentary on Virgil's *Georgics* and two short tracts describing the weather and its effects on crops and the food supply in two specific years, 1569 and 1570 (the latter describing a plague of black caterpillars and its effect on crops and trees).[47] A larger treatise, published in 1573, was devoted to a study of ancient Greek and Roman weights and measurements and their contemporary Roman counterparts. He wrote it after overseeing the reform and standardization of weights and measures in the city while serving as one of the three conservators.[48]

He also wrote a tract on the repair of the ancient aqueduct, the Acqua Vergine (discussed in chap. 3), and another on flood control. That Peto combined a practical approach with the study of ancient texts is evident in his tract on how to prevent Tiber River flooding. He attributed flooding primarily to the impediments to the flow of water caused by the piers of the bridges and especially the Tiber Island bridges. The river flooded in two ways—either it flowed into parts of the city, such as the Piazza della Rotonda (at the Pantheon), through the subterranean drains or conduits, or the river simply overran its banks, as at the Porta del Popolo and in Via Giulia. Peto described in detail his observations of the behavior of the river during floods in a number of locales throughout the city. The Tiber Island bridges, along with the many mills in the Tiber especially near the Ponte Sisto, constituted impediments that contributed to flooding. Peto advised that a ditch for drainage be created near the Castel Sant'Angelo. He also urged widening or removing one arch of the Ponte Sisto, removing one arch each from the island bridges, and redesigning the (still broken) Ponte Santa Maria or Ponte Rotto with one large arch instead of two.[49]

Andrea Bacci and the Del Tevere *of 1576*

In 1576 Andrea Bacci came out with a much expanded edition of his 1558 *Del Tevere*. This time he dedicated the book "to the Illustrious Senate and Glorious People of Rome," that is, to the Capitoline government. He hoped to demonstrate why the water was good for drinking, how it should be treated as a drinking supply, and how to remedy flooding, which he saw as the river's only defect.[50]

By this time the newly repaired Acqua Vergine was supplying to Rome abundant clean water, pushing Bacci to defend Tiber water in new ways. He argued—in contrast to his previous view that the rapid current of the river was sufficient to remove urban filth—that the closer one got to Rome, the more care must be taken to avoid drawing contaminated water. He made several recommendations for where in the river to throw sewage and other filth, and where to draw out drinking water. He also described his invention of a kind of floating dock held to the shore by ropes. He recommended that they be built so that people could walk out on them and collect their water from the center of the river, where it would be cleaner.[51]

Bacci's third book on flooding repeated much of what he had written in 1558, but he also informed the reader about subsequent events, such as the meetings called in 1566 by the newly elected Pope Pius V. The pope charged the participants to develop plans to prevent flooding of the river. At the beginning of the new pope's pontificate, Bacci explained, "his Holiness, desirous of the common health and good particularly of this dear city, issued a public edict to exercise ingenious minds toward this health-giving task of remedying the flooding of the Tiber." Bacci continued: "I found myself in some meetings that for this reason were convened in front of the very Illustrious and very Reverend Cardinals Sforza [Alessandro Sforza, 1534–1581] and Montepulciano, and the gentlemen masters of the streets, where, among many noble architects and engineers, and other men of merit in diverse professions, I understood a great diversity of opinions concerning this flooding and its remedies."[52]

Among all these, in Bacci's view, "very little resolution was heard." This was because most of the gathered assembly, "leaving the sciences aside, thought about it without method and by use of the empiricists, having regard only to remedies," and not knowing the "causes of the evil" lacked the ability to come to a solution.[53]

According to Bacci, some people proposed remedies already tried by the ancients but without noticing the difficulties they had had. Some gave in to the impossible without design (*proposito*). Others held "their designs secret." Others were frightened by the cost. Others offered secret new inventions. Of such inventions, Bacci believed that "every time that they depart from the method that the ancients ordinarily used in this concern, they will always be utterly mistaken." He mentioned these problems not because he wanted to actually try the inventions of others or because he wished to make anyone despair of ever being able to find a remedy. Rather, the diversity of plans came out of the difficulties that all the most ingenious ancients and moderns had confronted.[54]

In his discussion of remedies that he actually recommended, Bacci undoubtedly endeared himself to the cash-strapped communal officers with some of his far less radical and less costly recommendations: "I would begin with remedies more common and easier, conforming to the causes that have been said to concur with these inundations, thus as to enlarge and clean the bed of the Tiber, to fortify the banks, to straighten the course, and to lift from it every impediment." Bacci also suggested proceeding cautiously—the work itself would show the way. "I would like therefore that it be judged where the greatest need was."[55]

Bacci suggested several additional remedies. One was planting the banks in certain places with field maples (*opij*) and silver poplars (*albucci*), which "would improve the earthen banks so that they would not wash away in river floods [and] thus would procure supplies of wood for certain needs that occur, and would give some beauty to the river."[56] Another remedy was to clean the sewers (*chiaviche*) on an ongoing basis. And a third enlarged on an already useful remedy—to expand the ditches started by Pius IV.[57]

Bacci was well rewarded for his ongoing efforts to study the Tiber River and to advise on the issue of flood control. A record in the Capitoline Archive shows that he was granted Roman citizenship in 1576. Here he was praised for his writings on the Tiber River and on flood control. Perhaps ironically, because of his preference for Tiber water over aqueduct water for drinking, he received another prize commodity six months later—an *oncia* (i.e., the amount that flows through a 1.86 cm–wide pipe) of fresh, continuously running water from the newly repaired Acqua Vergine. This water, which was a variable amount because of the variable rate of flow, would be delivered by a pipe into his own home, which was situated directly over an Aqua Vergine conduit in Via Condotti in the *rione* Colonna.[58]

Paolo Clarante da Terni

Perhaps Bacci's reward for his advice and writings encouraged others to write their own tracts. One of these was Paolo Clarante (fl. 1560s–1570s) from the town of Terni in Umbria. Clarante was a physician who worked as an *archiatro* (physician to the pope) in the employ of the family of Pius IV. Most of what little is known about him comes from his writings of the 1570s. He addressed to the healthy and to the sick, brief (one-page) instructions about what to do in the time of plague (probably in 1576), and he wrote a treatise on the calendar, published in 1576, for consideration in Gregory XIII's process of calendar reform. Finally, in 1577, he published a tract on how to prevent Tiber River flooding.[59]

Clarante dedicated his tract on flooding to Gregory XIII. He assured the pope that he had been thinking about the topic for a long time. Before considering remedies, he believed it crucial to consider the causes of flooding, the most important of which was extraordinary rains. Further reasons for flooding included the melting of snow and the winds, especially the south wind that holds back the water of the river at the sea. These causes, Clarante concluded, point to a single remedy, namely, "a new mouth [of the river], and that [requires] a new riverbed." This new trench or canal would begin at the coast east of Ostia, go to Ponte Milvio, and then go through the city, ending at the Ripa, the main river port of Rome. The idea was that it would fill only during floods. Clarante systematically itemized his reasons for rejecting the many other remedies that had been proposed over the years. But he (like Luca Peto and Andrea Bacci) would also open the arches of the bridges (causing less blockage by piers) and remove other impediments from the river. Clarante insisted that such remedies were "not a task for the simple architect but [for] even the philosopher to find the causes and at the same time, the physician to adapt remedies to them, and a good physician to treat not the small world that is man but the great, whose head is Rome." Clarante revealed that under Pius V, with the authority of the Holy See and the Camera Apostolica, he had worked on a canal near the coast that would go to the sea, but the work had ceased at the death of Pius V. He ended with a detailed discussion of how the new river mouth would be constructed—with piles and a wall—to prevent continuous silting.[60]

Lorenzo Parigioli

In 1579, Lorenzo Parigioli, a priest from Perugia, wrote a tract on Tiber River flooding, an unusual topic for him in that his other writings concerned grammar and poetry. He dedicated his exposition on flooding to Giacomo Boncompagni, Gregory XIII's son, who was castellan—in charge of the Vatican fort, the Castel Sant'Angelo—and served as captain general of the church. The tract focused on the causes of flooding. Although Parigioli himself was a bookish man (as his references to ancient writers reveal), he admonished that "one does not learn everything from his books." To the contrary, "actual practice and observation of things" came "before theory and the arts." The former is often very different from the latter—"more secure, more intelligible and better." Parigioli outlined the causes of flooding. They included rain and snow; encumbrance of bridge piers and the narrowness of the river bed; and wind from the direction of the Tyrannean Sea, the sea itself, and the sand at the sea. Framing his discussion of each of these causes with references to ancient historians and poets, he argued that the main cause of flooding was excessive rain and that the common opinion that the south wind and the sea caused flooding could be observed to be untrue. The Tiber did not flood during days of high wind unaccompanied by rain.[61] Despite Parigioli's stated interest in experience and skill, he did not venture to suggest practical solutions to the problem.

The Floods of 1589

Despite all the writings advising on flood prevention, flooding continued to disrupt the city. Two major inundations devastated Rome during the last year of the papacy of Sixtus V. The floods arrived a week apart, one on November 4, 1589, and the second on November 10, 1589, at sixteen meters above sea level. A vivid description of the November 4 flood exists, written by an unknown visitor with the initials I. G. and addressed to an unknown highborn patron (*Molto Mag. Sig. mio*, i.e., My Very Great Signore). I. G. informed the addressee that among the reports (*avvisi*) that he could give was "the great flood made by the terrible Tiber River." He explained that Wednesday night, the river rose in such a way that all of Rome was flooded. An infinite number of families were trapped in houses, and it was necessary to bring them food in boats to prevent them from dying of hunger. The cries heard throughout the city frightened everyone, and it seemed that Judgment Day had arrived. People were moved to compassion, especially hearing the cries of the children. Because of the difficulty of grinding grain, bread could not be found; however, the police officers (*bargelli*, the police of the Governor of Rome) found grain and bakers and saw to it that bread was distributed. At the Ripetta port, lumber and barrels of wine were carried away by the floodwaters with damage valued at more than four million gold scudi.[62]

I. G. recorded the palaces and houses destroyed, especially at the turns of the river and especially the houses behind the church of San Rocco (near the Ripetta), where lived "the gentle women of this city" (prostitutes). The water "almost had the height to carry all to the sea." I. G. remarked that "it was a pleasure to see them [the prostitutes] running away, some without slippers, some barefoot, others disheveled and some [running] one way, others another, screaming for mercy in high voices and running through those streets and through that water." I. G. also reported how the water came up through the sewers, making a loud sound and flooding the area of the Pantheon, and in other places, such as the ghetto, Trastevere, and the Borgo. Indeed, for the morning mass, the cardinals were compelled to arrive through the corridor of the *castello* (i.e., the elevated passageway from the Castel Sant'Angelo to the Vatican).[63]

The indefatigable Andrea Bacci also took note of the floods of 1589. His observations can be found in a ten-page handwritten letter possibly composed around 1592. This missive appears in the flyleaves of a beautifully bound copy of the 1576 edition of *Del Tevere* in the Vatican Library. Addressed to Pope Clement VIII (ruled 1592–1605), it seems likely that Bacci wrote the letter in the fly leaves and presented the book to the pope close to Clement's papal election, which occurred on January 30, 1592.[64]

In his letter to Clement, Bacci emphasized that "in our own times" the Tiber flooded more often than in the past. Although the two floods of 1589 involved midlevel flooding, nevertheless, "one navigated through Rome for three days in many places not without notable ruin and damage." Bacci specified his own contribution to solving the problem: he had collected various judgments, remedies, and histories from the past, but "as far as the execution, it seemed to me to leave the burden to the engineers and [those] more practiced than me in the work."[65]

Yet Bacci also emphasized his own past experience and work on the problem, reiterating his participation in Pius V's commission headed by cardinals Montepulciano and Sforza. He again itemized ancient remedies for flooding and repeated his opposition to the idea of creating a long diversionary trench through the city. He added a lengthy discussion of the idea of adding two additional channels to the river outside of the city beginning several miles from where it flows into the sea, creating a trident-shaped mouth that would presumably allow more water to flow out more quickly.[66]

This attractive book with its handwritten letter addressed to Clement seems to have ensured Bacci's continued involvement in trying to solve the problem of flooding. And the problem proved to be ongoing. On December 24, 1598, came the worst flood ever recorded in Rome, a catastrophe memorialized by nineteen flood markers around the city.[67] After this terrible flood, Bacci served on another papal commission created by Clement and put out yet another edition of his book on the Tiber River, this one dedicated to the conservators of Rome and including an additional book that was an Italian translation of the Latin treatise by Ludovico Gómez on the flood of 1530 (the first treatise entirely devoted to Tiber River flooding).[68]

When Bacci died in 1600, the problem of flooding was far from solved. Proposals continued to be written throughout the next century and beyond. The problem of Tiber River flooding in the center of Rome was not effectively solved until the high embankment walls enclosing the river (and cutting it off from the city) were completed between 1876 and 1910.[69]

◆ ◆

The range of individuals concerned with Tiber River water for drinking and Tiber River flooding suggests a kind of public arena in which flooding was seen as a legitimate topic of discussion by both practitioners and university-educated men. The meetings or congregations called to consider the problem of flooding, which were attended by both papal and city officials, heard many opinions from numerous people. Such meetings, as well as the diverse writings on the topic, point to the public nature of the discussions.

It is notable that many of the writings considered here tied practical engineering solutions to two aspects of learned culture. The first was natural philosophy, and included the assumption that the causes of flooding needed to be understood before practical solutions could be undertaken. Second, knowledge of the ancients, including ancient hydraulic practices related to the river, was often considered essential to devising a workable solution for the present. What it did not do was develop mathematical analyses of water flow and other aspects of flooding related to the development of the new sciences, which for Rome occurred in the next generation (importantly, with the work of Benedetto Castelli [1578–1643] a student of Galileo).[70]

A solution to Tiber River flooding was not found in the sixteenth century. Nonetheless, the discussions about both the potability of the river water and solutions to the problem of flooding had important cultural consequences. The perceived connections between natural philosophy, antiquarian studies, and flood prevention were instrumental in bringing issues of practical engineering into learned culture.

2

The Streets and Sewers of Rome

Archival and other primary sources reveal a Rome rarely described in modern scholarship—a city whose streets and alleys often functioned as depositories for mud, manure, and sewage. Such passageways, especially in rainy weather, presented quagmires that would have emitted a powerful stench. The Tiber River flowed through an urban fabric including streets paved and unpaved, alleys, and drainage conduits including sewers—some open trenches visible to all and some underground. The lives of the people of Rome were closely tied to these structures and to the processes of maintaining them, and they were directly affected by failures to maintain them.

Street traffic included horses and their riders, horse-drawn carriages, carts pulled by oxen carrying heavy loads such as blocks of stone, donkeys carrying water barrels led by their masters (the *acquaroli*), pedestrians including beggars and prostitutes, vendors of all kinds, men and women dressed in finery, shrouded widows, and animals—pigs, sheep, cows, buffalo, and goats. Street cleaning and waste disposal, taken for granted in modern cities, were ongoing, never-solved problems in late sixteenth-century Rome.

There is a long history of laws, edicts, and papal bulls on the Roman streets—an administrative structure, at least at the top, was in place. Yet this structure, which included cardinal-led congregations (*congregazioni*) and the masters of the streets (*maestri delle strade*), was not the equivalent of a modern bureaucratic structure. There were no permanent employees, and there was no permanent funding. "Sanitation professionals" in the modern sense did not exist.

Masters of the Streets: An Evolving Office

The masters of the streets were officials who, beginning in the thirteenth century, were elected annually and often reelected. They were Roman patricians from venerable Roman families who enjoyed substantial local influence and power. The duties and functions of these urban officers underwent a development and expansion from their first appearance in the communal records in 1233. They oversaw street cleaning and paving and adjudicated urban property disputes that arose between neighbors. They also dealt with bridges, sewers, and urban structures such as walls and porticoes.[1] The legal instruments by which their duties were defined and gradually expanded are important sources for understanding the condition of the streets of Rome and the ongoing attempts to maintain and govern them.

The masters of the streets began as officers of the Roman Commune. The document of 1233 calls them *magistri aedificorum* and specifies that there are three of them assisted by two submasters. In the next century, a Roman city statute of 1363 contains a clause on the masters showing that they were provided with two submasters (usually skilled workers such as bricklayers or masons [*muratori*]), a notary to record their decisions, and an assessor (*assessore*). In this statute, the masters were to decide on all questions concerning walls, streets, squares, and property divisions in and out of the city. Their duties included adjudicating conflicts between proprietors with adjoining properties, making sure the streets were clean and unobstructed, overseeing the fountains and watercourses, preventing sewage and other filth (including animal viscera from butchers) from being dumped into the street, and maintaining bridges.[2]

The first extant statute devoted entirely to the masters of the streets is dated 1410. The twenty-six clauses of the statute specify the range of authority of the masters and their responsibilities, many of which resembled previous regulations and would remain in place for more than two centuries. Key provisions included those authorizing the masters to remove fences and other obstructions from the street, to have the streets cleaned, to demand money from abutting houses for street paving, and to repair sewers and drains.[3]

During the fifteenth century, this 1410 statute remained influential, but a gradual change in the office of the masters of the streets came about. Their duties expanded. In addition, a change in the authority that oversaw them occurred—from the Capitoline government to the Camera Apostolica, or papal bureaucracy. These changes coincided with the return of the papacy to Rome in 1420, starting with the Colonna pope, Martin V. Gradually, throughout the fifteenth century, the popes reasserted their authority over the city. At the same time, the roles of the masters of the streets expanded. Their responsibilities came to include granting licenses for building construction and demolition. This enabled them to have great influence on the development of the huge urban expansion that began in fifteenth century Rome.[4]

The Urban Bull as Mirror of Urban Conditions

Although this book focuses on the late sixteenth century, it is useful to review earlier bulls and edicts on the streets of Rome because late sixteenth-century legal measures were dependent on and included substantial reiterations of earlier bulls and edicts. Also, it is instructive to read edicts about the condition of the streets decade after decade for a century and a half.[5] It becomes clear that Roman streets often were filled with sewage and other filth—as much or more by the end of the sixteenth century as at its beginning. When read together in chronological order, communal laws and papal bulls and edicts provide graphic evidence not for the actual disposal of waste and street cleaning (the extent of which cannot be known in complete detail) but for the condition of the streets at particular moments in time. The remarkable similarity of descriptions of filth from one decade to the next should give pause to scholars who tend to read laws enacted about public waste disposal as evidence of progressive improvement.

Pope Martin V's papal bull *Etsi in Cunctarum*, issued on March 31, 1425, restored and reinvigorated the office of the masters of the streets, which office, the bull claimed, had been in disuse for a long time. The bull stated that the city and its districts, streets, bridges, and other parts "had suffered great disfigurement, or rather, abominable ruin and deprivation." Some inhabitants of the city, "such as butchers, fishmongers, shoemakers, tanners, and other diverse artisans" throw "viscera, intestines, heads, feet, bone, blood, and also skins, rotten meat and fish, and other fetid and corrupt things" into the streets, piazze, and other places both public and private. Another very different kind of abuse was that some Romans "with temerity and sacrilegious audacity" took over and usurped public and private spaces for their own personal use.[6]

About twenty-seven years later, Pope Nicholas V promulgated new statutes for the *maestri di li edifiti*. The most important change was that the masters of the streets were now placed directly under the authority of the papacy rather than (as formerly) under the Capitoline government. The pope also gave the masters an expanded range of responsibilities. They were specifically authorized to destroy structures (such as porticoes) protruding into the streets and to give licenses for private building. They must ensure that the streets were paved—especially the major streets—at the expense of the owners of adjacent houses. If owners refused to cooperate, marshals and police could forcibly collect the taxes. (This tax, collected by the masters or their subordinates door-to-door, was called a *gettito*. It remained a standard way for collecting funds for street work throughout the sixteenth century.) The masters were to prohibit the disposal of manure, fetid material, and dead animals into the streets. They must ensure that the people clean the streets on a regular schedule, that carters collect the waste every Saturday and dump it into the Tiber River, and that the "filth of the Tiber" (probably referring to sewage on the riverbanks) be cleaned twice a year at the expense of the same carters.[7]

The statutes delineating the duties of the masters of the streets were *attempts* at regulating the streets and keeping them clean—not necessarily accomplished achievements. Failure becomes particularly clear in 1467 when Paul II (ruled 1464–1471) had one of his own trusted men, Girolamo Giganti (ca. 1520s–ca. 1574) supervise a special commission to clean the streets, which had become intolerably filthy. Extant is a register of 679 citations made by Giganti's men between July 21 and October 12, 1467, for violations, primarily those involving throwing human and/or animal waste into the street in front of individual houses. Paul's attempts signal the failure of the masters as supervisors of street cleaning.[8]

Under the subsequent pope, Sixtus IV (ruled 1471–1484),[9] on January 8, 1480, the *camerlengo* of the Camera Apostolica, Guillaume d'Estouteville, issued an edict forbidding an array of abuses that provides a graphic description of city streets filled with the waste of humans and animals and obstructed by debris left over from building construction. Romans were prohibited from throwing excavated dirt (*sterro*), mortar (*calcinaccio*), rubble (*rempeture*), and clay (*luto*) into the streets. When it rained, no one was allowed to throw waste into the public streets. Waste on the streets should be raked to prevent drains from filling up with it. No one should throw fecal matter (*fecia*) into the drains or sewers (*chiaviche*). Nor should they throw garbage (*spazatura*) or stinking water or other fetid matter into the drains. Everyone who had paved in front of their houses must clean every Saturday and throw the sewage (*immunditie*) into the river. Those whose houses were fronted by an unpaved street must throw the clay and waste (*luto* and *immunditie*) in front of their houses into the river once a month. No person could make drainage spouts (*sciacquatori*) from their houses through which either water or human waste flowed into the streets and public places, nor could human waste be placed in small alleys (*vicoli*) where rain caused it to run into the street and create bad odors. Every Saturday certain named piazze (such as the Campo de'Fiori) had to be cleaned. Barber-surgeons (practitioners who both cut hair and performed surgery in their shops) could not continuously throw their water and waste materials into the street—which then becomes disgusting and muddy (*brute et fangose*). The list included clauses prohibiting porticoes and other structures extending from private buildings into the street.[10]

Sixtus IV did not stop there. On June 30, 1480, he issued a bull, *Etse de cunctarum,* that would influence urban policies and development through the sixteenth century. The bull gave the masters of the streets wide-ranging powers to expropriate private property in the interest of opening up streets and piazze. It was difficult, the bull complained, for the multitude of citizens, members of the curia, and others who live in the city—as well as pilgrims—to walk easily in the narrow streets. Two horses could pass only with difficulty. The most important streets needed to be widened, porticoes and other constructions that obstructed the street had to be removed, and when suitable, the streets were to be paved with brick.[11]

The bull went further—it gave (to modern ears) startling powers to owners of houses and palaces to appropriate (with the assistance of the masters of the streets) the property of their neighbors if that property was unoccupied or in a state of ruin. The owner would be paid a price agreed on by two experts, and if the experts disagreed, the decision would go to the *camerlengo*. The bull addressed many contingencies—if a house was uninhabitable, for example, and the owners of houses on each side wanted it, it would go to the one "most in need," or if both neighbors had need, part would go to one and part to the other. In clause after clause, the masters were given authority to appropriate houses and other properties for the benefit of enlarging the houses of neighbors or widening streets and enlarging piazze.[12]

Sixtus IV's novel bull, designed to assist the elite classes in expanding their palaces, was enforced to a greater or lesser degree throughout the sixteenth century. At the same time, a new officer, the president of the streets (*presidente delle strade*), became increasingly important. The president was an officer under the *camerlengo*. He was a member of the clergy—often a bishop, and presided over the secular patricians who were the masters of the streets. The authority of this cleric in the Camera Apostolica emerged gradually, at first with little more than formal authority, then with greater decisiveness and with the clear will to compete with the prestige of the secular masters.[13]

Organizational changes did not make the issue of fetid, debris-filled streets go away. Particularly notable at midcentury was Julius III's 1551 edict that initiated taxes for street cleaning. It was directed at merchants, store owners, artisanal workshops, inns, wine shops, and all other commercial establishments. These proprietors were required to pay four *giuli* per year for street cleaning, a tax to be collected by the masters of the streets. The edict emphasized how incommodious the filth of the streets was for everyone—cardinals, prelates, members of the curia,

Roman citizens, residents, and visitors. Because of the resulting "exhalations and stench of the air" there were many "infections and pestilent diseases."[14]

Julius III's edict contained provisions for implementation. Officers were appointed—a collector, a treasurer, and a supervisor of the cleaning—and their duties specified. Provisions were made for the workers to be paid and to have the necessary cleaning tools. An overseer was appointed who was to record all payments. The edict authorized penalties for those who disobeyed. But this tax for street cleaning was soon diverted for other purposes and finally was discontinued. The next year, in 1552, Julius did not even nominate officers for the positions of master of the streets.[15]

Portrait of the Streets: 1560–1590

Despite the long tradition of edicts and papal bulls directing the masters to maintain clean streets and remove obstructions, Rome in the early 1560s was far from the magnificent city that had been imagined for it. In her astute description of the urban landscape under Pius IV, Elena Bonara describes the city in 1564 as "still a profoundly medieval city." The region of the hills was largely abandoned. The population was for the most part amassed on the left bank of the Tiber, but even there the countryside invaded. Vineyards, fields, and orchards interrupted the continuity of streets and buildings. Wolves roamed around the walls of the Vatican, cows grazed on the Palatine. In populated areas, porticoes and external stairs hindered the passage of carts and horses. Few streets were paved. Many were mired in dirty water, sewage, skins, and viscera thrown in the street. Small piazze and alleys were often filled with refuse.[16]

An edict issued at a public meeting of the Capitoline Council on July 22, 1564, confirms Bonara's portrait of the city. It orders that "the city be cleaned from all filth and putrid things"; that pigs and goats not be led through the city; that "filth not be thrown into the streets from windows"; that if filth is thrown into the Tiber, it should in no way be thrown on the banks; and "that infected animals not be led through the city." The *caporioni* are instructed to look out for violators in their districts and report them at once to the conservators and prior.[17]

The Streets under Pius IV

The edict was issued four years into the reign of Pius IV, a year before his death. Yet the pope cared deeply about the magnificence of the city. His bull issued a year later, *Inter multiplices curas* of August 23, 1565, assessed the provisions of previous bulls concerning the streets and the masters of the streets and reinstated their provisions.[18]

Pius IV's bull was very much shaped and supported by the writings of Marc Antonio Bardo (fl. 1560s), a legal scholar who also served as an officer in the Presidenza delle Strade—the governmental department in which the president and masters of the streets served. Bardo delineated the legal foundations for the new bull in small treatises that described (and in some cases reprinted) prior papal bulls. Bardo, from Siena, was in Rome by 1557. He was the right-hand man of the president of the streets and the person who actually oversaw the magistracy and the masters.[19]

In 1565 Bardo wrote and had published three tracts on the streets of Rome. The first tract treated the legal basis for preemption of property, for construction, and for the redesign and remaking of piazze and streets, providing the legal foundation for *Inter multiples curas*. The volume began with twenty pages of indexes and tables. Clearly, it was intended for use by authorities such as the masters of the streets and the president to locate relevant legal precedents and procedures. Within the work, Sixtus IV's 1480 bull *Etse de cunctarum* was reprinted in full and accompanied by Bardo's extensive commentary. He also recounted the history of the office of master of the streets.[20]

Comparing the present pope to his fifteenth-century predecessor, Sixtus IV, Bardo extolled Pius IV's urban activities, pointing to the many buildings constructed under his rule. These structures added to the beauty of the city and were also useful. Further, public streets and the gates of the city had been newly constructed or enlarged and decorated. Bardo cited numerous law codes and texts in support of the clauses of *Inter multiples curas*. He followed his extensive commentary with twenty-one *quaestiones* concerning issues that were not verbatim from the Sixtine bull (Sixtus IV's *Etse de cunctarum*) but had arisen since in the mind of the pope and others and as a result of lawsuits.[21] By using the word *quaestiones*, a term from medieval scholasticism referring to the disputed questions to be discussed, Bardo elevated his discussion of the laws concerning streets to a philosophical discourse of the kind that occurred in contemporary universities.

Bardo's second publication on streets began with a description of the jurisdiction of the magistracy of the streets and of the masters themselves, including references to the ancient office and to the bulls of Martin V and Sixtus IV. He then devoted a chapter each to the other officers of the magistracy. They included the president, an assessor or auditor (i.e., his own job), a notary, the skilled architects and "geometers" called the submasters, the *depositarius* or treasurer, and commissioners to oversee particular places (such as the new street being built, the Via Pia). Also included are reprints of bulls and edicts relevant to the Roman streets beginning with the 1425 bull of Martin V and including Pius IV's bull *Inter multiplices curas* issued on August 23, 1565.[22]

In *Inter multiplices curas,* Pius IV reinstated the policies of his predecessors from the fifteenth and early sixteenth centuries (Sixtus IV and Leo X) that encouraged the building of elegant houses and palaces along new and newly straightened (and paved) streets, and he discouraged empty lots and ramshackle or even small houses. Yet the bull, issued on August 23, 1565, little more than three months before the pope's death, was as much a record of the failure of these policies as it was their renewal. Clauses of the bull revealed egregious abuses made under the auspices of the old laws. Clause four, for example, noted that "not without discontent of our mind," some had forced their neighbors to sell their houses or other property to them with the stated intent of expanding or building newly on the property and then had actually failed to build anything.[23]

The bull addressed disputes that had arisen and provided measures to prevent them. It specified the privileged position of neighbors in the sale of a property (they had first option to buy) and the rights of owners over tenants.[24] Despite this evidence of abuse, the bull reiterated that people could force their neighbors to sell their property to them. Although such property was usually occupied by a tenant, the bull remarks, the neighbor could also force the sale of a property occupied by an owner, or a property owned by a pious institution, as long as that property was outside the walled area of that institution. The essential context of this clause was the ever-expanding need for palace remodeling and construction on the part of Roman elites and the difficulty of finding space to appropriately expand and remodel existing structures.[25]

The bull called for the demolition or forced sale of humble, half-ruined, and half-built structures and vacant lots to prospective builders. Many such structures had become depositories for "dirt and filth."[26]

The disposal of human and animal waste was a particular concern. "In the cherished city of Rome," one clause declared, "there are a great number of alleys, two or three palms wide, or thereabout [ca. 2 ft.], in which dirt and filth are thrown repeatedly." Moreover the flanking walls, deteriorated by putrefaction, "bring a foul stench with the production of pestilence and bad air." Eventually the walls collapsed. No one was allowed to build other structures on such walls. If they did, they would be exiled from the Papal States and their goods would be confiscated. In order to eliminate said walls and alleys, the first among neighbors that wanted to build there (referring presumably to building an extension to their own house), after having obtained a license from the masters, even without having paid the price, could lay boards, quarry stones, and obtain other construction materials in order to do so.[27]

The bull admonished that "because very many houses of this city of Rome have certain small drains, commonly called *clavichettas,* discharging water and sewage from those houses into streets and public places," it ordered such drains to be closed within six months. Whatever other "latrines or water pipes and sluices, commonly called *sciacquatoria,* and cisterns" that flowed into public streets and alleys and squares also must be closed. People could build such water pipes in their own courtyards but not so that they flowed into public streets or squares.[28]

Overgrown vegetation was a problem. *Inter multiplices curas* admonished that the city was marred by "many thickets of rushes," which spoiled the urban landscape, prevented pure air from circulating, obstructed a clear view of the river, and provided opportunities for the commission of crimes. Such rushes were to be uprooted within six months and replaced with vineyards and orchards. The bull also commanded that the Via Angelica (a new street that ran from the Vatican Porta Sant'Angelo to the north) could not be obstructed by thickets of rushes or by vines and orchards. Indeed, along this street only meadows were allowed. Especially in the Borgo and Borgo Pia (the expansion of the Borgo created by Pius IV), irrigation of gardens created bad air. Thus, all gardens except for artichoke gardens (which did not require irrigation), were to be destroyed and the land cultivated back into vineyards and meadows. Further, in the city itself, vegetable gardens and orchards inside of apartment blocks must be enclosed by walls.[29] Thus, the pope legislated for clean (and dry) air following the theory of miasma that viewed bad air as the major cause of illness. (From our own modern point of view and framed in different terms, Pius IV was correct. Irrigation systems produced water—which enabled more mosquitoes to breed and thus more *mal aria* [bad air].)

The bull legislated that street obstructions such as stairs and porticoes be removed, and it penalized construction workers who built such obstructions with exile and the confiscation of their property. Streets must be wide enough for two coaches to pass.[30]

Complaining that artisans were spread throughout the city, the bull demanded a reform in which each craft occupy a particular area. Especially crafts such as candlemaking, tanning, and butchering that produced noxious odors (miasmas) and waste materials (such as animal guts) must be confined to the area between Ponte Sisto and Tiber Island.[31]

Had its provisions been fully carried out, *Inter multiplices curas* would have caused tremendous disruption not to mention hostility among the population. The destruction of gardens, the forced relocation of artisans, the compelling of families and individuals to move or sell their houses, and the destruction of stairs, balconies, porticoes, and street bridges all pointed to the desire for unobstructed streets but also amounted to an autocratic disregard for neighborhoods and various groups of people within the city. The extent to which the provisions of the bull were carried out is very unclear. Certainly the death of Pope Pius IV on December 9, 1565, little more than three months after the bull's promulgation, would have radically diminished its effect. In the end, it undoubtedly stands as a detailed source for understanding the urban conditions of Rome as they were rather than as an indication of changes actually put into effect.

Pius V's Revisions of Street Law

Pius IV's successor, Michele Ghislieri, called Pius V, achieved a successful act of delegation when he created the Congregation on Streets, Bridges, and Fountains (Congregazione super viis, pontibus et fontibus), led by cardinals Montepulciano and Sforza. The Congregation superimposed another layer of oversight over the president and masters of the streets. It usually met monthly from its first meeting on April 4, 1567, until its last on February 3, 1588. During Cardinal Montepulciano's lifetime (he died in 1574), it always met in his palace on Via Giulia. Montepulciano did not miss a single meeting, and there is reason to believe that he was the congregation's effective leader. The minutes of the meetings of the congregation are extant and represent a highly informative document relevant to over twenty years of work on streets, sewers, and other urban infrastructure issues.[32]

Pius V ended the policies of Sixtus IV, Leo X, and Pius IV that allowed owners of houses to expropriate the houses and other real property of their next-door neighbors in order to enlarge their own palaces. His bull *Reductio ad terminus iuris,* issued on April 10, 1571, rescinded the laws because "several lawsuits, questions, controversies about plots, and various grievances and hurts, disturbances and inconveniences were generated and brought forward by several persons." Further, daily complaints were brought to the pope concerning the injustice of these laws.[33] Pius's revocation of these policies had important exceptions. Expropriations needed to enlarge and remodel places of worship, and public buildings, streets, and piazze could still force owners to relinquish their property.[34]

Gregory XIII's Return to Pius IV's Policies

Pius V's reinstitution of the property rights of neighbors did not last long. The next pope, Gregory XIII, soon returned to aggressive policies of urban expansion and beautification, reversing Pius V's reversal. Gregory's bull of October 1, 1574, *Quae publice utilia,* gave the *camerlengo* of the Camera Apostolica and the masters of the streets full authority to authorize (and force) the sale of property both for the widening and straightening of streets and for the construction or expansion of private houses or palaces. As in Pius IV's *Inter multiplices curas,* neighbors were to be given first choice of purchase, various contingencies (such as two neighbors wanting the same property) were provided for, and in cases of conflict, the advantage and ornamentation of the city itself was to take first priority.[35]

Perhaps as important, in 1580, under Gregory XIII, a revised law code for the city of Rome was put into effect. The code—the work of the Roman magistrate Luca Peto—regularized procedures to be followed by the masters of the streets in matters of construction and renovation. The architect of the Roman people must inspect the work to be done and was required to create a drawing or plan and carefully calculate expenses. Specified members of the public (such as affected neighbors) as well as relevant officials must be notified of all renovations or planned constructions including explanations of the work to be undertaken and the expense. The statute detailed how taxes to finance the project were to be collected and distributed and how contractors were to be chosen. In the sale of houses, neighbors were given priority.[36]

The 1580 law made street paving mandatory. Throwing of refuse onto the streets was prohibited. Owners of houses on the street had to be informed at least eight days before scheduled paving work and had the opportunity to state why such paving should not be carried out. (Clearly they might not want paving because they did not want to pay for it.) If owners wanted to do the work themselves, they were required to complete it within a certain amount of time. Workers who paved the street must not work at night, must not mix old, worn-out stones with new ones, must use well-fired bricks, and must use excellent lime and well-mixed sand for mortar. If workers used substandard materials they would be forced to redo the work at their own expense and pay a fine. Further, the masters must in no way allow manure and sewage to be thrown on the street. Fires of malodorous things were forbidden, especially in summer. Neither the masters of the streets nor the submasters were permitted to accept materials or even food or drink from the masons and bricklayers who paved the streets (the law seems to be saying that the masters could not accept bribes in exchange for paving contracts).[37]

Five years later, on January 30, 1585, the masters of the streets issued a "General Edict" with sixty-one clauses to ensure that "the roads and streets of this city of Rome and its territory are maintained clean and repaired for the health of the inhabitants and the universal welfare." The comprehensive regulations seemed to cover all bases, each clause laying out a regulation and establishing the penalty for violation. Some of the clauses dealt with sewage and animal manure. Carters were forbidden to carry manure in their wagons without a net underneath and a cover on top. No one could throw manure or filth into the street in front of their houses or into other streets or courses of water. It was forbidden to throw straw into the street or burn it there, and it was forbidden to throw filth from windows or balconies. Pipes carrying sewage must not discharge into the streets or into alleys that opened onto streets. *Muratori* (masons) and others could only empty cisterns and sewers at night and only with a license from the masters of the streets.[38]

Another group of regulations dealt with artisanal production and sales. Artisans must not throw rubbish in front of their shops, nor could their counters extend into the street farther than two palms (about 17½ in.). They must have a license to build such a counter. Butchers could not strip flesh from skins or do anything that put putrid material in the street. Candlemakers could not cook (animal fat) in their houses where their shops were but only in designated locations. People in general were forbidden from washing clothes in the street or stringing clotheslines across the streets even high up.[39]

Debris and obstructions from building construction received much attention. Sellers of wood and pozzolana, carters of stone, muleteers selling lumber and other materials were all regulated in terms of where they could place their materials (not in the street). They were forbidden to sell without a license. Stonecutters, lime burners, rope makers, and other workers must remove their materials from the streets. Digging in the street was strictly forbidden without a license; existing holes must be filled. There could be no bricklaying or other building in a public place without a license, and bricks for all public building must be well made. Carters could not throw stone or gravel or manure into the street.[40]

Familiar regulations dealing with animals were repeated. Butchers and cattlemen could not lead cows or oxen through the densely populated areas of the city. Letting pigs or goats run in the street was forbidden. Huge wagons pulled by oxen could not be led through populated areas without express license from the masters of the streets.[41] These comprehensive regulations for the street provide a window into the chaotic Roman urban landscape. The extent to which they were promulgated and enforced is unknown, but it is relevant that shortly after they were articulated, the pope died.

The masters of the streets worked under the *camerlengo* and the pope, but they also enjoyed tremendous discretionary powers. What is emphasized in most accounts is that they now operated under the authority of the popes rather than the Capitoline Council. This was the case, but they should not therefore be seen as administrators in a modern bureaucracy. Often masters of the streets would have served in the communal government in other capacities—as members of the Capitoline Council, for instance, or even as one of the three conservators. Eleonora Canepari especially has shown how the great power and authority of urban elites such as the masters was firmly tied to intricate and often long-standing ties to the artisan classes in the *rioni* in which they lived. The masters are recorded as virtually always attending both the secret and the public council meetings of the Capitoline government. They were formally controlled by the pope, the president of the streets, and after 1568, overseen by the cardinals heading the Congregation of Streets, Bridges, and Fountains. Nevertheless, they would have enjoyed great local power and autonomy, particularly in the many cases in which the pope and cardinals did not have direct interest, sufficiently detailed knowledge, or adequate local ties.[42]

Sixtus V's Organizational Reforms

Sixtus V, famously known for his extensive and effective efforts at urban reform, extended his activities beyond physical reconstruction to administrative reorganization. On December 1, 1586, less than a year after his inauguration, he issued the bull *Ut primum potestas,* making the Borgo (the district around the Vatican) into the fourteenth *rione.* Civic Rome and papal Rome were now, at least on paper, one entity.[43] The pope thereby attempted to place the entire city under more effective papal control by creating a more unified administrative structure.

Sixtus V also abolished the Congregation of Streets, Bridges, and Fountains that had functioned more or less continuously since 1567. In its place, on January 22, 1588, he issued the bull *Immensa aeterni Dei,* which created fifteen permanent congregations, including the Congregation for Streets, Bridges, and Waters (*Congregatio decimatertia pro viis, pontibus, et aquis curandis*), which was in effect a continuation of its predecessor. In so doing, he made the congregations, which previously had been created as temporary expedients (although often quite long standing) into permanent governmental bodies, each headed by one or more cardinals—an organizational structure that remains to this day.[44]

In very much the same spirit, on March 7, 1587, Sixtus V changed the office of the masters of the streets from two to fourteen, one for each of the fourteen *rione.* This reform can be seen as part of his effort to exert more effective control over the city and was undoubtedly a response to what must have been the wide-ranging power and local autonomy of the patrician masters. The master of the street for each *rione* would be responsible for keeping the district in good condition and clean. At the same time, "we entirely dismiss and remove both [of the previous masters] from this office." The new masters would have full authority without need of petitions over both public and private places. Their responsibilities included the care of buildings, walls, wooden structures, water, channels, and drains. These new masters would pursue their offices honestly and intelligently and would be in charge for a year. Further, they would elect two from their number to act as judges.[45]

Each of the fourteen new masters of the streets would be paid a salary of ten scudi a month from the *gabella dello studio,* often called the wine tax. This tax had been established to fund the university (the *studio*) by the Pope Eugenio IV (ruled 1431–1447) through a papal bull, *A supremo paterfamilias,* on February, 7, 1432. It was called the wine tax because it was a levy on foreign wine imported into the city.

Its proceeds could be used directly but were more often used to fund the interest on a bond issue called a *monte,* in this case the *monte dello studio.* Since its initial enactment, its proceeds had often been diverted from the university to other uses.[46]

Despite its apparently secure funding, Sixtus V's new organization of the masters and the general authority of the new congregation for the streets was not long lasting. The new congregation did not maintain its authority over the streets but soon exercised power only over ecclesiastical disputes. After the death of Pope Sixtus V on August 27, 1590, there were three successive conclaves and three popes. Urban VII (September 1590) ruled for twelve days, Gregory IV (1590–1591) for almost a year, and Innocent IX (1591) for sixty-two days. After this rapid succession of conclaves and short-lived popes (during which time we can safely assume that little work was done on the streets), Clement VIII Aldobrandi (1592–1605) was elected. By 1594, the office of the masters of the streets had reverted to two men—two Roman patricians—each elected for the term of a year.[47]

Sewers, Drains, and Waste Disposal

A close-up examination of a single issue—waste disposal and sewage—during a few years allows a better understanding of the decisions made and practices carried out on a daily basis. This is possible because of the extant archival documents, including contracts given to skilled workers for particular projects as well as the minutes of the Congregation on Streets, Bridges, and Fountains, which, because it functioned from 1567 to 1588, allows the investigation of a twenty-year period from the perspective of one particular kind of urban problem.[48]

In the late sixteenth century, much of Rome's waste was not disposed of through sewers. Yet as Katherine Rinne has pointed out, the great number of terms used in this period for sewers and drains can lead to confusion. Those terms—including *chiavica, chiavichetta, chivicone, canale, cloaca, fogna,* and *fossa*—could refer to a variety of drainage structures, such as closed pipes for drainage of fountains or large pipes that were sewers or simple open ditches that served as drains or sewers (or both).[49] It is not always clear to which kind of structure various documents refer. A further complication is that drains that funnel water (e.g., from an aqueduct) and sewers that convey sewage are distinct, but they could and often did function as both.

A sewer could be an open trench, creating a tremendous stench as its putrid matter descended to the Tiber. Even closed sewers could and did become clogged with dirt, debris, and fetid waste. During floods, as Luca Peto observed in his 1570 description of the flood of 1557, the river water flowed up the sewers. Such backflow would carry massive amounts of sewage onto the streets of Rome.[50]

In the late sixteenth century, even when the existing sewers were not clogged

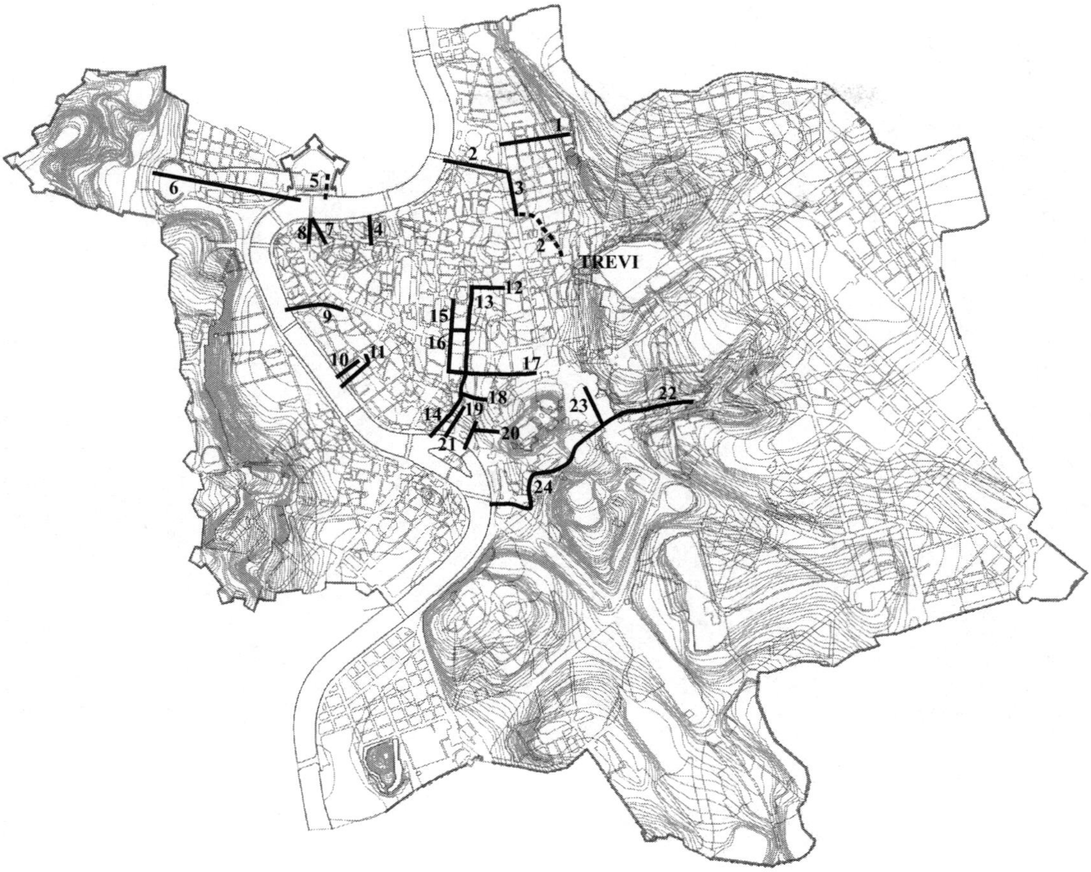

Figure 2.1 Most important documented sewers in late sixteenth-century Rome: *1,* Cloaca di San Sebastianello; *2,* Chiavica di San Silvestro; *3,* Chiavica dell'Arco di Portogallo; *4,* Chiavica di Tor di Nona; *5,* Chiavica di Castel Sant'Angelo; *6,* Chiavica del Borgo; *7,* Chiavica di Panico; *8,* Chiavica di Ponte; *9,* Chiavica di Santa Lucia; *10,* Chiavica di Palazzo Farnese; *11,* Chiavica di Campo de' Fiori; *12,* Chiavica della Minerva; *13,* Chiavica dell'Olmo; *14,* Chiavica della Giudea; *15,* Chiavica della Rotonda; *16,* Chiavica della Ciambella; *17,* Chiavica dei Calcarari; *18,* Sewer in front of Santa Caterina dei Funari; *19,* Cloaca Maxima di Sant'Ambrogio in Pescheria; *20,* Cloaca sotto Piazza Montanara; *21,* Cloaca di Portico d'Ottavia; *22,* Chiavica della Suburra; *23,* Chiavica della Spoglia di Cristo; *24,* Cloaca Maxima. Diagram © 2017 by Katherine Wentworth Rinne with kind permission.

or otherwise damaged—and therefore functioned as conduits for the discharge of waste—as is clear from the sewer map shown in figure 2.1, they hardly provided sufficient coverage for the entire city or for even half the city. Rome was not a modern networked city—to use a concept developed by historians of technology and urbanism. As Leslie Tomory aptly described in an important study (on London's hydraulic works), such a city has "a materially integrated infrastructure, with a particular focus on the built urban environment." Tomory has suggested that the hydraulic system in London in the late seventeenth century might make that city an example of the first networked city. He refers to the hydraulic system gradually

Figure 2.2 Cloaca Maxima looking east, Augustan-era tract, below Forum Transitorium. Photograph © 2017 by John North Hopkins with kind permission.

created by London's New River Company after 1660, in which a huge expansion of customers forced the company to design each new component in light of the rest of the network. Tomory's analysis moves the timeline of the networked city back two hundred years, from the nineteenth century to the seventeenth.[51]

But not to sixteenth-century Rome. Here, a common type of drain was a surface-level channel running down the middle or sides of a street to carry away rainwater often clogged with debris, despite prohibitions to the contrary. Although water still flowed through the famous ancient drain, the Cloaca Maxima (fig. 2.2), most other ancient drains were filled with dirt and rubbish deep underground and had long since ceased to function. They were often too deep to rehabilitate, if they could be found at all, since the ground level had risen at least ten meters in the intervening centuries.[52]

Nevertheless, sewers and drains did exist alongside other means of waste disposal. Using a combination of archival documents and archaeological reports, I have been able to document twenty-four, more or less functioning sewers and/or drains in late sixteenth-century Rome.[53] It can be seen that with the exception of the two short sewers draining into the Chiavica del Olmo, and of the Chiavica della Suburra running into the *Cloaca Maxima,* these conduits were not connected to each other or to any other integrated system. They consisted of limited separate drainage systems that served discrete parts of the city or particular palaces.

The Chiavica di San Silvestro

The issues and problems surrounding urban sewers in sixteenth-century Rome can be underscored by a closer look at one of the longest sewer/drains in the city, the Chiavica di San Silvestro (fig. 2.1, 2), which ran from the Trevi Fountain through the Campo Marzio to the river. For most of the sixteenth century, it was an open sewer about 1.24 miles (2 km) long. A tax record of 1538 shows that taxes were charged to 407 houses along the sewer for its repair or cleaning, including the Monastery of San Silvestro. Some of these houses were as far away as the Piazza Colonna and connected by an open drain that ran down the Via del Corso.[54] Yet nearly thirty years later, on December 3, 1567, the Capitoline Council recorded that in the year 1566, "the same water going outside of the conduit of the Acqua Vergine of Trevi, through a ditch" and "passing through the gardens of the Monks of the Convertite [Santa Maria Maddalena delle Convertite, a convent for reformed prostitutes] and of San Silvestro, generating very bad air," caused many people to be infected and as many died as died of the *peste* (plague). But now, the record continued, it had been repaired. The council, "knowing that for the future it is necessary to diligently care for it so that it does not fall into similar disorder," made one Vincenzo Bellincino (who lived nearby and who had worked hard to help the affected families) "custodian of the said sewer, ditch, and water." The council further commanded that "all owners of the houses, gardens, and lands where the sewer ditch (*cloaca fossa*) passes or will pass must not throw or cause to be thrown any sort of filth or other thing" into the drain.[55]

On September 18, 1568, it was noted that vaulting for the Chiavica di San Silvestro had been completed and that money designated for the Acqua Vergine was used to pay for it.[56] This record seems to suggest that the problem had been fixed and that the formerly open sewer was now covered. Yet problems remained. On October 12, 1568, the congregation instructs "the architects" that in their visit to the hospital of San Giacomo degli Incurabili (on other business), they could also make a decision about the sewer (which ran next to the hospital and was used by it).[57]

It is not clear what condition the sewer was in, but whatever had been done was only partial, because on January 11, 1569, the Congregation on Streets, Bridges, and Fountains considered the outcry of the neighbors "about the very great stench, cause both of deaths and of insupportable damage in the future."[58] In April 1569 the issue of open sewage was again on the table. A magistrate, Paolo del Bufalo

(d. ca. 1588), and Alessandro Uberti, secretary of the governor of Rome, informed the congregation "that if the sewage is not cleaned in the channel under the Trinità and also the sewer of San Silvestro not completed, the usual pestilent influence will come."[59] In July a request was made for the "systematization" of the *cloaca* by Paolo del Bufalo and neighbors because it was unsafe. A tax for the work was imposed for those affected.[60] By 1570, the open sewer of San Silvestro had been cleaned and closed as part of the completion of the Acqua Vergine and served as the drainage conduit of Acqua Vergine water into the Tiber River—a great amelioration of a long-standing problem of sanitation in the center of the city.[61]

So there were improvements. Still, late sixteenth-century Rome was far from achieving a modern, networked system of sewers and drains. Rather, the city operated under an eclectic system that included enclosed sewers and drains, open sewers, cesspits (*pozzi neri*) that frequently overflowed, and the dumping of garbage and sewage directly onto the street or into narrow alleys. Some of this fetid material was removed by carters and street cleaners and dumped into the river, but clearly much remained in the streets and alleys, creating putrid, muddy quagmires, especially during and after rain.[62] On September 22, 1574, the masters of the streets, it was noted, should collect a tax for the Cloaca di San Silvestro for the payment of newly hired workmen.[63] No funds for ongoing maintenance meant that a new house-to-house tax collection had to be made for this purpose.

On the basis of literary sources, Douglas Biow has demonstrated the Italian Renaissance obsession with cleanliness. (Although Biow's study shows in equal measure an interest in, almost an attraction to filth.) The very different evidence of the present study shows not that Romans were indifferent to filth in the streets and alleys, but indeed that they made (at least some of the time) valiant efforts to remove sewage and other kinds of waste.[64]

The minutes of the Congregation on Streets, Bridges, and Fountains adds to the evidence of the fundamental concern for waste disposal and clean streets. In the twenty years that the congregation met on a monthly basis, rarely a meeting went by without dealing with an issue involving sewage or street cleaning. Examples include removing mud and manure from the city walls; waste removal in general and the collection of taxes to effect it; the punishment of carters for throwing sewage into the Tiber; prohibitions against throwing sewage, manure, and waste into the streets; contracts given to street cleaners; prohibition of pipes draining sewage from houses and palaces into the street; edicts against illegal dumping of sewage and debris into the Colosseum and elsewhere; orders to proprietors to put waste in front of their houses on certain days for pickup; orders to proprietors to pay their waste-pickup taxes or remove their own waste without using city carters; and discussions about whether the number of carters should be increased.[65]

The congregation's efforts to remove mud, sewage, and debris from the streets and alleys of Rome required an ongoing effort. But it was also piecemeal and not highly successful—largely, in my opinion, for structural reasons. The financing for street cleaning and for paving was contingent upon collecting taxes from the proprietors of shops, houses, and palaces on the streets affected. Tax collection and street cleaning were not sharply separated operations as they are in modern cities. In sixteenth-century Rome, street cleaners could be carters with contracts with the city or contractors who supervised their own carters. These contracts had to be renewed and taxes collected door-to-door, again and again. There were no permanent funds. Worker requests to the congregation for pay owed to them were common. There was no permanent agency with paid employees who cleaned the streets and collected waste. The result was that the high-level congregation again and again occupied itself with particular streets, individual disposal issues, and particular contracts. And virtually every meeting of the congregation was taken up in part by requests for exemption from the required street cleaning (and street paving) taxes by poor widows and others claiming poverty—requests that must have taken up an inordinate amount of time.[66]

◆ ◆

Rome in the late sixteenth century was not widely different from many other cities in the same period—it was an early modern city, not a modern one. It would be anachronistic to measure Roman sanitation in the late sixteenth century from the point of view of modern standards. Ideas about hygiene and sanitation have varied greatly from one historical culture to another. Numerous premodern cities in addition to Rome struggled with similar problems related to street cleaning and waste removal. In premodern times, structures and practices differed from modern ones, as did ideas, attitudes, and beliefs about such practices. It is well to recall that before the 1880s, there was no understanding of the germ theory of disease or bacteriology, and therefore little of the modern understanding of the connection between human waste and the transmission of disease. The miasmatic theory of disease viewed visible filth and smells as disease-causing entities, but if water appeared clean then it usually was considered clean.[67]

Describing Rome a hundred years after this period, Richard Krautheimer takes account of the new palaces on the outskirts but adds that "the inner core of the city retained its old narrow, dark and dank streets, winding and crooked, many unpaved and muddy, filled with refuse and impassable after a rain."[68] Renato Sansa, whose important work on Rome's sanitation focuses on the seventeenth and eighteenth centuries, is surely correct in rejecting the stereotype of preindustrial cities, including Rome, as indifferent to issues of sanitation. But his work also explicates a mixed bag of methods for achieving the elimination of urban waste materials as well as a lack of success. This failure in Rome is evidenced by the eighteenth-

Figure 2.3 Eighteenth-century street sign prohibiting the throwing of garbage and sewage into the alley. Vicolo della Luce at the corner of Via della Lungaretta, Trastevere, Rome. "By order of the Very Illustrious and Reverend Monsignor President of the Streets, it is forbidden to throw, carry, and dump filth into any part of this alley, and it is forbidden to let a rubbish heap accumulate here under penalty of ten scudi and other, corporal [punishment] according to judgment as [established] by the edict of December 30, 1763." Photograph May 2017 © Bob Korn with kind permission.

century incised stones that can still be seen on many walls in the central area of the city, signed by the masters and/or the president of the streets, ordering citizens not to throw *immondizia* into the streets (fig. 2.3). Such signs indicate continued efforts to create a cleaner city but also provide evidence for the continuation into the eighteenth century of the centuries-long practice of throwing human waste and garbage into the streets.[69]

In sixteenth-century Rome, sanitation was clearly a concern, but a fully integrated urban system of sanitation supported by public funding was not even envisioned let alone approached in practice. Sanitation efforts often had as much to do with the aesthetics of a magnificent city as with notions of public health. (Plagues and outbreaks of disease were certainly a concern, but they were not necessarily closely associated in the minds of contemporaries with sewage and other filth on the streets or in the water.) Whatever its motivation, efforts to keep the streets of Rome free from filth on a permanent basis were not successful. The streets of late sixteenth-century Rome were often filled with mud, sewage from humans and animals, the putrid remains of animals, and rubble.

3

Repairing the Acqua Vergine

Conflict and Process

By the autumn of 1566, thousands of scudi had been paid out to the architect in charge of the massive undertaking that was to restore fully the only ancient aqueduct that still (barely) functioned—the Acqua Vergine. Yet an evaluation in December found the project to be in utter disarray. The pope who had initiated the visionary project, Pius IV, had died a year before, no doubt deeply disappointed that his farsighted undertaking had come to nothing. Antonio Trevisi, the architect in question, was taken to prison. He reportedly died there, probably shortly thereafter.

Yet four years later, in 1570, this ancient aqueduct to the Trevi Fountain, most of which ran underground, had been repaired. A plentiful new supply of water from the Salone Springs—about 6.5 miles (10.5 km) east of the city near Tivoli—now flowed into Rome, supplying fresh water to the burgeoning population for drinking and washing. By the mid-1580s, nine fountains flowing with aqueduct water had transformed the center of Rome. The Acqua Vergine also provided the luxury of running water to the palaces of those wealthy individuals who lived in the low-lying Campo Marzio and to a few nonwealthy beneficiaries as well. Giacomo Della Porta (1532–1602), Bartolomeo Grippetto (also known as Bartolomeo Gritti or Griptus [1510–1584]), and contractors working under them constructed new fountains that still grace the center of Rome and that are still fed by the waters of the Acqua Vergine. They remain today as elegant and characteristic features of modern Rome.[1]

The actual project of reconstruction was begun in 1561 under Pius IV and finally accomplished under Pius V in 1570. It is a story of years of frustration, bitterness, failure, conflict, and eventual success.

The Ancient Infrastructure

Rome's location was ideally suited for aqueduct construction. The city is surrounded by hills and mountains shaped by prehistoric volcanoes that contain large volcanic lakes—Lake Bracciano to the north and Lakes Albano and Nemi to the northwest in the Alban Hills. These lakes along with numerous springs and streams are located at altitudes higher than the city itself. Thus, the city's physical geography facilitated the construction of gravity-powered aqueducts, which must be designed with a downward slope and be graded to allow a regular, even flow from the source to the destination (neither too steep, which would erode the aqueduct walls and destabilize the structure, especially at bends, nor too shallow, which would inhibit the flow of water). To maintain the gradient, the ancient Romans built the aqueducts underground or above, sometimes on high arches, depending on the terrain.[2]

Between 312 BCE and 226 CE, the ancient Romans had constructed eleven aqueducts that supplied fresh water to Rome. At that time, during the reign of Emperor Augustus (ruled 27 BCE–14 CE), Rome had a population of one million. The ancient Romans needed the aqueducts to supply their enormous and culturally important bath complexes as well as fountains throughout the city. The construction of an aqueduct constituted an immense undertaking that provided economic and social benefits to rulers and patrons.[3] Contrary to popular belief, what was not high on the ancient Roman list of priorities was the public health of ordinary people. The very low average life expectancy in ancient Rome—about twenty-five to thirty years due to the high rate of infant mortality—resulted in part from grossly unsanitary conditions of the streets and ordinary dwellings. The often idealized view of ancient Rome as a water-cleansed, sanitary city is highly romantic and far from historical reality.[4]

Once an aqueduct was built, the channel required constant maintenance to keep it clean, to remove the calcium carbonate encrustation (sometimes called sinter) that formed inside the conduits, and to repair cracks and leaks. This work required that the underground sections of the aqueduct be built with intermittent vertical access shafts called *putei*, which were usually covered at the surface with stone or wooden slabs. In ancient times the workforce for maintaining the aqueducts ranged from 240 to 700 slaves supervised by the curator of the aqueducts

Figure 3.1 Remains of the Aqua Claudia in the Parco degli Acquedotti, Rome. The channel, or *specus*, of the Claudia is visible, above which can be seen a remnant of the ruined channel of the Anio Novus. Photograph May 2017 © Bob Korn with kind permission.

(*curator aquarum*). Julius Frontinus, who wrote a tract on aqueducts in the first century CE (*De aquae ductu urbis Romae*), was a *curator aquarum*.[5]

As the population of Rome declined in the early medieval period, the aqueducts deteriorated. Despite sporadic papal efforts to carry out repairs, by the eleventh century only the mostly underground Acqua Vergine remained a (partially) functioning conduit. Nicholas V saw to the most notable repair of the Acqua Vergine in the mid-fifteenth century. This low-lying aqueduct ended at the Trevi Fountain, which in the sixteenth century looked far different from the eighteenth-century extravaganza we see today.[6] Nonfunctioning remnants of the other aqueducts remained, including spectacular arched structures, as strikingly visible in the sixteenth century as they are today (fig. 3.1) and reminders of ancient Roman engineering skill as well as of the slave labor that built and maintained them.

The Acqua Vergine (called the Aqua Virgo by the ancient Romans) ran from its origins in the Salone Springs east of Rome and then took a sharp turn and entered the city from the north by way of the Pincian Hill. The ancient Romans cut much of the channel in a course that ran as deep as 131 feet (40 m) underground to create the correct gradation for the gravity-powered flow of water from the springs to the city.[7]

Where the channel traversed sections of gravel or clay, the ancients built a solid floor, walls, and ceiling. Elsewhere, they tunneled through tuff, the volcanic rock characteristic of the area. Instead of settling tanks, they constructed an occasional bend or zigzag in the channel to slow the flow of the water, which would allow sand and rock to settle out. At each bend and periodically elsewhere along the channel, they constructed shafts to allow access for cleaning, repair, and removal of sand, stones, and sinter. The Acqua Vergine was unusual because it received its water supply not only from the Salone Springs but also from other springs and streams along its entire course, thereby serving as a channel into the city for many sources of nearby potable water.[8]

Agostino Steuco: Learned Humanist and Urban Reformer

By the sixteenth century, the ancient conduits within the city walls were almost entirely destroyed, and the actual course of the Acqua Vergine had become a mystery. Because of the hidden, underground course of the conduit for most of its route, the location of its primary source springs had been forgotten. The first person in the sixteenth century to solve the mystery was Agostino Steuco (1496–1549), a learned theologian and humanist highly skilled in Greek and Latin. Paul III appointed him librarian of the Vatican Library in 1538. Steuco vacated his library post for several months in 1545 to go out into the Roman countryside to search for the lost conduit.[9]

Early in his career he had written biblical commentaries and polemical tracts against the Protestants, but when he arrived in Rome in 1535, he turned to problems of urban renewal and hydraulic engineering. As Ronald Delph has shown, Steuco began thinking about the city as a whole and how it could be transformed to better reflect the glory of the popes and the Catholic Church. He adopted the humanist view that prevailed in Rome in the 1530s and 1540s—that the popes were the legitimate heirs of the Roman *imperium* and should rightly work to return the city to its ancient splendor. In small printed tracts he proposed urban improvements. One of these was the construction of a wide avenue from the river port of Ripetta to the Porta Flaminia (the gate at Piazza del Popolo) and from there to the Capitoline. This avenue would be ornamented with dramatic fountains and renamed the Via Pauli, after Pope Paul III. Finally, he wanted the Acqua Vergine to be repaired and made fully functional so that water might be provided for the proposed new fountains.[10]

Before it could be repaired, the actual course of the ancient structure had to be located. Although the conduit appeared to be fed by a spring about a mile from the city boundaries, Frontinus (*De aquae ductu urbis Romae,* 1.10) had stated that the source was, rather, at the eighth (ancient) Roman milestone from the city to the east.[11]

Steuco tracked the course of the entire aqueduct by following extant vestiges of the ancient structure, including the trail of airshafts located about every 100 yards. It was a difficult task made more difficult because farmers had frequently filled the shafts with dirt to prevent their animals from falling into them. But by finding the shafts, one by one, Steuco arrived at the Salone Springs. He discovered that the Romans originally had surrounded the springs with an enclosure lined with waterproof hydraulic concrete (*opus signinum*) that funneled the water to the aqueduct but that the concrete had disintegrated, and now the water flowed into a tributary of the Aniene River.[12]

After his successful search, Steuco emphasized the importance of his own experience in relocating the ancient aqueduct and then (so he hoped), having it reconstructed. He assured readers that "not only reason itself and the contemplation of things, but also experience and practice and evidence examined with my own eyes" inspired him to think that the aqueduct could be restored "with minor expense."[13] Steuco wrote a treatise about his discovery, establishing himself as a serious student of Roman archaeology and of classical scholarship.

In his tract, he referred to the skilled architects who accompanied him: "And according to the architects themselves who were in charge [and] who often had meetings to examine these things and evaluate them with care, the whole expense would come to the sum of 15,000 scudi of gold." He further referred to "consultants and experts" who had advised that once the water was brought back to the city, the expenses could be recouped immediately from selling it to those wishing to pipe it into their houses, as once had been the Roman custom.[14] Steuco's forward-looking desire to restore the aqueduct was not one that he would live to see accomplished. Paul III's desire to repair the Acqua Vergine remained unfulfilled, but the project would be taken up again in the early 1560s.

Pius IV's Urban Failure

Giovanni Angelo Medici, like Agostino Steuco, was a protégé of Paul III Farnese. Giovanni Angelo traveled to Rome from time to time in the 1530s and 1540s and may well have been acquainted with Steuco and his exploration of the aqueduct. After he became Pius IV in 1559, he seems to have immediately taken up the task of reconstructing the Acqua Vergine. He raised money for the project and created an oversight commission, including Cardinal Montepulciano.[15]

The repair of the Acqua Vergine turned out to be more difficult than Steuco had envisioned. It represented a large-scale and immensely costly public-works project of a kind that was uncharted territory for late sixteenth-century Rome. The undertaking produced much interest, bitter conflict, and for Pius IV, a devastating failure. Some aspects of the project during Pius IV's reign are obscure. Although some contracts are extant, account books are not. Some reports about what happened were written years later, and it is not always clear when events referred to took place.

The protagonists included Pirro Ligorio, whose interest in waters is evidenced in his encyclopedia volume on water (located in Naples) discussed in chapter 1;

the magistrate Luca Peto, who had written about flood control; the architect/engineer Antonio Trevisi, who had promoted a large-scale canalization project in his treatise on the Tiber River and on the Bufalini map and who ultimately was given the contract to repair the aqueduct; the Capitoline Council, which sent its own representatives to inspect the project (including Luca Peto); and the Camera Apostolica and the pope.

Pirro Ligorio's Scathing Account

In the early 1560s, Pirro Ligorio served as architect of St. Peter's and the Vatican palace (fig. 3.2). He oversaw the construction and decoration of Pius IV's loggia, which was part of that palace. He also guided the design and construction of the beautiful building in the Vatican gardens called the Casino and had it decorated with ancient sculptures and busts, many of which he had discovered during his archaeological digs. Ligorio was an avid investigator of ancient Roman ruins and antiquities and one of the most knowledgeable antiquarians in Rome. He was extremely well placed during the rule of Pius IV and seems to have avidly promoted the repair of the Acqua Vergine.[16]

It was Ligorio's misfortune that Pius IV's successor, Pius V, actively opposed the humanist and antiquarian pursuits of his predecessor and thus of Pirro Ligorio. In light of this unfavorable new atmosphere, Ligorio left Rome to return to Este family patronage, continuing his work on the renowned gardens and fountains at the Villa d'Este in Tivoli and then moving on to Ferrara to work for Alfonso II d'Este (1533–1597).[17]

Ligorio wrote his account of the Acqua Vergine after he was forced to leave Rome. It appears in his voluminous encyclopedia of antiquities housed in the Archivio di Stato of Turin — a many volume, handwritten work that he created primarily in the 1570s. From his account, it is clear that he had urged the repair of the Acqua Vergine and had been rebuffed. In his encyclopedia, under the headword "Acqua Vergine," he provided a short history of the aqueduct and then announced that "the same Pirro Ligorio who has written this work had made a proposal to Pope Pius IV so that its own spring [the Salone Springs] might be channeled."[18]

Here Ligorio lambasted the Roman magistrate Luca Peto, who had become one of the overseers of the Acqua Vergine project and who wrote an account of its ultimate success without any mention of Pirro Ligorio. According to Ligorio, Luca Peto prevented his (Ligorio's) idea from going forward. Indeed, "the great doctor Peto was against this work," Ligorio raged, and "almost caused such a worthy project to fail, denying that water could come to Rome as if it had not already been in Rome." Finally, "having seen that the thing succeeded, he wrote the work in which he boasts that he did everything, so great were his power, persuasion, and lies."[19] It

Figure 3.2 Portraits (or self-portraits?) of Pirro Ligorio as an old man and as a younger man, "Disegni di Pirro Ligorio." Turin, Archivio di Stato di Torino, Ms. J.a.II.17. Su concessione del Ministero dei Beni e delle Attività Culturali e del Turismo.

is known that Ligorio made a trip from Ferrara to Rome in 1570, the year in which the Acqua Vergine project was finally completed. He must have come across Luca Peto's tract on the project (which Peto published in 1570 just as the repair of the aqueduct was completed). There Peto does take a great deal of credit, and Pirro is not mentioned at all.[20] Evidently Peto's account filled Pirro Ligorio with rage.

Ligorio's account is supported by a large, detailed map of the Acqua Vergine (fig. 3.3), assignable to him on the basis of its handwriting and now in the Uffizi in Florence.[21] His deep knowledge of the aqueduct is also evident from his accurate description of the conduit in a later volume of the encyclopedia under the headword "Piscina." Here, within a discussion of various kinds of ancient settling tanks, cisterns, and *castella* (distribution tanks) for both rainwater and aqueduct water found among the ruins of Rome, Ligorio described the aqueduct.[22]

He noted that "even though we have investigated the Acqua Vergine with curiosity, both inside and outside the city and in its springs," he had not discovered cisterns for cleansing along the conduit; however, distribution tanks—some of which were highly decorated with triumphal subjects—had been discovered in the area of the Via Lata (the present-day Via del Corso) and in the Campo Marzio.[23]

Figure 3.3 Pirro Ligorio. Detail of drawing of the Acqua Vergine—section at the Bocca di Leone, ca. 1560. Florence, Galleria degli Uffizi, Gabinetto dei Disegni e delle Stampe, Uffizi n. 4236A. The small circles on the aqueduct indicate *putei*, or shafts. Ink and wash on paper. Su concessione del Ministero per i Beni e le Attività Culturali e del Turismo.

Here again, Ligorio angrily castigated Luca Peto: "we have wished to place here some few words that although they are revealed belatedly, I hope that (even though late) they would be pleasing to the curious to hear the truth in recognizing the lies." He mentioned an "Onophrio"—probably Onofrio Panvinio (1529–1568)—and Luca Peto. Peto "has particularly advanced himself as the inventor and conductor of the Acqua Vergine into Rome." He had praised his own accomplishment in a treatise, believing that Ligorio would not take offence. Ligorio explained that he had investigated every part of the aqueduct including the springs at his own expense and had made a representation of it in design (undoubtedly a reference to the Uffizi map) and had showed Pope Pius IV the cost of reconstruction. However, "Luca Peto along with another scoundrel" opposed his plan and acted in a way that his own (i.e., Ligorio's) costs and time remained hidden. Recounting the disaster of Antonio Trevisi's contract, Ligorio concluded that the popes Pius V and Gregory XIII brought the thing to conclusion, whereupon Luca Peto claimed to be the author of it "in order to get his hands on new money."[24]

Resuming his archaeological description of the aqueduct, Ligorio described the structure in detail, including its course and the materials used to construct it. His account included a history of the aqueduct based on ancient texts as well as on physical descriptions. He described and illustrated an aboveground arched section, one of the few outside of Rome, in the region where the conduit crosses the stream of the Marranella not far from its juncture with the Aniene (fig. 3.4).

Ligorio then followed the structure into the city, detailing arches, pilasters, and other structures.[25]

Ligorio was deeply unhappy with the work that had been completed in 1570. He exclaimed on "the calamity of the work made in throwing together and patching" that came about "through defect of the usurpers." Because insufficient money had been spent on the structure of the walls, "the work will easily be short lived and frail." He claimed to have instructed Pius IV on how the task should be carried out. Most of his recommendations followed ancient building practices, which he himself had investigated thoroughly, and which he greatly admired. First, he advised, at the beginning of the aqueduct one must not fail to use *signinum*, the fine lime mortar used by the ancients to waterproof the interior surfaces. Second, sluice doors must be created at appropriate places so that the water can be diverted and the aqueduct cleaned. Third, for security, the marble blocks (*marmi*) that had been placed in the aqueduct by Belisarius in 537 to prevent the Goths from entering the city through the aqueduct should be left in place. Fourth, the area around the shafts should be cleared, or the shafts covered to prevent leaves and dirt from falling in. Fifth, the inferior water from the Herculaneum River (which derived from a marsh) should not be let into the Acqua Vergine. Finally, the walls of the aque-

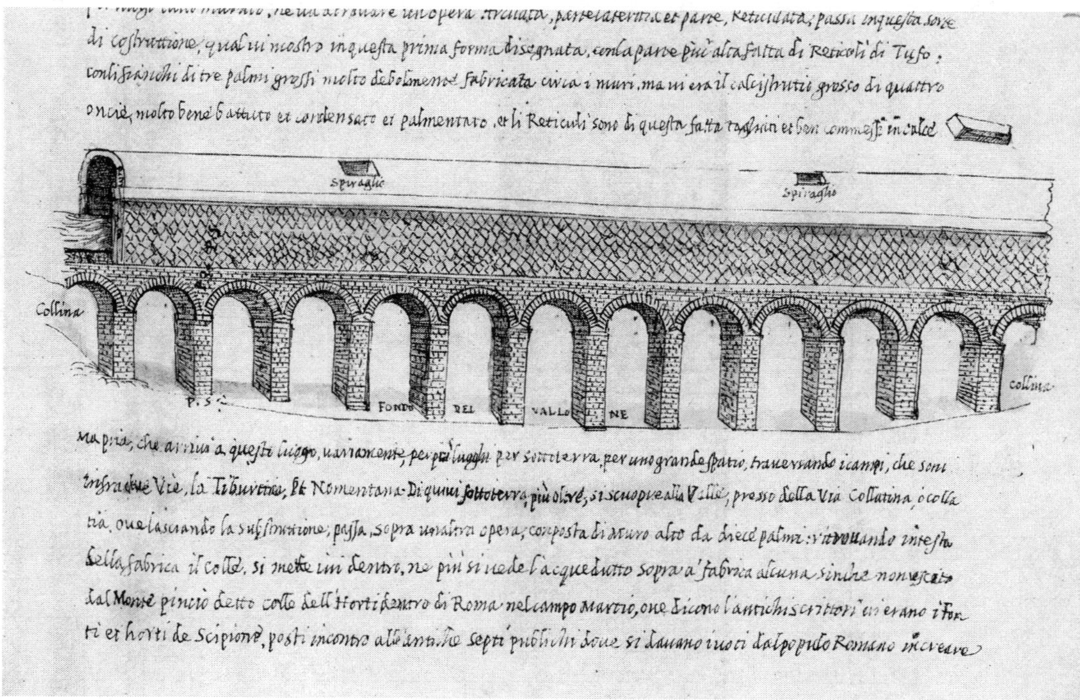

Figure 3.4 Detail of folio 15r from Pirro Ligorio's encyclopedia with a drawing of a section of the Acqua Vergine near the Marranella. Turin, Archivio di Stato, Ms. J.a.II.1, fol. 15r. Su concessione del Ministero dei Beni e delle Attività Culturali e del Turismo.

duct should be made four palms thick, rather than the three palms of the ancients, especially in parts built over land without arch work, such as at Bocca di Leone.[26]

Ligorio claimed that his proposals were ignored, but Luca Peto's own tract on the aqueduct repair suggests that at least some of the procedures Ligorio outlined had been followed.

Luca Peto and the Search for the Aqueduct

The Roman magistrate Luca Peto wrote his own tract (the one Ligorio had scathingly criticized) on the Acqua Vergine in 1570 after the conduit had been completed. In it, Peto discussed the persistent confusion (on the part of the Capitoline Council) about the actual course of the ancient aqueduct. Clearly Steuco's tract from the 1540s, which explains the route of the conduit, was not available to the council, nor presumably was Ligorio's clear map now in the Ufizzi. (It is possible, but undocumented, that Ligorio took part in the exploration described by Peto and that his map in the Uffizi was one result.) Peto explained that the matter was discussed at a public meeting of the Capitoline Council on June 3, 1561. It seemed that no one could describe the work to be done, since it was believed that the springs that provided water to the aqueduct were not the Salone Springs but rather those at the salt works near the (no longer extant) Porta Trigemina, between the Aventine Hill and the Tiber River. The confusion, Peto suggested, lay in the corruption of the word *Salone*, referring to the Salone Springs, with *saline*, referring to the salt works, which were located at the terminus of the Appian aqueduct. Here he referred to the frequent confusion, which caused some to identify the Aqua Appia with the Acqua Vergine, both of which ran underground. Peto continued: "The charge was entrusted to me, I do not know why, that I look at the springs and places carefully and report back to the Senate." (The Capitoline Council had appointed him, its minutes show, on June 3, 1561.) With the texts of Frontinus and Pliny in hand, he reported that he had thoroughly investigated the area around the eighth milestone (at the Salone Springs), which was where Frontinus said the origins of the Acqua Vergine could be found. "And indeed," he wrote, "having inspected the arched work and the substructures that were attributed to the Vergine by Frontinus, I reported boldly, even if the underground canal in certain places did not appear, that this was the Vergine and could be restored."[27]

In his discussion of the aqueduct, Peto combined the display and use of ancient learning with detailed accounts of his own personal investigation. He transcribed for his readers the two ancient inscriptions found on the markers or boundary stones in Cardinal Montepulciano's garden and pointed to another inscription discovered by the antiquarian Andrea Fulvio (ca. 1470–1527) at the foot of the Pincian Hill. Emphasizing that his own archaeological explorations had closely followed Frontinus, he described his exploration of the subterranean parts of the structure, especially those areas where Frontinus had mentioned "acquisitions," that is, the inflow of new streams. He related that his investigations were accomplished by various means: "some by horse where there was not much water, some by foot where it is permitted to walk with dry feet, some truly by small boat accommodated to it, drawn by men where it was a navigable stream."[28]

Peto's tract directly contradicts Ligorio's account that states that he and a friend (probably Onofrio Panvinio) discouraged any attempt to restore the aqueduct. Possibly, if Ligorio's account is true, what they were discouraging was Ligorio's involvement in the project rather than the project itself. Clearly there was conflict and competition over repairing the Acqua Vergine in the early 1560s. The ultimate victor—granted a lucrative contract to carry out this major urban project—was Antonio Trevisi.

Antonio Trevisi and the Acqua Vergine

Pius IV personally chose Antonio Trevisi undoubtedly in part because of Trevisi's tract on flood control and his publication of the Bufalini map, with its letters proposing canalization. It is also probable that Trevisi had provided a plan for the reconstruction of the aqueduct. In any case, the Camera Apostolica gave Trevisi the contract on April 18, 1561, arranging to pay him 20,000 gold ducats. The contract specified the sections of the aqueduct and the work to be done on them. It described the course of the aqueduct, beginning east of the city at the Salone Springs, running west and north for most of its course, and then turning sharply south to enter the northern end of the city at the Pincian Hill (fig. 3.5).[29]

The contract stipulated that Trevisi must clean and repair and reconstruct the entire aqueduct "with good and suitable materials," whether underground or above. He must put "all the other waters" (i.e., the other streams) that flowed into the aqueduct in ancient times back into the conduit. He must remake all the broken and ruined parts of the aqueduct (such as an aboveground section at Bocca di Leone), creating sloping sides in the walls so that the rain runs off. He must clean the entire underground conduit. In clause after clause the contract reiterated the same points for the various sections—clean, repair, construct with good materials, and make with sloping walls and exit holes for rain water "as one sees it had at the time of the ancients."[30]

The work, which was divided among masons and other workers, was to be "inspected by three architects": one chosen by the Camera Apostolica, one by the People of Rome, and one by "this Antonio." Further, Trevisi was obliged to have the work done within a year and entirely at his own expense (i.e., he would be paid by the Camera, but then would have to buy materials and pay wages with

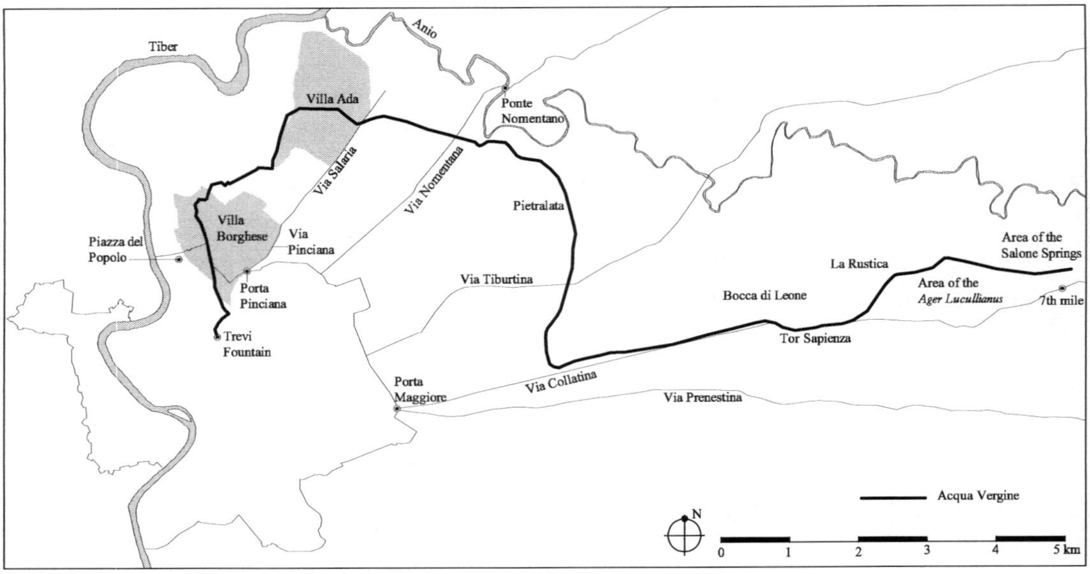

Figure 3.5 Course of the Acqua Vergine from the area of the Salone Springs (*far right*) to the Trevi Fountain at the center of Rome. Courtesy Chiara Bariviera with kind permission.

that money). And he was given authority to excavate stone and pozzolana (a volcanic ash that strengthened and added hydraulic qualities to concrete) in convenient, nearby places, paying "honorable and reasonable prices according to the judgment of the Camera." Finally, Trevisi's schedule of payments was established: 8,000 ducats at the beginning and then 1,000 a month until the project was completed (twelve months later).[31] Archival records show a steady flow of payment authorizations by the Camera Apostolica between 1561 and September 1566—years longer than the single year first projected.[32]

Trevisi failed to complete the project—or (some said) he barely started it—even though the time allotted stretched four years beyond his initial one-year contract. Some expressed outrage. In his tract on the aqueduct, the magistrate Luca Peto fumed that Trevisi was a man "of enough rashness," who, along with "reckless skilled masons," defrauded the pope and the Roman People. Peto reported that these men had divided up the work places and the work among themselves and promised that they would complete the work for 18,000 scudi. But they cheated the People by submitting cost overestimates with the plan of keeping the unspent balance.[33] The work, Peto explained, was to be carried out "with money collected by that man Treviso and distributed to masons." The work was inspected by the president of the project, illustrious Gabrio Serbelloni (cousin of Pius IV), who found that it was barely started. So many disagreements and quarrels had arisen among the contractors and subcontracted artisans that finally it was reported to us that "the work could not be completed unless Trevisi was removed."[34]

The pope refused to fire Trevisi "with the excuse that it was fair that Trevisi, the author of the work, should be the one to complete it, but we should just pay attention that the work was attended to quickly and correctly."[35] (What Pius meant by calling Trevisi the "author" of the work is unclear, but suggests that Trevisi had proposed a detailed plan.)

In his Turin encyclopedia of antiquities, Pirro Ligorio also reported that the work of the aqueduct was given to one who underhandedly played with the money (i.e., Trevisi). Ligorio's advice had not been taken in order to give "by piecework or combined labor" a contract to "Antonio [Trevisi] from Lecce, Charmer of Snakes." In the pontificate of Pius IV, Ligorio complained, the work was done "with so much confusion of master masons and with so many quarrels" that it was not completed and "the contractor Psyllaro [i.e., the snake charmer, Trevisi] died in prison because of it." [36]

The Role of the Capitoline Council

During the early 1560s, the Capitoline Council discussed the aqueduct frequently, becoming increasingly mistrustful of Trevisi as time went on. All began well in a Public Council meeting that took place on October 12, 1560. There it was reported that the pope (Pius IV) had called the magistrates to a *congregazione* for which "about ten cardinals" had been deputized. The pope had made it clear that "for the beauty and benefit of this land," he very much wanted water from the Salone Springs to be conducted to Rome. Two cardinals—Della Cueva (Bartolomé de la Cueva y Toledo, 1499–1562) and Crispi (Tiberio Crispi, 1498–1566)—were directed "to hold a council and understand the opinion of the people." The council decided to deputize four gentlemen (unspecified) who together with the conservators and prior would go to thank the pope for his liberality and kindness and tell him that "the People very happily have accepted this his offer and that thus His Holiness should have such a very praiseworthy task started and carried out." This was approved unanimously by voice vote and without dissent.[37]

It perhaps dawned on the Capitoline Council only slowly that the People actually would be required to make major financial contributions to the new project. At the latest, that realization came on June 3, 1561, about six weeks after the Camera Apostolica had given Trevisi his contract. On that date, the conservators called a Public Council meeting and informed the council that the pope believed that conducting the water of the Salone Springs into Rome "will result in grandness to the city and benefit to this people." The pope therefore had informed the people that he had made a contract with the architect (Trevisi) for 20,000 ducats from the Camera Apostolica. His Holiness wanted to pay 6,000 himself, he wanted the College of Cardinals to pay 5,000, the Cancelleria to pay 3,000, and the remaining (6,000) to be paid by the People (the Capitoline Council). The Public Council was informed that this was discussed in the Secret Council of the day before, but there had not been a quorum to pass it.[38]

The Capitoline Council was not pleased. It created a committee, which included Luca Peto, to investigate whether the Salone Spring water really could be conducted to Rome and whether it could be conducted through the same channel that the Camera Apostolica had told master Antonio (Trevisi) it had passed through in ancient times.[39]

Despite its conflicts with the pope, the Capitoline Council intermittently expressed alarm at the paucity of water produced by the unrepaired Acqua Vergine, especially given Rome's growing population. The already inadequate supply of water was occasionally made worse or stopped altogether by people diverting water from the aqueduct for private use. At a Public Council meeting on June 20, 1561, it was reported that "the water was impeded at times in a way that for two days it does not come to the public." The report continued that it was necessary "to make an edict against those who impede the said water in any way whatsoever."[40] Possibly they were glad to hear at the Secret Council meeting of July 5 that the project of conducting the water of the Salone had already been started.[41] But the task of raising their share of the cost had not begun.

On August 23, 1561, the Public Council considered ways to raise this money. The first idea was to take a "census" (*censo,* an annuity) from some of the officers of the city. Most officers of the city purchased their offices and made a profit by keeping a portion of the income (such as taxes or custom duties) generated by the office. The decision of the council meant that the officers chosen would be asked to give a sum of money to the Camera Apostolica for the repair of the Acqua Vergine, for which they would receive an annual interest payment. The next month, on September 6, the Secret Council deputized four noblemen along with the three conservators and the prior to raise the money.[42] At the next Public Council meeting, it was reported that the money would be collected from the notary of the Ripa (a customs officer at Rome's main river port), who would be given interest at 7½ percent. These *pecuniae ad censum,* as they were called, were in essence loans that paid an annual interest.[43]

The *gabelle,* or taxes on certain products, were another source of possible revenue. The Capitoline Council always resisted the meat tax, *gabella della carne,* sometimes called the *quattrino della carne.* This hated tax directly affected the mouths of a Roman population—often threatened with near starvation or worse—with a levy of one-fifth of a baiocco, also called a quattrino, for every pound of meat sold. On August 23, 1561, the cardinal *camerlengo* suggested the meat tax as a source of funding, an idea that the council rejected on this occasion in favor of the *censo.*[44]

But the income from the *censo* proved insufficient. On April 9, 1562, a representative of the Camera Apostolica, Cardinal Santa Fiora (Guido Ascanio Sforza di Santa Fiora, 1518–1564), urged the Secret Council that the People pay the other third of the installment for the aqueduct so that the work could continue.[45] At the Public Council meeting of April 28, it was decided that since the notary of the Ripa had been sufficiently taxed, they would have to resort to other officials.[46] Meetings of the next months brought suggestions of other officers who might be taxed, but no action.[47]

In the end, the Capitoline Council objected to the unfairness of the situation, suggesting that the financing of the aqueduct had been administered inequitably. At the Public Council meeting of December 15, 1562, it was noted that the Commune had paid two-thirds of what it owed and that the cardinal *camerlengo* many times had solicited the final third. However, "it did not seem to us honest, the two thirds not being paid up to now by the others who are obligated [that is, the pope and the cardinals], that the People must pay at present the last third." Instead, the council suggested, it would like to pay 500 scudi now and then the remaining simultaneously along with the payments made by the others. Further, it would like the money to come from the imported wine tax (*gabella dello studio*).[48] The council's protest was to no avail. At a Public Council meeting on February 26, 1563, it was noted that the reverend *camerlengo* had insisted that the Commune pay the remaining third and that further, the pope did not agree that the money should come from the wine tax, but rather it should be taken from communal officials.[49]

The argument about this tax on imported wine was a familiar one in struggles over funding between the papacy and the Capitoline Council. As mentioned previously, the *gabella dello studio* was a tax originally intended to fund the Roman university but often diverted. The Capitoline Council favored this tax on imported wine over other taxes. The pope almost always resisted the suggestion. The pope wanted the income from the *gabella dello studio* for the *studio* (university) or for some other purpose. Seemingly because there was plenty of wine made within the Roman walls and more or less cheaply available, a tax on imported wine would not have a huge negative effect on the Roman population. The Capitoline Council thus preferred it over the meat tax or other forms of taxation.

The Papal Curia continued to pressure the Capitoline government. In a letter of August 28, 1563, Giovanni Serbelloni (Cardinal San Giorgio), nephew of the pope, sent a letter urging the conservators to pay up.[50] Some months later, on February 17, 1564, the minutes of a Secret Council meeting recorded that the pope, "with very great insistence," urged the people "with every speed" to give what was owed for the aqueduct, "a work so honorable and necessary."[51] Eight months later (October 31, 1564), the order of business of a Secret Council meeting included "a new order for bringing to conclusion the work of conducting the water of the Salone to Rome." Several council members, including the magistrate Luca Peto, had discussed the matter with the pope. Again the council proposed using the wine tax to fund it, and the Public Council of November 23 approved this source.[52] But the minutes of a Secret Council of January 15, 1565, report that the pope had again disagreed with this source of funding because he wanted to use the wine tax for the construction of a new university building (what would become the Roman University, La Sapienza).[53]

Parallel to the struggle over financing was the concern about the competence of the man whom the pope deemed a "sufficient architect," namely, Antonio Trevisi. On January 14, 1562, the minutes of a Secret Council meeting report that the council has chosen its own architect—Bartolomeo Grippetto—to inspect the work. This was done in compliance with Trevisi's contract, instructing that an architect from the People should inspect the work. Grippetto was from the town of Caravaggio, twenty-five miles east of Milan in Lombardy. He was deputized "to review the work of the water of the Salone and to visit it continuously so that the work is done suitably." The council also recommended that two gentlemen be appointed to whom Grippetto could report. These gentlemen (who are not named), along with the conservators and prior, were also to oversee the entire operation so that the work might be carried out "to the satisfaction of the People."[54]

On April 9, 1562, at a Secret Council meeting, the proposal was made that some suitable provision (i.e., a salary) be made for Grippetto's work of oversight. Further, not two but four noble men (*nobiles viri*) were chosen to serve in the next year—Mario Frangipane, Rutilio Alberini, Orazio Naro, and Luca Peto—a hefty and undoubtedly necessary backup for Grippetto's work of inspecting the work of a rival architect—and one employed and defended by the pope.[55]

Pius IV died in 1565 with the aqueduct still in disrepair. On September 13, 1566, the Camera Apostolica gave Trevisi his final payment.[56] Then, on December 6, 1566, it ordered an evaluation of his work.[57] Peto noted that now the supervision for finishing the aqueduct was given to Cardinal Politiani (i.e., Montepulciano), who began by investigating how much was completed, "since meanwhile that wretched Trevisius had quite met his death." Peto confirmed that he and another Roman magistrate, Orazio Naro, were appointed, along with two architects, Giacomo Della Porta and Bartolomeo Grippetto, to see that the work was carried out.[58]

A New Attempt

The new pope, Pius V (Michele Ghislieri), elected on January 7, 1566, occupied himself more with piety and less with urban reconstruction than had his predecessor. Pius V had been promoted to bishop, cardinal, and then grand inquisitor by Paul IV Carafa. He emphatically disapproved of and undid some of Pius IV's works. He ordered, for example, that the "pagan" statues from the Casino of Pius IV and in the Belvedere (placed there by Pirro Ligorio) be removed.[59]

Early in his pontificate, the Acqua Vergine did not appear to be one of Pius V's priorities. Indeed, in a reversal of the usual scenario, we see the Capitoline Council urging the pope. The minutes of the Public Council meeting on March 2, 1566, reported that the council insisted that the aqueduct project had to be completed. "The channeling of the Acqua di Salone being very important work where a great sum of money has been spent," it noted, nevertheless, "not only did the result for which the money was spent not follow, but it can be said that the expense and work already done was lost and thrown away." The council supplicated the pope that the work be continued and completed, and that to this effect, the tax already enacted be exacted.[60]

The council's request was soon supported by a formidable ally, Cardinal Montepulciano. The record of the June 8, 1566, Secret Council meeting indicates that the cardinal had made the pope understand that the People wished to run the Salone Spring water through the city and distribute it to the houses of various gentlemen. The pope now undertook this major public-works project and gave the Commune permission to tax those houses that would benefit, following a suggestion that Agostino Steuco had made years before.[61] Here a new solution to the problem of how to finance the repair and maintenance of the aqueduct had been put forward, namely, the taxing of houses that would benefit from the new supply of water.

Another notice pointed to the new leadership role played by Cardinal Montepulciano. On July 6, 1566, the minutes of a Secret Council meeting recorded that a congregation concerning the aqueduct was held in Montepulciano's palace on Via Giulia. In attendance were the gentlemen appointed by the Capitoline Council to attend to the aqueduct (their names not mentioned). It was now understood that "His Holiness was resolute that the said work in every way be carried out to its proper completion."[62]

The pope was undoubtedly motivated further by an outbreak of illness in August of 1566 brought on by polluted water. Well water became contaminated in the *rioni* of Colonna and Trevi. At first about twenty workers laboring at a canal, who drank from a fountain, became ill. The disease spread to four thousand families in the area. The pope became alarmed, fearing its spread to the rest of the city.[63] (Retrospective diagnosis suggests that the pathogen was typhus. It would not have been understood this way in the sixteenth century, but given the number of workers taken ill simultaneously after drinking, they could and did deduce that the cause was contaminated water rather than their usual culprit, bad air.)

Pius V, albeit somewhat slowly, created a new standing committee—the Congregation on Streets, Bridges, and Fountains (Congregatio super viis, pontibus et fontibus) — to deal with the aqueduct and other urban issues. The new congregation held its first meeting on April 4, 1567, in the palace of Cardinal Montepulciano. It was headed by the cardinal until his death in 1574. Pius V created a second congregation in 1570, also headed by Cardinal Montepulciano to deal specifically with the urban conduits and fountains of the Acqua Vergine.[64]

The Efficient Cardinal Montepulciano

It was auspicious and probably crucial that the practical-minded Cardinal Montepulciano was one of the two cardinals who effectively oversaw the work of the new congregation (the other being Flavio Orsini, 1532–1581) and the Acqua Vergine repair (fig. 3.6). Montepulciano was uniquely suited for the job and held a personal interest in its success. He was born Giovanni Ricci into a noble family from Montepulciano (south of Florence) in 1497. To escape an unsympathetic stepmother, he traveled to Rome in 1513, arriving at the age of fifteen under the protection of the Tarugi family from Montepulciano. His rising career in Rome can be attributed to his efficiency and his ability to carry out complex diplomatic and administrative tasks. He became the master of the household (*maestro di casa*) in charge of the complex affairs of the residence of Cardinal Del Monte (Giovanni Maria Ciocchi del Monte, 1487–1555). Then he was drawn into the service of the Farnese pope, Paul III. In 1536 the pope sent him to Spain as papal *nunzio* (papal ambassador). He also served as papal *nunzio* in France, Germany, and the Low Countries, and in 1544 he was made *nunzio* to Portugal and archbishop of Manfredonia. In 1545, he became administrator of the Tuscan town of Chiusi. In February 1550, his old protector and employer, Cardinal Del Monte, became Pope Julius III. Julius called Ricci back from Spain to manage the papal finances, which were in complete disorder, and in 1551 made him a cardinal.[65]

Giovanni Ricci was not a deeply spiritual man, nor did he exhibit any interest in canon law or theology. Rather, he was a highly competent administrator and overseer of practical affairs. Such talents were desperately needed and highly valued at a time when the renovation of Rome was considered an important key to the presentation of a reformed Catholic Church. After Julius III died in 1555, his successors continued to appreciate the cardinal's talents and use his services. Pius IV appointed him president of the streets and in 1561 elevated his home town, Montepulciano, to cathedral rank and put him at its head, from which time he was known as Cardinal Montepulciano. At the death of Pius IV in 1565, he came close to being elected pope. However, Carlo Borromeo, the influential nephew of the late pope, found him to be reprehensible because of his lack of spiritual qualities,

Figure 3.6 Scipione Pulzone, called il Gaetano, portrait of Cardinal Ricci (Montepulciano), 1569. Oil on canvas. 26¼ × 20¼ in. (66.7 × 51.4 cm). Harvard Art Museums/Fogg Museum, William M. Prichard Fund, 1934.66. Photo: Imaging Department © President and Fellows of Harvard College.

his love of extravagance, and the fact that he had an illegitimate son (Giovanni Ricci) born to a noble Portuguese woman, Francesca d'Andrada, during his term as papal *nunzio* in Portugal.[66]

Throughout his cardinalate, Montepulciano utilized his wealth to its limits to further his projects of building construction and collecting antiquities, making him one of the major builders of his era and one of the foremost collectors of antiquities in Rome, equal to the Farnese and Medici despite his far fewer family resources. Shortly after he was made cardinal in 1552, he purchased the palace on the Via Giulia (Palazzo Sachetti) from Orazio da Sangallo, the son of the renowned architect Antonio da Sangallo. The purchase included Antonio's impressive collection of antiquities. The cardinal continued to expand that collection, and he was especially well positioned to do this after Pius IV appointed him president of the streets, a position that allowed him to claim antiquities discovered during excavations connected with construction and street repair. On May 30, 1564, Cardinal Montepulciano purchased from the Crescenzi family the villa on the Pincian Hill next to Trinità dei Monti, now known as the Villa Medici.[67]

The purchase of the Crescenzi property provided the occasion for extensive construction and renovation of the buildings and the creation of magnificent gardens — at least a third of the expenditure went to the gardens, fountains, and water conduits. The Acqua Vergine passed directly under Montepulciano's new property on its way to the Trevi Fountain. When Pius V assigned Cardinal Montepulciano, along with Cardinal Flavio Orsini, to oversee the congregation charged with completing the reconstruction of the aqueduct, he undoubtedly recognized that Montepulciano was not only a highly effective supervisor of complex projects but also that he had an enormous personal interest in seeing a fully restored aqueduct. Only then would the cardinal have a sufficient supply of water for his magnificent new gardens on the Pincian Hill.[68] It was also helpful that his codirector, Flavio Orsini, owned extensive gardens in the Campo Marzio (between Via del Babuino and the present-day Corso).[69]

From Failure to Success

The first meeting of the Congregation of Streets, Bridges, and Fountains was held on April 4, 1567, but the Acqua Vergine project at first received no attention. The minutes of the congregation indicate that early months were focused on cleaning and paving streets, on taxes levied for that work, and on requests for exemptions from those taxes. The congregation, headed by Montepulciano, turned its attention to the aqueduct only more than a year later in the summer of 1568. One signal that it had done so is a notice in the minutes of a secret Capitoline Council meeting in July that the pope wished to have the aqueduct completed. The People would need to contribute more money. Further, they would need to have a drain (*chiavica*) completed from the Trevi Fountain to the river, so that the expected abundance of water from the Trevi would have an exit and so that stagnant water would not cause pestilence. Then, on August 14, an *avviso* reported that "it is resolved that the construction already begun by Pius IV to conduct here the Salone water be finished according to plan, for public benefit."[70]

This notice suggests a hands-on approach that is borne out by subsequent minutes of the congregation. By September 1568, Luca Peto and Oratio Naro had compared the prices and offers of various masons and had made contracts. Further, for unknown reasons, the head architect, Nanni di Baccio Bigio (ca. 1513–1568), a major architect in Rome who had done much work for Cardinal Montepulciano, was replaced by "Giacomino," that is, Giacomo Della Porta (1532–1602).[71]

Giacomo Della Porta was born in Porlezza in Northern Italy near Como. He was the son of a sculptor, and in his youth, he was apprenticed as a stucco worker (*stuccatore*). He was in Rome by 1559. In his early Roman years, he is always called a sculptor (*scultore*) and was also occupied with buying and selling excavated objects such as statues, coins, and other antiquities in addition to pieces of marble to be sold for making lime—a common Roman occupation. In the early 1560s, he began to be awarded contracts to design and build structures, including buildings, facades, and gates. In 1563 he was recorded as the *misuratore* (surveyor, appraiser, measurer) of the repair work of the Senate palace on the Capitoline Hill, working under Michelangelo. After Michelangelo's death in 1564, he was named architect of the Roman People (*architetto del Popolo Romano*), that is, of the Capitoline Council. As such he closely supervised construction on the Capitoline Hill (the palaces of the Conservators and of the Senate), as well as numerous other civic projects. When the architect Giacomo Barozzi da Vignola (1507–1573) died in 1573, Giacomo Della Porta became both architect of the Farnese family and the architect of Saint Peter's. He can be found everywhere in late sixteenth-century Rome, his hand in numerous projects. Yet he does not seem to be involved in conspicuous efforts to gain patronage, nor did he undertake writings or other "higher profile" activities. His low-key but highly successful career suggests that he possessed the combined characteristics of trustworthiness and competence.[72]

The congregation itself dealt with structural issues pertaining to the aqueduct. In the meeting of September 21, 1568, a concern was raised that the inferior water of the Marranella stream should not run into the new aqueduct. The new architect in charge, Giacomo Della Porta, was ordered to go as quickly as possible along with Bartolomeo Grippetto to find out what was needed to clean the channel. (Little is known about Grippetto beyond his being mentioned as employed in a variety of projects and being a trusted architect of the Roman magistrates.) The congregation lastly turned again to the issue of money, instructing that security was to be taken from three contractors who had been put in charge of parts of the aqueduct.[73]

As the November minutes reveal, the congregation closely supervised the entire project. The architects were instructed that they must give reports on the work done by the masons every fifteen days in writing to the overseers of the work (i.e., to Luca Peto and Orazio Naro). They must report on money already spent as well as on money needed. A channel dug through the country estate of Camillo di Rustici had done damage, and he was to be compensated. Finally, Cardinal Montepulciano noted "that master Giacomino [i.e., Giacomo] Della Porta, appointed architect, is very diligent and has gone and done much in the business of the aqueduct of Salone." He suggested payment (to the architect) "worthy of his labors."[74]

Money was an unending problem. At the next meeting Luca Peto and the treasurer were instructed to go to the pope "to make him understand that difficulty [unspecified but undoubtedly about money] concerning the business of the aqueduct in order to know what His Beatitude will command."[75] On December 23, 1568, the issue of money was raised at a public meeting of the Capitoline Council. This time the use of the *monte dello studio* (a bond issue, the interest of which would be paid by the wine tax) was proposed as a way of raising the needed 10,000 scudi. A committee of four men were chosen to go with the conservators and prior to discuss the matter with the pope.[76]

Monti, or bonds secured on a funded debt were increasingly important in sixteenth-century Rome. In a *monte,* a single share was called a *luogo;* multiple shares were *luoghi di monti.* Bonds were one of two kinds: either *monti non-vacabili,* that is, long-term bonds that could be passed on to heirs or exchanged by acts of living persons, or *monti vacabili,* which conferred rights on the shareholder only until his or her death, at which point the bonds were returned to the Camera Apostolica. The disadvantages of the latter were compensated by interest rates that paid about 2 percent higher than the former.[77]

The operations of a *monte* or funded debt began with its creation or "erection" (*erezione*). The foundational document of a *monte* would state the reason for its creation; the number of shares or *luoghi;* the value (usually 100 scudi per *luogo,* but it could vary); whether the *monte* was *vacabile* or *non-vacabile;* the amount of interest to be paid per *luogo* (stated precisely in scudi, and giuli or baiocchi); how often the interest would be paid (usually every two months); the type of currency to be used (gold or silver); the endowment of the *monte,* that is, the source of the interest payments (in this case the endowment suggested was the *gabella dello studio* or wine tax); the conditions under which the bonds could be redeemed; and finally, the officials who were to administer it.[78]

The erection of a *monte* could result in a large amount of ready cash but would also increase the Capitoline Council's highly burdensome debt load. Nevertheless, on June 15, 1569, two deputies charged with finding money for the Acqua Vergine, Prospero Boccapaduli and Pietro Antonio Bandini, met with the conservators and the prior to discuss raising some of the money from the *gabella dello studio,* taxing foreign wine, and some from a bond issue (a *monte*) with an interest rate of 7 percent.[79]

Beyond money, decisions had to be made about the actual repair of the aqueduct. Peto's detailed discussion included an account of improvements in the ancient structure. For example, the ancient aqueduct had received water from the Herculaneum Stream (Rivus Herculaneus), which supplied very poor-quality water, lowering the water quality of the aqueduct as a whole. In the restoration, this marshy section was bypassed by changing the route so that the gradation could be maintained. To do this, new arches were constructed at the place called Bocca di Leone. Workers constructed a tunnel through a mountain and added a sluice so that if the water became flawed at this point it could be diverted. They also strengthened the walls of the aqueduct and made other repairs.[80]

In September 1570, the channel from the Salone Springs to the Trevi Fountain was at last completed. After ten years of difficulties and conflict, abundant water finally flowed into the Trevi Fountain. The city as a whole immediately benefited, as did particular individuals who were rewarded with their own supply of water. A papal edict granted one *oncia* of water from the aqueduct to Montepulciano for his villa on the Pincian hill.[81] Luca Peto received an *oncia* of water from the aqueduct for his own home, for his use and that of his successors.[82] Cardinal Orsini was granted three *oncie* of water for his vast garden in the Campo Marzio, including water for the nearby Hospital for the Incurables (Ospedale degli Incurabili).[83] Bartolomeo Grippetto was awarded a quarter *oncia* of water for his private use in his house in the Campo Marzio.[84] (The lack of a recorded water reward at this time for Giacomo della Porta suggests that his residence at this time was located beyond the reach of the Acqua Vergine.)

On October 25, 1570, Pope Pius V issued an edict signaling his desire to have Acqua Vergine water distributed widely in the low-lying parts of the Campo Marzio and displayed in elegant fountains as was the case in ancient times. The pope lauded the completion of the great undertaking that had been celebrated "with the great applause and joy of the whole population for public utility and the embellishment of this city." He mandated that fountains be constructed throughout the city and that individuals be permitted to buy water for their individual houses.[85]

Conduits, Pipes, and the Price of Water

Now that plentiful water poured into the low-lying area of the city, it became possible to envision the distribution of that water to strategically placed fountains. But wider distribution of Vergine water required the construction of an additional distribution tank or *castellum* (a tank into which the aqueduct water flowed and then exited through pipes leading to diverse locations). What was needed was a new system of underground pipes, and new fountains (beyond the single Trevi Fountain). Such new constructions would enable the distribution of water to public squares as well as to private palaces and homes in the low-lying Campo Marzio. Because it was powered by gravity, water from the Acqua Vergine could only be piped to areas that were lower than the terminus of the aqueduct at the Trevi Fountain, which is nineteen meters above sea level. The water must flow continuously downhill.[86]

A new congregation, also headed by cardinals Montepulciano and Orsini, was created to deal solely with the Acqua Vergine and its fountains. It met from

November 4, 1570, until February 10, 1571.[87] Its first meeting outlined a plan to construct eighteen fountains strategically located around the Campo Marzio.[88] A few months later, on January 16, 1571, the minutes to a public meeting of the Capitoline Council record how the large public project of the fountains was to be paid for. A Florentine merchant, Bartolomeo Bonamici had purchased 180 shares for a new bond issue (*monte*), the interest to be funded by the meat tax at a rate of 7 percent per share (*luogo*). Bonamici would not pay for his shares all at once but would pay out gradually as needed for the construction of the fountains.[89]

The Castellum of Sebastianello and Underground Conduits

To enable a network of Vergine pipes throughout the low-lying Campo Marzio, a new *castellum*, or distribution tank, had to be built. Giacomo Della Porta and Bartolomeo Grippetto supervised the construction of the new tank. They had it built beneath the Pincian Hill at San Sebastianello (just north of present-day Piazza di Spagna) at the beginning of a new road, Via del Sebastianello, that led up to Cardinal Montepulciano's villa. As Rinne has noted, the *castellum* was built underground and was—crucially—seventy centimeters higher than the Trevi fountain, located 750 meters to the south, increasing the available pressure by about 15 percent, a necessary increase for further distribution into the Campo Marzio.[90]

On July 28, 1571, a contract was awarded to two masons to build a conduit from the distribution tank at San Sebastianello to Piazza del Popolo, at the north entrance to the city. The contract mandated specific and detailed instructions. The masons were to begin the conduit at the garden of the Roman patrician Orazio Naro and to build it straight to Piazza del Popolo. The conduit was to be made of terra cotta sections, with the size specified. The joints between pipe sections were to be made with the lathe and be well seasoned (*ben staggionati*) and well fired (*cotti*). Every ten *canne* (about 72 ft. or 22 m) there was to be a manhole over a chamber attached to a settling tank, which would allow an individual to descend into the chamber to clean the tank. The exact materials and dimensions of the apparatus and its foundations were specified. The settling tank was to be made of peperino stone, and it would be connected to a shaft to the street (to release air). The opening of the chamber (on the street) was to be covered by a lid made of travertine (fig. 3.7).[91] The building of this section of the conduit infrastructure seems to have gone smoothly.

Another section of the conduit line provoked a heated disagreement. This conflict came about as a result of a proposal made by the sculptor, Guglielmo Della Porta (1515–1577, not to be confused with Giacomo Della Porta, who was a relative). Guglielmo's proposal for the middle and southern branches of the conduit and their accompanying fountains was ultimately accepted over the objections of

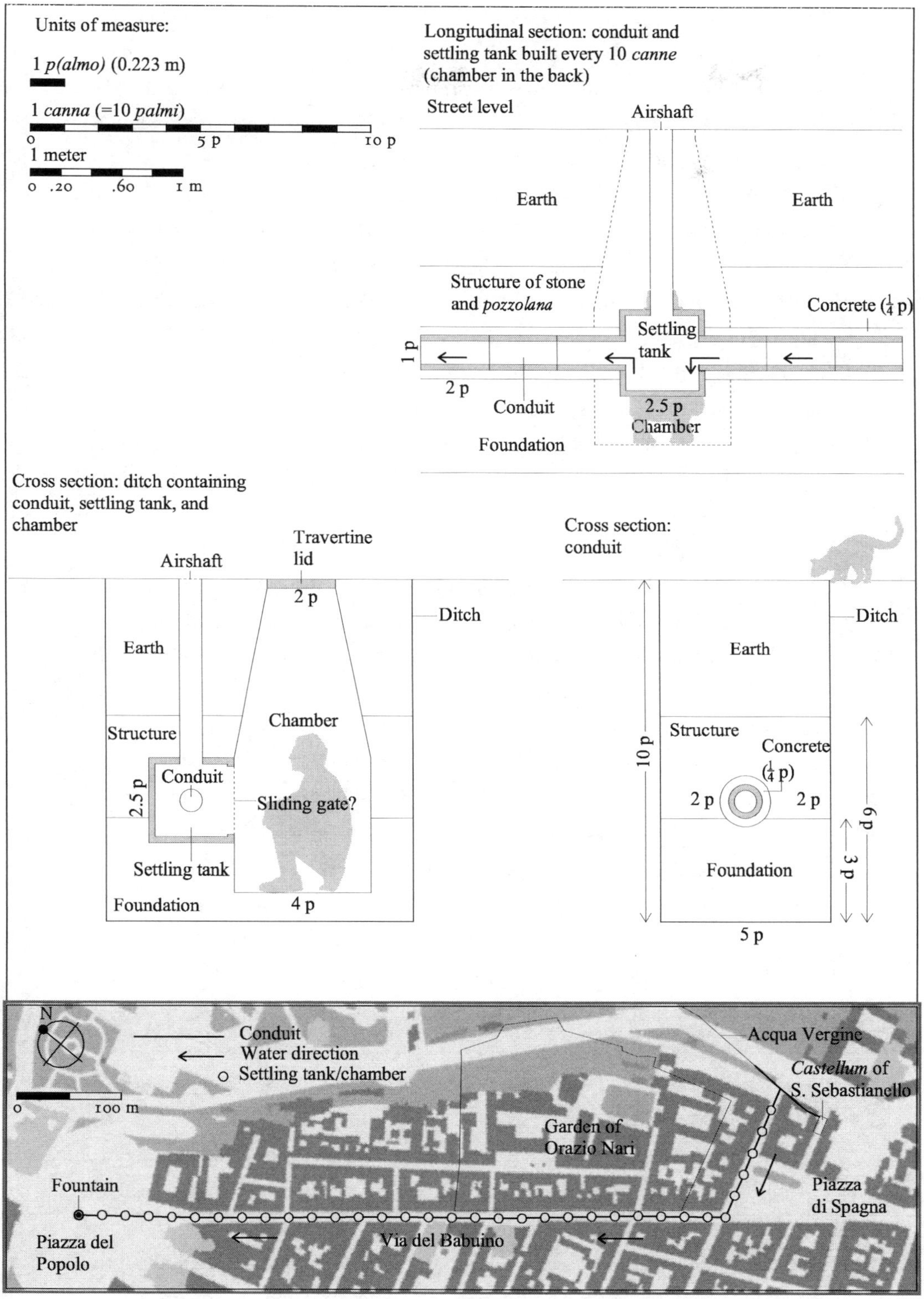

Figure 3.7 A manhole/settling tank on the conduit from San Sebastianello to Piazza del Popolo. Reconstructed on the basis of a contract, ASR, Notai del Tribunale delle Acque e Strade, b. 6, prot. II, fols. 269r–271r. Reconstructed by author and Chiara Bariviera. Drawing © 2017 by Chiara Bariviera with kind permission.

the Capitoline Council. Guglielmo was a well-known sculptor from northern Italy. He was born in Porlezza near Como and worked in both Milan and Genoa before coming to Rome, where he is first documented in 1546. He was called "del Piombo" because in 1540 he was named the *piombatore apostolico,* that is, the official in charge of the lead used for sealing papal bulls.[92]

The Capitoline Council's dissent from the choice of Guglielmo appeared first in the minutes of a Secret Council meeting of March 5, 1571. The council expressed a "foreboding" (*presentito*) that Guglielmo del Piombo sought a contract for the construction of the fountains "without informing either us or the deputies of the People [i.e., the deputies of the council who sat on the congregation]." The council insisted it should have "a consultive vote." They continued, "further lamenting" that Cardinal Orsini had issued an edict commanding that the Capitoline Council could do nothing without the approval of the masters of the streets. This seemed difficult (*duro*) and against the statutes and ancient sovereign power (*possessiones*) of the conservators.[93]

Despite the council's "foreboding" and unusually direct protest, two months later, on May 17, 1571, Guglielmo Della Porta (del Piombo) received the contract. It referred to Guglielmo's plan (*disegno* and later *disegno de carta,* indicating that it was a design drawn on paper), which had to be deposited in the Capitoline so that the overseers could refer to it. It specified exact measurements of the pipes and foundations and specified elements such as drainage plugs and bolts, Guglielmo's pay, and the time schedule. The key provision specified that the material to be used would be stone from the quarry of Orte (near Viterbo), similar to marble.[94] Why exactly Guglielmo was able to get a contract for constructing this part of the conduit is unclear. Possibly his close association with the pope (from his position as *piombatore* in charge of papal seals) provided leverage from the pope himself.

The contract was approved by the cardinalate Congregation on Streets, Bridges, and Fountains. The notary, Ottavio Gracchi noted that one member, Ludovico Mattei, a representative of the People (the Capitoline Council), "in stipulation and obligation of this kind of contract does not wish to take an oath nor to oblige himself willingly to the enactment of the [contract] written below on the Acqua Vergine."[95]

Yet the next day, May 18, 1571, at a public meeting, the council capitulated. The previous day its representatives had been told "on behalf of Our Lord" his Holiness, that the decision to give the contract to Fra del Piombo (i.e., Guglielmo) was final and "an end had to be made" of this matter. The pope had judged that the conduits would be better made and more durable if made with the stone from Orte that Guglielmo had proposed using. In the council's capitulating decree, it wishfully suggested "that if within the next eight days some master had come to do the same work better and less expensively for the utility of the People," the illustrious Lords (i.e., the Cardinals) might give it to this master "and have the stipulated contract [with Guglielmo] canceled."[96] The Capitoline Council made another last-ditch effort to prevent Guglielmo from getting the contract. Together with the appointed deputies, they went to the pope and told him that they were not satisfied with the decision about Guglielmo Della Porta and that "he [i.e., the pope] might have wanted again to take advice from men skilled in the art." The pope replied "very quickly with brusque words" that the matter at hand had been given to the cardinals and that the project was being overseen by "expert and judicious men," and that he only wished "that they have made the ditches well."[97]

Guglielmo's work on the conduits began with the hiring of three stonemasons (*scarpellini*), who would be making the conduit sections out of the stone from Orte.[98] As was usual, after the first trenches had been dug and three *canne* (about 22 ft.) of conduit had been laid, a test was carried out to see whether the course had been surveyed properly, to see whether a constant gradient had been created, and to assess the water pressure in the newly laid pipes. The first test, in October 1571, was an utter failure—not because of the gradient but because the chosen stone from Orte, which was a travertine and which was porous, leaked water everywhere. Giacomo Della Porta (an overseer of the entire Acqua Vergine as well as of the conduit from the *castellum* of San Sebastianella to the Piazza del Popolo) wrote a scathing report. It was said many times, Giacomo emphasized, that "the design and invention of the *Frate* [Guglielmo Della Porta] would not have been good and that it would always cause damage that would never be able to be remedied." This was proven, the report continued. A test was made after some of the conduit was built—it did not hold the water and "caused damage in the same way as a sieve [*crivello*]."[99]

Instead, Giacomo Della Porta noted, pipes made of terra cotta that go through Rome do not leak, "except a conduit of terra cotta where some are found broken because of the master that put it into the work." In contrast, "the sound of water in the cellars" indicated "how much the said conduit of travertine holds." Giacomo remonstrated that the invention of the travertine conduit "was disapproved of by the infinite judgments of intelligent people," judgments made before the conduit was constructed, which concluded that "it [the conduit] would prove to be dangerous and then would be unfixable and of this there are infinite witnesses."[100]

Giacomo Della Porta reported the happy conclusion (for him, anyhow) of the debacle. Finally, "the work of the conduits of terra cotta" was given to the masters to put them into the trench (*chiavica*). An overseer, Vincenzo Della Fontane, was put in direct charge of the work and was given 4 scudi per month. Further, "the two architects who had made the water of the Salone springs come to Rome"—namely Bartolomeo Grippetto and he himself Giacomo Della Porta—were put in charge of soliciting (*sollecitar*), that is, hiring and acquiring materials and making the accounts, measurements, and cost estimates of the work (*conti, misure, stime*) and of paying out the money spent. As a result, "with their judgment they have continually helped and repaired all the errors and damages" that had happened—"all with great advantage to the People."[101]

Pricing the New Water

In addition to building a new physical infrastructure for the distribution of the Acqua Vergine water in the Campo Marzio, an administrative system for selling and distributing the water had to be devised. The new distribution system would include a schedule of prices for the sale of water to particular institutions or individuals who would run a pipe from the new conduit to their dwelling or other kind of structure. It also included regulations for making such pipes and for their use.

On June 25, 1571, the conservator Tommaso Cavalieri presented a memo to the council laying out his pricing plan. Cavalieri's plan was based on the assumption that purchasers of water would lay their own pipe from the main conduit. There should be consideration, Cavalieri explained, of the distance that the pipe has to travel from the main conduit to its terminus. For every *canna* (a Roman *canna* at this time was 2.23 m or 7.33 ft.) from the main conduit, after the first three *canne*, the price (40 scudi per *oncia*) should decrease half a scudo. However, if the distance was very great, so as not to give the water as a gift, the price should be at least fifteen scudi. A pipe of one-half an *oncia* would cost half (i.e., 20 scudi), so that, Cavalieri continued, "many would be able to enjoy this benefit." Further, no one except cardinals should be allowed to purchase more than one *oncia* of water. Cavalieri provided two reasons for having the price diminish according to the distance of the receiver from the main conduit. First, the work (of the aqueduct as a whole) was done at public cost and everyone must participate equally. Those who lived close to the main conduit didn't have as much work (and thus as much cost) to lay their own pipeline, whereas those farther away would spend much more. Further, "the beauty consists in the many fountains throughout the city." Setting a high price for the water added to its cost, meaning that some would choose not to buy water, and the city would lack beauty with fewer fountains. Cavalieri concluded that with his low pricing, many would buy water and build fountains. The city, with its abundance of Acqua Vergine water, would make thousands of scudi.[102] This forward-looking piece of urban planning, with its understanding that Acqua Vergine water was plentiful and its consideration for both individual users and the city as a whole, was rejected.

Instead, at the Secret Council meeting of October 16, 1571, Cavalieri's proposed price of forty scudi per *oncia* was doubled to eighty scudi per *oncia,* forty scudi per half *oncia,* twenty scudi per fourth *oncia,* and so forth. An *oncia* was the amount of water that flowed through a pipe (*canella* or *fistula*) that was 1.86 cm in diameter. The actual amount of water would have been variable, since the rate of flow varied with conditions and could not at that time be measured accurately.[103] What people purchased when they purchased an *oncia* of water was not a fixed amount of water but the inflow of water from the aqueduct from a pipe of a particular diameter.

The Capitoline Council also devised regulations for the sale of Acqua Vergine water to private individuals (who could run a pipeline from one of the conduits to their house or palace). Prospero Boccapaduli (1505–1585)—who often served as one of the masters of the streets and was one of the deputies of the Capitoline Council who attended the Congregation on Streets, Bridges, and Fountains—drew up the regulations. There could be only one pipe per house or palace no larger than one *oncia*. The price of the water varied depending on the distance of the house from the main conduit (this part of Cavalieri's plan was accepted). The pipe had to be made of metal; pipe and drill (for boring into the main conduit) were specified; digging in public streets was protected as long as the digger repaired the street. Careful records of sales were required.[104]

◆ ◆

So the Acqua Vergine aqueduct repair was finally completed. It was not without problems, as an edict issued almost twenty-five years later (in 1598) makes clear. The edict discusses broken pipes, leaks, flooded cellars, and pipes laid too low—enabling more water to the owner but reducing the output for neighbors. The edict prescribed a uniform pipe level and steep fines for violations.[105] It suggests that Giacomo della Porta's glowing report of his and Bartolomeo Grippetto's work on the infrastructure may require some qualification.

Still, fresh Acqua Vergine water provided enormous benefits to cardinals and other elite persons living in the Campo Marzio as well as to ordinary Romans. The nine Acqua Vergine fountains, designed and built by Giacomo della Porta and Bartolomeo Gippetto, provided a real alternative to Tiber River water. They were pleasing embellishments of the city then and remain so today.

4

Contested Infrastructure

Broken Bridges

In 1561 a construction disaster occurred on the broken Ponte Santa Maria. This bridge, located just downstream from Tiber Island, had been damaged and rendered useless by the flood of 1557. We learn of the disaster from a comment published twenty-seven years later by Girolamo Ferrucci in his annotations to the 1588 edition of Andrea Fulvio's *L'antichità di Roma*. The mishap occurred during a repair effort initiated by Pius IV.[1]

The pope and Capitoline Council had agreed to repair the bridge with wood rather than with the more expensive but more durable stone "such that it might serve for the commodity of the carts and horses that bring much merchandise from *Ripa* [*Grande*] to Rome." A great sum of money was spent on the repair, growing to 8,000 scudi. Then one day, while a huge machine (*machina*) was being pulled "with capstans and ropes from one end of the bridge to the other, one of the heavier hawsers that sustained most of the weight broke, and the machine, falling headlong into the Tiber, was broken, and the whole went into ruin with very great damage to the builder who made it." The structure, Ferrucci repeated, "crashed precipitously into the Tiber, annulling everything and going into such ruin and perdition that nothing could be recovered."[2]

The Ponte Santa Maria—in 1561 a ruined structure with a huge gap at the center—was the first stone bridge to be built by the ancient Romans. The longest bridge in the city from ancient times, it possessed six arches and five piers. During its history the bridge was called by a variety of names, including Ponte Senatorio. Constructed in 179 BCE, its core is composed of tufa from the Aniene River. Subsequent repairs were made with the Roman technique of *muratura a sacco*—a rubble core conglomerate of lime mortar, stone rubble, broken brick, and other materials faced with stone.[3]

The bridge had been repaired several times in the fifteenth century. Then in the sixteenth, during the last years of the papacy of Paul III Farnese, in 1548 and 1549, signs of instability prompted another major repair effort. Initially Michelangelo (Michelangelo Buonarroti, 1475–1564) headed the work of stabilizing the bridge. After the death of Paul III in 1549, the new pontiff, Julius III, gave the job to the Florentine architect Nanni di Bartolomeo Lippi, called Nanni di Baccio Bigio. Baccio Bigio's repair of the Ponte Santa Maria lasted until the flood of 1557. The flood destroyed two arches, the bridge deck, and a pier (the one that had been repaired by Michelangelo and again by Baccio Bigio).[4]

The Ponte Santa Maria under Pius IV

Before the *machina* crashed into the Tiber River, the repair of the Ponte Santa Maria was a concern of the Capitoline Council as well as the papacy. In the summer of 1561, the minutes of the Capitoline Council reveal two worrisome issues: the materials to be used in the repair and how to pay for it. On July 11, 1561, the council agreed with the pope that using wood rather than stone for the repair, although less durable, would be far cheaper. A Public Council meeting formally approved the use of wood on August 14. The next month the council took up materials again. This time, on September 16, 1561, a Public Council decided that although the bridge would be repaired with wood, it was necessary to repair "some parts in stone" so that the wooden reconstruction could be placed thereupon.[5] These "parts" must refer to the pier and perhaps the arches as well.

But how to raise the funds to pay for the repair? At the July 11 meeting, the councilors heard that the pope had admonished them that whoever did the work should give security and that the masters of the streets should raise taxes for the project (i.e., collect a *gettito* door-to-door). On August 6, the Public Council decided that "the illustrious Lord conservators, the prior, the masters of the streets, and four chosen gentlemen [not named here] go to His Holiness and make him aware of the misery of the people." This delegation would urge the pope to allow money for the repair of the bridge to be taken from the *gabella dello studio* (the tax on foreign wine). On August 14, the Public Council was informed that the pope had graciously accepted the council's request and that the money would indeed come from the wine tax. On September 16, 1561, the Public Council estimated that the stonework would cost 400 scudi.[6]

For the rest of the year 1561, there is no mention of the bridge in the minutes. We can surmise that the large machine fell into the river shortly after, in October 1561.

The Island Bridges

Undoubtedly one reason the Capitoline Council members objected so strenuously to the collection of a *gettito* for the Ponte Santa Maria is that one had just been collected for the repair of one of the Tiber Island bridges, the Ponte Fabricio (called the Ponte Quatro Capi), which crosses from the island to the left bank.[7] A document dated July 30, 1560 (the summer before the Ponte Santa Maria fiasco), reveals that a submaster of the street—none other than Antonio Trevisi—went door-to-door collecting a *gettito* from surrounding dwellings and other structures, for the most part in neighboring Trastevere. Trevisi wrote in his own hand that the taxes were collected "by me Antonio Trevisi, submaster of the street," including from "some houses, warehouses, mills, and barns, for the repair of the Ponte Quattro Capi, which is very necessary and much frequented because of the ruin of the Ponte Santa Maria." The document recorded taxes levied on about 465 households and other entities in the vicinity of the island bridge. Most of the taxed entities were houses (306), for which the owners are named as well as non-owner residents. Other structures taxed included warehouses (*magazeni*, about seventy), workshops (*botteghe*, about fifteen), barns (*fenilli*, nine), grain storage facilities (*granari*, ten), two palaces, a monastery, a church, six bakeries, a hospital, two butcher shops, and seven mills.[8] The bridge was successfully repaired under the aegis of the next pope, Pius V.

The Ponte Sisto

Another crucial bridge, the Ponte Sisto (fig. 4.1), located upriver from Tiber Island, seemed on the verge of collapse. Almost a hundred years earlier, between 1473 and 1475, Pope Sixtus IV had presided over the construction of this beautiful bridge with its distinctive oculus. Masons had built the Ponte Sisto on the foundations of a ruined ancient bridge, the Ponte Aurelio.[9] Now, ninety years later, one of the piers was leaning dangerously. Pius IV, who had not yet given up on the Ponte Santa Maria, seems to have become alarmed. At a Secret Council meeting on August 26, 1564, the magistrates heard that the pope had insisted that the repair of the Santa Maria, but principally the repair of the Ponte Sisto, had become "very necessary because it threatens ruin as has been reported by skilled men."[10]

Figure 4.1 Ponte Sisto, Rome. Photo taken from the south side of bridge on the right bank. May 2017 © Bob Korn with kind permission.

From this date in August 1564 until Pius IV's death in December 1565, there is an ongoing and contentious exchange about how to pay for reconstruction of one or both of the bridges interspersed with papal exhortations about the danger of the Ponte Sisto's imminent collapse. As usual, the Capitoline Council preferred to use the *gabella dello studio*, but the pope insisted that money from these taxes from imported wine "were depleted"[11] and that the meat tax be used. A delegation from the council to the pope argued that the bridge was necessary "not only to the People, but to the clerics and the Camera Apostolica," and that they should contribute to the repair as well.[12]

The council tried other money-raising measures. A bond issue (*monte vacabile*) was created, but then the pope appropriated the income for other projects while admonishing the council "not to delay more" on the bridge.[13] The issue of both bridges became regular topics of discussion. This included a discussion on October 8, 1565, about the use of brick arches for the Ponte Santa Maria and the choice of "six or more" gentlemen to figure out how to do this and how to raise the money. At a special bridge-repair meeting on the same day, the Ponte Sisto restoration was postponed until April or May, when the wood could be acquired for 500 scudi; an architect would be sought for the Ponte Santa Maria to do the job for less than 5000 scudi; and the committee would beg the pope to contribute.[14] At a Secret

Council on November 21, 1565, the pope's (apparent) offer to help repair both bridges was reported, and the council resolved to find a way to raise the money.[15]

On December 3, 1565, the pope again urged that the council attend to the repair of the bridges. This time he had put two individuals in charge—Monsignore Di Macerata (someone from the city of Macerata in the Marches) and *signor* Gabrio, that is, Gabrio Serbelloni. He also had decided, with the advice of architects, how to best carry out the restoration. For the Ponte Santa Maria, "it is necessary to make two arches and a buttress [*sperone*] or pier, because the work would thus be much more trustworthy, secure, and durable, and in the case of the Ponte Sisto, it is necessary to remake that pier that is in danger of collapse." The council agreed to give every thought "to the way to find some of the money in order to begin with said work, and for now to provide at least the purchase of the necessary wood."[16] This was the last recorded discussion of the council before the death of the pope on December 9, 1565.

Repair of the Ponte Sisto under Pius V

The new pope, Pius V, seems to have had a pragmatic approach and an ability to delegate practical tasks to men who could actually get things done. One sign of such pragmatism is that, if the records of the Capitoline Council are any indication, he appears to have dropped the massive task of repairing the Ponte Santa Maria altogether. Not a single mention of this bridge is made in the council minutes during his entire papacy.

The instability of the Ponte Sisto continued to be a matter of grave concern. Four months after Pius V's election on May 7, 1566, the minutes of a Secret Council meeting noted that the pope had many times solicited the council to restore the Ponte Sisto and that it would be worthy to "consult on the method" and "to satisfy the Holy desire of Our Lord and for the public benefit."[17] Over the next weeks, the pope was relentless—insisting that the repairs be made but also insisting that because of other expenses, the treasury of the Apostolic See was exhausted and could not contribute.[18] On June 19, at a Public Council meeting, the pope relayed that he was "not a little astounded that up until now there is not seen any preparation and that he wishes in all ways that it [the bridge] be restored."[19] Five months later, on November 21, 1566, Cardinal Sforza scolded the magistrates: "for many months and years His Holiness, just as his predecessor, has ordered the conservators" to organize the restoration of the Ponte Sisto. The task was urgent "because clearly one sees that the bridge worsens every day." If it were repaired today, one would spend little, but if it fell, "one would spend much more and with major inconvenience to the people."[20]

The Capitoline Council created committees, looked for sources of funding, and tried to resist papal pressure. One idea was to follow the precedent of Pius IV's Acqua Vergine funding—divide the cost equally between the Papacy, the cardinals, and the Roman People.[21] Another was to sell some offices, picking two in the depository (treasury) of the Camera Apostolica.[22] On September 9, 1566, the Secret Council heard that the masters of the streets had returned from consulting with the pope, and because no preparations had been made for repairing the bridges, he was threatening to take the council's meat tax and use it to repair the bridges.[23] At last, on December 3, 1566, the Secret Council decided to use money from the sale of its house of printing.[24]

For each of these attempts to find money, small committees were chosen to pursue the issue further. However, after the print shop decision, "gentlemen" were chosen to actually find men to do the work. On January 23, 1567, a Public Council meeting was held in the palace of the conservator Antonio Velli. Attending were the conservator Domizio Cavalieri, the prior Angelo Albertoni, and three magistrates—Luca Peto, Tommaso Cavalieri, and Marcello Negri—who had been chosen at a previous Public Council meeting as "assistants of the work." The group also chose the architect who would supervise the project—Matteo da Castello.[25]

Matteo Bartolini da Castello (ca. 1530–1589) was from Città di Castello in northern Umbria and was in Rome by 1558 working on various projects—the roof of the Sistine Chapel, the Church of Saint John Lateran, San Giovanni di Fiorentini, and the foundation of Pius IV's new Porta Pia. He was called *muratore* (bricklayer or mason) in some reports, but he would soon be called *architetto*.[26]

In the minutes of a Secret Council meeting on May 9, 1567, we get a glimpse of the state of the work at the Ponte Sisto. The 1,000 scudi earned from the sale of the Commune's house of printing "has already been spent for lumber and other necessary preparations." Beyond this, lime (*calce*) and volcanic rock (*pozzolana*) had been brought to the construction site, but there was no money to pay for it. After consulting with the elected *signori assistenti,* it was decided that 2,000 more scudi could be taken from the fund that had resulted from the sale of the house of printing.[27] A Public Council approved this six days later on May 15.[28]

Two years elapsed without further mention of the bridge repair in the council minutes. Then on April 28, 1569, it was reported at the Secret Council that the pope had ordered the magistrates to pay the bricklayers (*muratori*) and stonecutters (*scarpellini*) who had worked on the Ponte Sisto the 225 scudi that they were owed. It was decided that they would be paid from the *gabella dello studio* (wine tax) and that further payment of 1,000 scudi for work on the Ponte Sisto and the Tiber Island bridges should also be paid.[29]

At last the Ponte Sisto and the island bridges were repaired.

Repair of the Ponte Santa Maria under Gregory XIII

The Ponte Santa Maria, ignored during the pontificate of Pius V, again became an issue when Pius V's successor, Gregory XIII, was elected on May 13, 1572.[30] After his coronation, Gregory immediately focused on preparing for a jubilee in the year 1575. His first urban projects were aimed at readying the city for what was hoped to be a significant influx of pilgrims. The Ponte Santa Maria was important for such a yearlong event because the bridge was crucial not only for carrying goods from the Ripa Grande to the left bank but also as a conduit for pilgrims visiting the seven churches of Rome and particularly for walking from St. Peter's on the right bank to St. Paul's outside the walls on the left bank. This pious practice of visiting the seven churches—initiated some two decades before by Filippo Neri, founder of the Oratorians—was enthusiastically embraced by Gregory XIII.[31]

Gregory XIII, like his predecessors, was compelled to consult with the Capitoline government. The first mention of his interest in repairing the bridge appeared in the minutes for the Secret Council meeting of March 12, 1573. There it was proposed that "public edicts be affixed throughout the city" about the proposed work, the cost, and the contractors. The decision to do this was deferred to a Public Council meeting (but does not appear to have been brought up at the next Public Council meeting).[32] A month later, on April 4, at another Secret Council meeting, the councilors were informed that N.S. (*Nostro Signore*, Gregory XIII) "wished in every way" that the bridge be restored before the Holy Year. To do this the pope wished to take the money "from the augmentation of the 1,250 shares of the meat tax sold by the People."[33] (An existing *monte* or bond issue could be augmented by the issuance of a new group of shares [or *luoghi*]) for the same *monte*.) In this case, 1,250 more shares of the *monte della carne* were issued, with interest to be paid by the meat tax. In response to the pope's wishes, the council decided to create another committee at the public meeting on April 6, 1573,. Four magistrates were elected to find ways to raise the money.[34]

About a month later, on May 2 at a Secret Council meeting it was reported that

> according to the order given by the Public Council we have seen in many congregations the designs of the architects and heard their opinions on the restoration of the Ponte Santa Maria and that in sum, all concur that the bridge be reconstructed with two arches for greater strength, more beauty, and greater endurance, and all likewise agree that 25,000 scudi be spent, not more.[35]

Beyond the design of the repair, the council conceded that the money would be taken from the meat tax, and thus, it was reported, 25,000 scudi were taken from the tax and then 250 shares (of a *monte* or bond issue with the meat tax providing interest payments) were to be sold.[36] This was all confirmed in a Public Council meeting three days later on May 5 with a vote of thirty-four beans in the green box, or *bussula* (for the plan), and three in the white box (against).[37] Four gentlemen, who were eventually put in charge, purchased lime for the bridge construction and arranged for the transport of the needed lumber.[38]

Work on the Ponte Santa Maria progressed. More than a year later, on August 3, 1574, it was announced at a Secret Council meeting that the 25,000 scudi for the repair of the bridge had been used up and that 10,000 or 15,000 more scudi were needed. The conservators agreed and it was then approved by the Public Council that 10,000 more scudi be drawn from the meat tax.[39] Then, on October 15 of the same year, it turned out that there was a lack of sufficient travertine to finish the restoration of the bridge but that in the Colosseum there was a great quantity of travertine under the ruins (not in the structure itself) that could be used. The architect, Matteo (Matteo Bartolini da Castello), asked permission to use it.[40] The Public Council turned the proposal down on October 19, with thirty-four beans against in the white box and twenty-four beans for in the green box.[41]

Despite the travertine problem, the bridge seems to have been at least almost complete. On November 12, 1574, less than two months before the opening of the jubilee, the Secret Council issued an edict to recognize the work of "master Matteo Architect" in the work of the restoration of the Bridge of Santa Maria. He labored "valiantly and well," and "he worked not only with ingenuity but also in person" (meaning he was on site). And, most important, he made "the work that can resist whatever the force of the river."[42] In the Public Council meeting of December 2, it was again resolved to recognize "master Matteo, architect, for his labors." The appreciation, which involved a monetary reward, was read by the scribe, and the council voted to approve it—eighty beans in the green box were yays, but twenty-one beans in the white box were nays. The reason for the nays may well have been the cost, for Matteo was then given 2,500 *scudi monete* "for recognition and remuneration of his labors" in the restoration of the bridge, and in addition he was made a citizen of Rome, "with its privileges."[43] He was further awarded a mill on the right bank of the river near the repaired bridge as well as a small piece of land.[44]

But Matteo was unhappy about construction procedures the Commune had ordered against his advice. He took the highly unusual measure of sending a letter (undated, dictated to a professional scribe) addressed to the first conservator, Gaspare Sanguigni, and the prior, Virgilio Crescenzi. (Because these two Communal officers served together only in July, August, and September 1575, the letter can be dated to these months.) Matteo feared that the repaired part of the bridge would fall down and take the rest of the bridge with it. He referred to a previous meeting of a congregation and listed the Roman magistrates and officers who were present as well as the architects Bartolomeo Gritti (i.e., Grippetto), Giacomo Della Porta, and himself. Here, he reported, the "many arguments" that "I brought in as evidence to prove the evident necessity according to my opinion that one must strengthen that foundation that is between the pier that now is rebuilt toward Trastevere and the other pier toward Rome."[45]

Matteo had wanted to execute this reinforcement immediately. The other attendees at the meeting disagreed. It was said that "there was plenty of time and that it was possible to wait for some days"—that is, to wait before reinforcing the piers. "I, aware of the necessity, couldn't help saying," Matteo continued, "that I disagree and that I implored the *Signori* that if by any chance our Lord did not do it, there might be another intervention, because I thought for sure that the continuous eroding and undermining" that was caused by "the velocity of the water on the said foundation" meant that it was about to be destabilized and that "in a short time" the foundation of the new pier would collapse and "ruin and make fall down all the remaining that was made with such great cost of this illustrious People."[46] He seemed to be desperately attempting to protect his reputation in the event that the collapse he feared would occur.

The letter sheds light on Matteo's desire—refused by the Capitoline Council—to use stone from the Colosseum. Here he proposed another source of material for the reinforcement. "I have always said," he urged, "to reinforce it with a height of five to six palms (about 44–52 in.) with those blocks from the Temple of Peace." He was referring to stone from the Basilica of Maxentius in the Forum, which in the sixteenth century was erroneously (from a modern archaeological point of view) known as the Temple of Peace (Templum Pacis). Matteo's unusual letter shows him to have been deeply disturbed. He mentions the great benefits and honors that he has received from the council and notes that he was writing "with my very deep grief" and "begging" the illustrious Lords."[47]

Whether the council responded immediately or not, the successful repair must have felt like a great urban victory, undoubtedly a source of great satisfaction. The repaired bridge was covered in travertine and was decorated in all the spandrels with the Boncompagni family insignia, the dragon, as well as the pope's coat of arms placed over the lunettes, which effectively removed insignia of the previous popes who had made repairs. An inscription still visible on the parapet credits both the pope and the Capitoline Council for the repair. Calling the bridge the Ponte Senatorio, the inscription seems to nod to the work of the Capitoline government.[48]

Matteo Bartolini da Castello died in 1589. He was thereby spared the knowledge of the bridge's collapse in the flood of 1598 (fig. 4.2).[49] Its remnant standing in the river today, the Ponte Rotto, or Broken Bridge, is what remains after two

Figure 4.2 Gerard Ter Borch the Elder, Ponte Santa Maria (Ponte Rotto). 16[09?]. Drawing, ink on paper. Rijksmuseum, Amsterdam.

other arches were removed in the nineteenth century (despite public outcry) during the project to build the Tiber embankment walls. One of the piers of the remaining arch that stands today, it should be noted, is the one repaired by Matteo, and it appears to be heavily reinforced.[50] Matteo's repair of the Ponte Santa Maria was a great, if temporary, achievement that satisfied Gregory XIII's desire to have a functioning bridge in time for the jubilee of 1575.

Water for the Hills: The Acqua Felice

The Acqua Vergine, completed in 1570, brought plentiful water to the Campo Marzio. But the hills of Rome lacked sufficient water for further development. This scarcity became a thorn in the side of Gregory XIII, who was partial to the papal palace that he was building on the Quirinal and its accompanying gardens. The Quirinal and other hills in the eastern part of the city had been a focus for the development of magnificent palatial streets for aristocratic Rome at least since Pius IV's construction of the new street (Via Pia) and gate (Porta Pia) in the 1560s.[51]

The solution to the problem lay in the creation of a new aqueduct out of the remnants of ancient structures. One possibility envisioned use of the ancient Aqua Traiana to conduct water from Lake Bracciano, a plan that Paul V would use to make the Acqua Paola in the early seventeenth century. The easier solution, taken in the sixteenth century, involved a reconstruction of the ancient Aqua Alexandrina. This aqueduct had been and could again be fed by springs in the area of Pantano del Grifo (Panato dei Grifi) between Montecompatri and Colonna, about twenty-three miles from Rome. Using arches from other aqueducts as well, it could supply water (it was believed) not only to the Piazza delle Terme (now Piazza della Repubblica) at the Baths of Diocletian (at this time being transformed into Santa Maria degli Angeli) but also to other areas of the Esquiline and Quirinal Hills, the Capitoline, the Caelian Hill, the Roman Forum, and the Forum Boarum. The new Acqua Felice, as it was called, was constructed and had reached the Piazza de Terme on June 15, 1587, and by 1589 it had reached many of the other fountains that had been projected within the city.[52] But these accomplishments were not without difficulties and conflicts.

In an edict of February 1583, Gregory XIII approved a plan for three private individuals from noble families—Ortensio and Fabrizio Frangipane and Orazio Muti, along with their associates—to undertake the huge engineering project on their own and at their own expense. The project could be done, the edict pronounced, with "modest expense," and the water could then be sold for much more. The construction would be completed "for their [the Frangipanes' and Muti's] advantage, and with their own care and expense," and they could sell this water for "a price not greater than 500 scudi for a single *oncia*." Part of the conduit could be made using the existing Alessandrino aqueduct. The three men were authorized to deputize any officials that they needed; they could force the sale of land required for the conduit, but they must compensate the owners with prices determined by the masters of the streets. Gregory prohibited anyone from opposing or obstructing the project. After work was begun, the men were given three years to conduct the water to the Piazza, and then "for our apostolic palace [that is, the papal palace on the Quirinal] six *oncie* can be had for free." If they had not completed the project in three years, they would be constrained to pay the Camera Apostolica 6,000 ducats, two thousand per year.[53]

Gregory XIII's desire for the new aqueduct was reported to the Secret Capitoline Council three months later on May 27, 1583. The pope "very greatly desired that water of the Pantano dei Grifi" be conducted to Rome, the council minutes reported. Such an aqueduct was also greatly desired by merchants and others who wanted to conduct the water "all at their expense." The pope noted that the aqueduct would be useful in both public and private realms. It would promote habitation in other parts of the city and would allow the construction of four or five public fountains, such as one on the Capitoline Hill for public use. The council in turn noted that when the aqueduct was constructed, fountains should be constructed at the Baths of Diocletion in the *rione* of Monti (the Esquiline) and "for public use and public squares and must not be assigned and diverted to other things."[54] This seems to be an insistence on public use only and probably is a veiled objection to the private nature of the project and the likelihood that the nobles who had been given the go-ahead would do so mostly for their own palaces, taking water themselves and selling it for their own profit to other elite Romans.

A few months before Gregory's death, an *avviso* made it clear that the Roman People (i.e., the Capitoline Council) were expected to contribute a share of the cost of the aqueduct. The Commune would pay 25,000 scudi, the *avviso* revealed, "with the authority from the pope to impose great hardship on many things" in order to obtain the 500,000 scudi needed for the massive project.[55]

Gregory XIII died on April 10, 1585, before the end of the three years given to the Frangipanes and Muti. Felice Peretti was elected Sixtus V on April 24, 1585. With remarkable rapidity, on May 28, little more than a month later, the new pope had the land purchased around Pantano del Grifo. He put Cardinal Ferdinando de' Medici (1549–1587) in charge of the aqueduct, which he called Acqua Felice, and hired Matteo Bartolini da Castello (who had successfully repaired the Ponte Santa Maria ten years before) to be in charge of the work.[56]

The speed with which Sixtus jumped into the project suggests that the new pope's hand may have been manipulating things in the background through the nobles Frangipane and Muti during the rule of his enemy, Gregory XIII. Whether or not, it was relevant that his great new palace and garden complex, Montalto, would be a direct recipient of the new water supply (fig. 4.3). Like Cardinal Montepulciano before him (who had needed the revitalized Acqua Vergine for his palace and gardens on the Pincio), Sixtus needed a better water supply for the palatial new estate that he and his sister Camilla Peretti (1519–1605) were constructing on the Esquiline near the Basilica of Santa Maria Maggiore. (The Villa Montalto was destroyed in the nineteenth century to make way for the train station.)[57]

In his edict, Sixtus noted that the areas around the Lateran and adjacent places (i.e., mostly the hills) suffered from a scarcity of water and for that reason had few buildings "and absolutely is scarcely useful for the owners." He wished the new aqueduct "to be called Acqua Felice after me." The edict called for Ferdinando de' Medici and the "Commissary" of the Camera Apostolica (Giovanni Bernardino Piscina) to meet with the Colonna family (the noble family that owned the land) and agree on a price for the purchase of the water of Pantano del Grifo and lands around it. They were also to "negotiate and make an agreement with Matteo Bartolini da Castello as to what is necessary for bringing the water, whose charge, as we understood, he was about to receive." They were to establish Matteo's salary and attend to other details necessary to the success of the project. The water was to be brought to the Piazza (of Santa Maria degli Angeli) within two years.[58]

A document of June 1, 1585, stipulated the conditions of the sale of the water

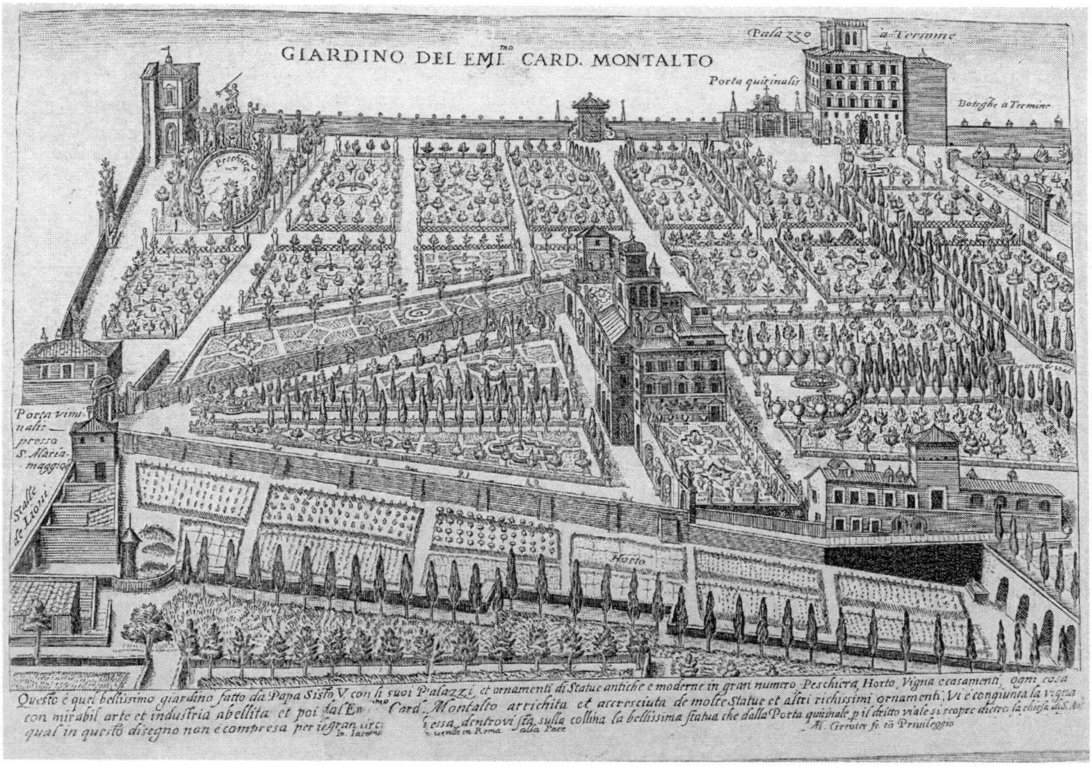

Figure 4.3 Matthäus Greuter, Villa Peretti-Montalto with gardens. In album *Giardini di Roma e dintorni*. Ca. 1623. Engraving/etching on paper. 9½ × 14 in. (242 × 351 mm). Rijksmuseum, Amsterdam.

at Pantano del Grifo to the Camera Apostolica. The Colonna family member who held the land, Marzio Colonna, sold all the water from Pantano del Grifo, including water used by the mill and the ironworks. Other stipulations were laid down that would allow the aqueduct to be constructed, including use of stone on the land, access for workers during construction and future upkeep, and prohibitions from impeding in any way the flow of the water. The *signor* Colonna would be paid 25,000 scudi for this sale. Further, ten *oncie* of water must be left to the seller either within the city or outside.[59]

An *avviso* on August 24 reported that a congregation concerning the aqueduct met in the palace of Cardinal Ferdinando de' Medici. This refers to the Palazzo Medici on the Pincio, which Ferdinando had purchased from Cardinal Montepulciano's heirs in 1576. The Medici cardinal's vast plans for the gardens of his newly acquired palace, including the spouting fountain known as the Parnassus, required a more plentiful and forcefully flowing source of water than the Acqua Vergine could provide. A successful Acqua Felice would fill the bill. The work on the aqueduct started less than a month later as evidenced by one of the extant account books.[60] These account books provide a detailed view of work on the aqueduct.[61] On the day that work began (September 28), an *avviso* reported that Sixtus desired to conduct to Rome "a head of water so great that it would serve not only Monte Cavallo [the Quirinale], where he plans to bring it, but many other places in the city," this showing "the great mind" of the pope.[62]

From September 28, 1585, until July 7, 1586, Matteo Bartolini da Castello directed the work. Payments went to men for diverse jobs, including demolition, construction, transport of supplies, and provisioning of workers. They included masons (*muratori*) and trench diggers (*cavatori*), carpenters and lime workers, men who constructed barracks (*capanne*) for the hundreds of men working on the project, carters of wood and other materials, water carriers (bringing water to slake the lime), and a surveyor. The account book recorded the hiring of cargo boats to carry materials (on the Teverone, i.e., Aniene River) as well as payments for metals such as copper, tin, and iron, and for nails. It recorded payments to provisioners of food and of horses.[63] On October 5 an *avviso* reported that three cardinals—Ferdinando de' Medici, Pierdonato Cesi (1521–1586), and Sixtus's grandnephew, Cardinal Montalto (1571–1623)—visited the Pantano del Grifo springs.[64] The great project was forging ahead.

To build such a gravity-powered aqueduct successfully, it had to be built with an appropriate gradient. This was not an issue if an ancient structure had been used (such as most of the Acqua Vergine), but for the Acqua Felice extensive new sections had to be built. The surveyors would use some kind of leveling and sighting device (and there were several varieties) and proceed section by section. The instrument would be placed along the course and had to be made perfectly level. Then one person made a sighting along a straight line to a distant object or assistant holding a calibrated rod upright. If the sighting were done correctly, a straight line could be established from one point to the next, and then the desired gradation provided in the construction. This procedure would be repeated over and over, section by section, until the entire length of the aqueduct had been built.[65]

Despite great efforts and expense, less than a year later, Matteo da Castello's work ended in failure.[66] A test was made of the part of the channel already dug, and the water flowed back toward the springs rather than down toward Rome. Matteo was immediately removed as supervisor of the project, which was handed over to Giovanni Fontana (1540–1614), the older brother of the more famous Domenico Fontana (1543–1607). On July 26, 1586, after Matteo had been dismissed, two cardinals, Ferdinando de' Medici and Alessandro Farnese, and Domenico Fontana persuaded the pope that for another 60,000 scudi, a plan to draw water from the higher springs of Zagarola would solve the problem. Sixtus V sent Giovanni Fontana to the springs to check them out.[67]

We know from an intriguing and little-known manuscript source—a manuscript book in the Biblioteca Nazionale of Naples by a fountain expert, Giovanni

Figure 4.4 Giovanni Antonio Nigrone, illustration of a leveling/surveying instrument for surveying an aqueduct. Biblioteca Nazionale di Napoli. Ms. XII G. tom. 59, fol. 274r (previously 266r). Su concessione del Ministero dei Beni e delle Attività culturali e del Turismo © Biblioteca Nazionale di Napoli.

Antonio Nigrone—that a survey was made as the project began again under Sixtus V. Nigrone, who was a client of Ferdinando de' Medici and had been designing his gardens at the Medici Palace on the Pincian Hill, went on the expedition to resurvey the springs. The expedition also included the sculptor Bartolomeo Ammannati from Florence, the shamefaced Matteo Bartolini da Castello, Domenico Fontana, and Raffaello da Sangallo. At issue was whether water could be pulled from the springs of Zagarolo, near the Pantano del Grifo but at a higher altitude. In his manuscript book, which was mostly devoted to fountains, Nigrone provided an account of the expedition. He first explained and provided vivid illustrations of the variety of level and sighting instruments that could be used to do the surveying for an aqueduct. He emphasized that it was essential that the "leveled rule" (*regholo a livelo*) was in fact perfectly level. He noted that Matteo's survey had been off by twenty-three palms (about 17 ft.). He had corrected the problem with an invention of his own—a small sighting device attached with straps to the level which, Nigrone claimed, gave a far more accurate reading (fig. 4.4).[68]

The beneficiary of Matteo da Castello's fiasco was Giovanni Fontana. Giovanni was from the large and successful Fontana workshop led by the papal architect, Domenico, who played a central role in Sixtus's numerous construction projects. The Fontana family had come to Rome from the small town of Melide in the canton of Ticino on Lake Lugano in present-day Switzerland. Giovanni's brother Domenico arrived in Rome in 1563 attracted to the work opportunities offered by the Roman construction boom.[69] Giovanni became known for his expertise in hydraulic projects.[70]

In his famous book, *Delle trasportazione dell'Obelisco Vaticano,* published in 1590, Domenico Fontana, without mentioning Matteo's fiasco, reported on the Acqua Felice project. He emphasized the difficulties of the project—"an undertaking certainly that cedes nothing to the ancients"—because the source of the water at the springs was a mere forty palms higher than the location (at Piazza delle Terme) where it was to be delivered, "so that it was necessary to use [in surveying the channel] an almost miraculous diligence." Most believed that it could not succeed.[71]

In his glowing account of his brother Giovanni's successful construction of the Acqua Felice, Domenico Fontana did not provide the nitty-gritty details of Matteo da Castello's debacle. However, it turns out that his brother Giovanni did write down the particulars of his survey and supervision of the project, and his notes were handed down within the Fontana family. These details of Matteo's failure appear in a tract on the Acqua Felice published more than a hundred years later in 1696 by Francesco Fontana (1668–1708), a descendent of Giovanni and Domenico and son of the better-known Carlo Fontana (1638–1714). Here Francesco describes the disgrace of Matteo da Castello in detail, reporting that Matteo, wishing "to make a test of the entrance of the water [into the canal] toward Rome, became aware, but too late for his own shame" that the water turned back, the reverse of what he had planned, "because it did not have enough force to climb where the architect firmly believed that it would."[72]

Giovanni Fontana, on the other hand, was "forced to use a wonderful diligence, combined with extreme labor, as he attested in some distinct records in manuscripts made by him concerning this work, which are conserved in my house, as coming from the same stock [i.e., he was from the same family]."[73] This is a remarkable record not only of the details of an engineering project completed a century before but as evidence that a family of architects kept and handed down family papers about their projects from one generation to the next.

Back in the sixteenth century, Domenico Fontana explained that although the springs had been found on the Colonna property, excavations also had been made two miles farther in order to find a greater water supply. He described the aqueduct as traversing twenty-two miles in a winding path. At times it was carried by tall arches, reaching a height of seventy palms (about 51 ft.) and a width of

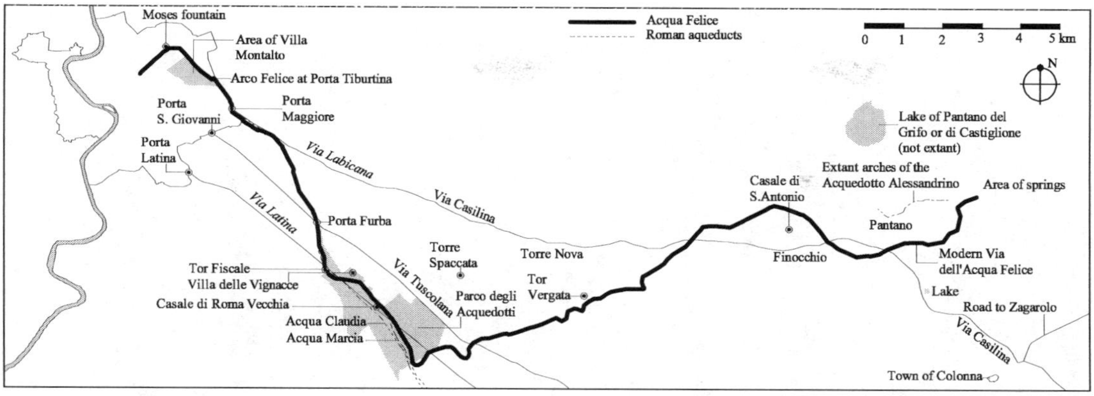

Figure 4.5 Course of the Acqua Felice outside of Rome from its springs (*far right*) into the city. Courtesy Chiara Bariviera with kind permission.

twelve palms (about 8¾ ft.). The aqueduct ran seven miles above ground and fifteen below. The underground section could be as much as seventy palms deep (about 51 ft.), and it had often been necessary to cut into hills. Two thousand men worked continuously, and at times even three or four thousand. It had taken eighteen months to complete and had cost 270,000 scudi, including the 25,000 given to Marzio Colonna for the water and land. The water was sent to all parts of Rome that previously had been abandoned because of dryness. Fontana provided an idyllic picture of the changes brought about by the new aqueduct. Sixtus, he reported, gave water to pious places, monasteries, cardinals, and other gentlemen who had gardens and vineyards in these places. The areas began to be inhabited again, and new palaces began to be built there. There was "almost a new Rome, and gardens receiving and drinking the moisture that they have so long desired, now grow fresh and verdant."[74]

The new aqueduct was a pastiche of old aqueduct sections and new constructions. Part was the ancient Alessandrino aqueduct. Using extant account books, Giancarlo Palmerio has analyzed the ways in which the builders used ancient aqueduct sections and in addition took materials such as stone from ancient aqueducts along the way to make the new Felice. The project was an enormous one, and it was also very spread out. Workers were organized into camps along the path of the conduit being constructed, and barracks were constructed for each camp. Extensive logistical support supplied provisions such as food and construction materials needed for the task at hand. As the project went along, the barracks were taken down and new ones constructed at the next section of the work by teams of carpenters. The trench diggers and masons probably worked together, the first group to dig and also survey ancient ruins for usable materials, the second to build the walls and other structures of the aqueduct (fig. 4.5).[75]

Surveyors, or "levelers" (*livellatori*), were a constant presence. They had to en-

Figure 4.6 A section of the channel of the Acqua Felice running along the surface of the ground with a stone manhole cover. Parco degli Acquedotti, Rome. Photograph May 2017 © Bob Korn with kind permission.

sure that the same gradation was maintained throughout. They were divided into squads and used various heavy leveling instruments that required a team of men to move. Trench digging was carried out in some places by creating open trenches, in which case the conduit had to be vaulted. At other locations the trenches were dug underground. Sometimes workers discovered ancient underground conduits that couldn't be used, and these had to be pierced through. Two crews of fifty men each dug vertical shafts along the pathway of the planned work. These shafts may have been used to establish the level of the underground conduit, the gallery being dug from one shaft to the next. Some if not all of these shafts were left to be used for future inspections and maintenance. During the work, more than fifty-two underground streams were discovered, which added water but slowed down progress (fig. 4.6).[76]

Some parts of the construction required entire subindustries, such as the preparation of lime for mortar. Huge quantities of lime were purchased and burned in lime kilns and combined with the volcanic rock pozzolana to make mortar. This included the major lime kiln close to the Porta Maggiore outside the city walls. Much mortar was required to bind the various found stones together, stones such as peperino, travertine, tufa, limestone, and also brick collected from the nearby remains of the ancient aqueducts, especially the Claudia and the Marcia. Thus costs were saved both from quarrying and transporting newly excavated stone. Near the Porta Maggiore as it entered the city, the new aqueduct was constructed on the arches of the Aqua Marcia and the Aqua Claudia.[77]

By December 1586, the Acqua Felice waters had arrived at Piazza di Santa Susanna (now Largo di Santa Susanna), where a terminal fountain or *mostra* was to be built (the future Moses Fountain). The water had been calculated to be able to be raised fifteen or sixteen palms (about 22 ft.) at the terminus. But a test on December 22 revealed that the water rose only 3.1 palms (about 4⅓ ft.), thereby preventing a spectacular or even minimally acceptable water display. No one was fired this time, but to remedy the problem, the Piazza di Santa Susanna was regraded and lowered as was the stretch of Via Pia (now Via XX Settembre—Via del Quirinale) to the Piazza del Quirinale. Domenico Fontana designed the Moses Fountain with three arches, the center one displaying a giant (and much disparaged) Moses sculpted by Leonardo Sormani (active in Rome 1550–ca. 1590) and Prospero Bresciano (or Antichi). Thereby, Sixtus tied his pontificate to the rule of Moses, who had brought water to the Sinai Desert just as he the pope was bringing water to the Quirinale Hill. Sixtus's connection to antiquity (and ancient wisdom) was reinforced by the Egyptian lions and medieval lions (four lions in all) from the mouths of which drinking water spouted. The triumphal arch, reminiscent of imperial Roman structures, bore a large inscription describing the new aqueduct and in praise of Sixtus.[78]

The triumphal display also served as a giant distribution tank. In a building attached to the rear of the fountain lay three large basins on the floor behind the back wall of the fountain. These basins received all the water from the conduit of the aqueduct and divided it into three parts, beginning its distribution to three different areas of the city. Then water poured into three additional outside basins (in front of the arches), flowing through horizontal openings in the exterior wall facing the piazza. These exterior basins were also fitted with pipes and thus acted as distribution tanks as well.[79]

Water from the Felice was a boon to the city—especially to the hills. Numerous fountains were built, both during the pontificate of Sixtus and in subsequent decades. The fountains on the Quirinal and other hills were sponsored and paid for by the Camera Apostolica and by private users, while others were "sponsored" by the Capitoline government. Conduits ran under the Via Pia and the Quirinale to another distribution tank on the western edge of the Quirinal. Additional conduits were run from this Quirinal *castellum*, including two siphons, one to private houses, monasteries, and convents, and then to a public fountain, constructed by Giacomo Della Porta, at the church of Santa Maria ai Monti (also called Madonna dei Monti) in a densely populated neighborhood. A second conduit from the Quirinal *castellum* ran to the new fountain on Capitoline Hill in front of the Senatorial Palace. This Capitoline fountain was designed (with some controversy and required revisions) by none other than Matteo Bartolini da Castello, who seems to have recovered from his dismissal as head of the Acqua Felice project as a whole.

The conduit from the Quirinal to the Capitoline was made with 1,480 terracotta pipes along with lead pipes (which withstand more pressure) for the siphon going up the Capitoline Hill.[80]

Understanding that the completed Acqua Felice would deteriorate if it were not maintained, Sixtus V issued a bull on February 19, 1590, *Suprema cura regiminis,* that provided for ongoing maintenance. The new Congregation on Streets, Bridges, and Waters was in charge. In his bull, Sixtus boasted that the construction of the aqueduct had been undertaken to restore the hills of Rome to their pristine splendor. An annual fund for the conservation of the aqueduct was established (of about 700 scudi). Two Roman citizens would be chosen as annual inspectors of the aqueducts and fountains and would receive an annual salary of 123 scudi to be taken from the *mons religionis* (a bond issue established by Pius V) and divided equally between them. They must check every part of the structure, report, and then repair all damage. If it was caused by their negligence, they must repair it at their own expense. Along the channel, a strip of twenty feet on both sides must be clear of structures and vegetation. It was forbidden to get water from the aqueduct other than at specific fountains. Penalties were established for those who polluted the water or caused damage to the structure and for those who illegally drained off water other than from the named fountains.[81]

◆ ◆

The infrastructure projects examined in this chapter at times were completed successfully and at other times they failed. The Ponte Santa Maria was a failed structure in 1557, a successful one in 1575, and a failed one in 1598. Beyond considerations of success or failure, however, the projects possessed certain things in common. First and foremost, their funding was not an agreed on matter nor one in which any prior provisions ever had been made. Further, the architects, engineers, and overseers of the projects, although in some cases old hands, were chosen in a process that was sometimes conflicted and often influenced by patronage considerations. There was a lack of standardized protocols for carrying out engineering projects, whether aqueduct construction, the laying of conduits, or the repair of bridges. This led to discussions; trial and error; changing contractors and architects; success or failure; and conflicts over money, personnel, and how best to approach the problems at hand.

5

Roman Topography and Images of the City

While Romans endeavored to repair and reconstruct their city, they also investigated the urban terrain for its ruins and antiquities. Usually by foot or on horse, interested parties searched for aqueduct conduits, explored and measured ancient ruins, and dug for ancient statues and coins. And while Rome was being rebuilt and explored, it was also being imagined in printed city views and maps.

City plans represented idealized images of the city, and they expressed diverse opinions about Roman topography—both ancient and contemporary. Making images of Rome went hand in hand with ongoing discussions and debates about ancient monuments and sites. Which ancient events happened around which of the ancient (now ruined) buildings? Where were sites mentioned in ancient texts located? How were known structures used in ancient times? Creating maps and images of the city was not simply a matter of visually recording the existing urban landscape or of accurately drawing carefully measured spaces.

Maps and City Views

The production of maps was a burgeoning activity in sixteenth-century Europe that was spurred on by expanding commercial interests in the

East, oceanic exploration and the discovery of new lands, the humanist investigation of ancient geographical texts, and the flourishing culture of print. As technical precision and accuracy increased, so did the intensity and variety of symbolic, political, and cultural functions of maps, including urban maps and city images.[1]

From midcentury this prolific production of maps and city views contributed to Rome's culture of print. Pilgrims and other travelers to the city eagerly snatched up souvenir prints depicting monuments within Rome and images of the city as a whole as it was in antiquity or in the present or a combination of both. From the 1540s the intense production of maps had a corollary in fervent archaeological investigation of Roman ruins. Discussions and arguments about ancient locations and structures could aptly be expressed in cartographical form. Maps also could demonstrate princely and noble authority by showcasing the territory they ruled. Printed maps could be displayed on walls, inserted into atlases, or gathered up into collections to be sold.[2]

By the second quarter of the sixteenth century, architect/engineers and antiquarians had begun to survey the third-century Aurelian Walls surrounding Rome—initially built between 271 and 282 CE by the emperor Aurelian and his successor Probus—and to record their results in cartographical form. Maps based on surveys sometimes displayed conclusions or arguments about ancient topography. Ancient monuments were also measured and drawn.[3]

In the sixteenth century, drawing maps to a fixed scale revolutionized mapmaking and brought together pictorial mapmakers and surveyors. Maps measured to scale had been produced in other times and places, such as in ancient Mesopotamia, in ancient and medieval China, and in ancient Rome, but these had not resulted in continuous cartographical practices. In late medieval Europe, surveyors for the most part measured land, drew boundary lines, and then described them in writing. Makers of city views did not create surveyed maps but rather depicted urban images without connecting distances on the plan proportionately to distances on the ground. Pictorial city views were not surveyed maps. Bird's-eye views of cities used pictorial perspective to establish a vantage point beyond the urban perimeter from a hill or an imaginary height. Most maps of Rome made in the second half of the sixteenth century used a combination of pictorial, surveying, and measuring techniques.[4]

Bufalini's famous woodcut map of 1551 represents an important exception to this combined pictorial and measured approach (see fig. 1.6). It is the first freestanding map of Rome since antiquity made true to scale by orthographic projection. (Orthographic projection is the projection—with lines perpendicular to the drawing surface—of an object such as a building. With orthographic projection the plan of a building can be made to scale with all its elements shown as accurately measured.) In Bufalini's map, each viewpoint was represented as though from directly above in strict orthogonal projection. The map is not totally accurate if judged by modern standards but was remarkably accurate for its time.[5]

Measurement, Surveying, and Mapmaking

Orthographic projection required the adoption of new mathematical techniques and the development of surveying instruments, both crucial to mapmaking in Rome. Mathematical practices included the taking of measurements on the ground, surveying with instruments, and the development of graphic techniques that could depict proportional dimensions on a flat surface. In contrast, most maps of Rome in the late sixteenth century were pictorial representations made by using perspectival techniques. Perspective was an essential technique for creating realistic-looking, recognizable cityscapes, but, because perspective required the use of orthogonal lines drawn to one or more vanishing points and resulted in foreshortening, such images could not reveal actual measurements of length, width, and distance.[6]

In contrast, Bufalini's map is an ichnographic plan made by orthographic projection. An ichnographic plan of a building is a ground plan, a horizontal section that shows the walls and other features measured to scale. Bufalini's map showed all the features of the city as in a ground plan. On a single horizontal plane he depicted the city as if from hundreds of viewpoints, each one directly above each topographical feature. The measurements of the streets, walls, and other features could be displayed and every element within it drawn to scale. Although much less recognizable than the everyday view of the city, ichnographic plans are more usable for locating structures and for understanding the actual size and dimensions of the city and the spatial relationships among the constructions within it.[7]

The development of surveyed maps was crucially influenced by the reception and wide assimilation of an ancient text—Claudius Ptolemy's *Geographia*. The *Geographia* was rediscovered by the Byzantine scholar Maximus Planudes in 1295 and was also known in the world of Arabic scholarship. In the early fifteenth century, the Greek scholar Manuel Chrysolaras brought a copy to Florence, where it was translated into Latin by his student Iacopo Angeli da Scarperia and became known by its Latin title, the *Cosmographia*. As Patrick Gautier Dalché has shown, the *Cosmographia* was in fact known in Western Europe before its arrival in Florence. But with Angeli's Latin translation, it began to exert enormous influence on the development of cartography. Ptolemy had located approximately eight thousand sites in the known world and listed their coordinates, and he showed how to project the spherical earth onto a flat surface. Fifteenth-century scholars created and corrected numerous maps based on their study of the *Cosmographia* and also supplemented its information.[8]

Alberti's Survey of Rome

A century before Bufalini, the fifteenth-century humanist and polymath Leon Battista Alberti (1404–1472), influenced by the *Cosmographia,* first elucidated techniques for measuring and mapping Rome. While residing in the city between 1443 and 1455, Alberti developed a method for creating a measured map of the city, which he explained in 1450 in his *Descriptio urbis Romae.* This tract, combined with what is known about Alberti's surveying methods from his small treatise on practical mathematics, the *Ludi rerum mathematicarum* (written in the 1440s), reveals how he may have proceeded in his survey of the city. No fifteenth-century map created by Alberti's method survives, and scholars have suggested that he wrote the *Descriptio* and distributed it without an accompanying map because he was mindful of the poor ability of the manuscript tradition within which he worked to accurately transmit complex drawings and other visual materials.[9]

Alberti described his method of creating a map from a high, more or less central point in the city, namely, the Capitoline Hill. First he constructed an instrument that consisted of a calibrated disc that he called a horizon (*horizonte*) and a movable rule or radius. He divided the perimeter of the disc into forty-eight degrees (*gradi*), each subdivided into four *minuta.* At its center he attached the moveable rule or radius that reached to the circumference of the circle and that was itself divided into fifty degrees, each also subdivided by four (fig. 5.1). The degrees on the radius were proportional to measured feet on the ground. The procedure required standing on the Capitoline Hill with the disc fixed and flat and the first degree unit turned toward the north. Walking around the instrument, the surveyor pointed the radius at a structure he wanted to survey. The point of the radius on the perimeter gave him the horizon degree and minute, indicating the direction or bearing of the structure.[10]

The distance of specific buildings and other structures from the Capitoline Hill also had to be determined. Although it is not known precisely how Alberti found such distances (which he listed in the *Descriptio*), he was familiar with several methods of surveying and could have used one or more likely a combination of them. He may have paced the distances. He may have used a hodometer, described in the *Ludi rerum mathematicarum,* which was a wheeled instrument with a little hole bored into the axle so that at every rotation of the wheel, a pebble would fall from a container attached above the axle into a small pouch below, the distance then being calculated by counting the pebbles. The *Ludi* also described a primitive form of the surveying instrument known as a transit, and it outlined the principles of triangulation whereby a surveyor could discover the position of a building without direct measurement.[11]

Alberti determined the relationship between a degree of the radius and a mea-

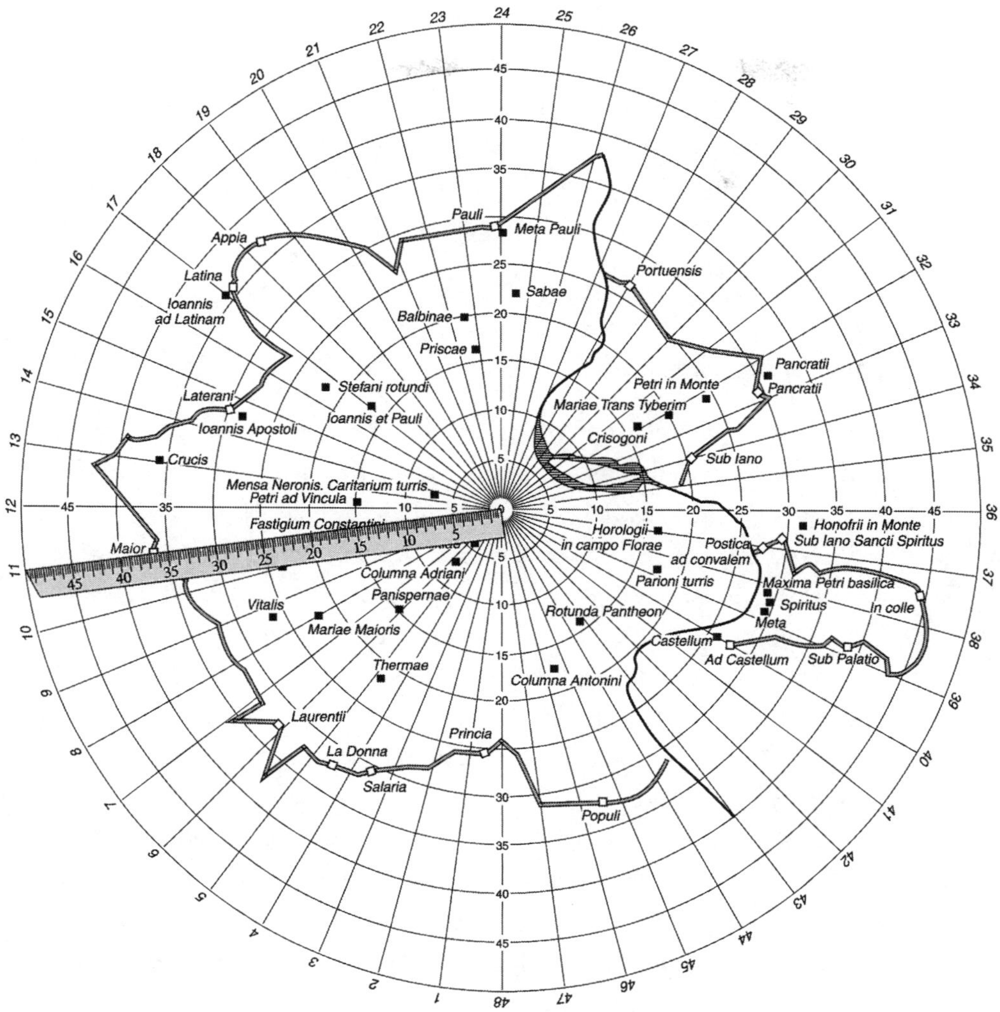

Figure 5.1 Photoshop construction with the radius placed on a map of Rome constructed with Alberti's horizon, north at bottom, east at left. The map (without radius) was constructed by Vasilij P. Zubov and published in Leon Battista Alberti, *Desât' knig o zodčestve* [*De re aedificatoria libri decem*] (Moscow: Izdatel'stvo Vsesoûznoj Akademii Arhitektury, 1937), 2:129. Zubov followed the text of Alberti's *Descriptio urbis Romae* as established (with some errors) by Girolamo Mancini (1890). At the center of the map is Capitoline Hill. The radius shows the bearing of Porta Maggiore (*Maior* on map; 11°) and its distance from Capitoline Hill (36°2′). Alberti did not give the scale of radius degrees to measurements on the ground, but given the known distance between the Capitoline Hill and Porta Maggiore (1.667 mi. or 2.68 km), we can calculate that each degree on the radius was valued at about 241.14 feet or 73.5 meters.

sure on the ground. Knowing the distance from the Capitoline Hill to the structure in question would then have given him the degree and minute indicated by the written calibrator on the movable radius. Most of the *Descriptio* is made up of tables giving the location of the object on the perimeter of the *horizonte* and its location on the radius. To create the map, Alberti would have placed the horizon on a drawing surface, and put the Capitoline Hill at the center. Taking the information from his tables and using the horizon, he would then have been able to trace the course of the city's walls and gates and the course of the Tiber River (connecting the dots between measured locations such as bridges over the river and gates in the walls), and he would have been able to place significant churches and monuments on a drawn map. The result would have been an orthogonal plan giving precise spatial locations of buildings and reference points such as walls and the distances separating them.[12]

Measuring and Depicting Sixteenth-Century Rome

After Alberti, surveying and measuring techniques, instrumentation, and map-making techniques developed apace, creating the foundations for the flowering of Roman mapmaking. In his map of Imola, a town near Bologna, Leonardo da Vinci created the earliest extant ichnographic map from the Renaissance (fig. 5.2). Leonardo's approach was the one later described by the painter Raphael (1483–1520) in his letter to Pope Leo X written between 1513 and 1520.[13]

Raphael wrote his letter in the context of his intense study of Rome in collaboration with humanists such as Angelo Colocci (1467–1549). The letter contained the first definitive statement of the three types of drawings made with orthographic projection, namely, plan, elevation, and section. Raphael described his surveying instrument, which he called a *bussola,* in detail. It was similar to Alberti's "horizon" with the addition of a magnetic compass at the center to facilitate the establishment of a baseline on the north-south axis. Although Raphael described it for the first time, it was undoubtedly in fairly common use among surveyors, and it was the instrument that Leonardo used for creating the Imola plan.[14]

Surveying and the creation of measured drawings are increasingly in evidence from the 1530s. Particularly significant for these activities was the extensive work on Roman city walls and fortifications ordered by Paul III in the 1530s and 1540s. The work was headed by the architect Antonio Sangallo the Younger (1484–1546) and included the repair of the Aurelian walls.[15] In addition, architect/engineers began the construction of new defensive walls around the Borgo and strengthened the Vatican fort that was the Castel Sant'Angelo (a project taken up again in the 1550s and 1560s), which required the construction of bastions. Extant is a collection of hand-drawn and surveyed maps in the Vatican Library including a drawing of the walls circumventing the city and a number of maps showing measured plans for fortification of the Borgo. In the Uffizi in Florence, drawings by the architect Antonio da Sangallo the Younger show his measured survey of a section of the Aurelian wall (fig. 5.3).[16]

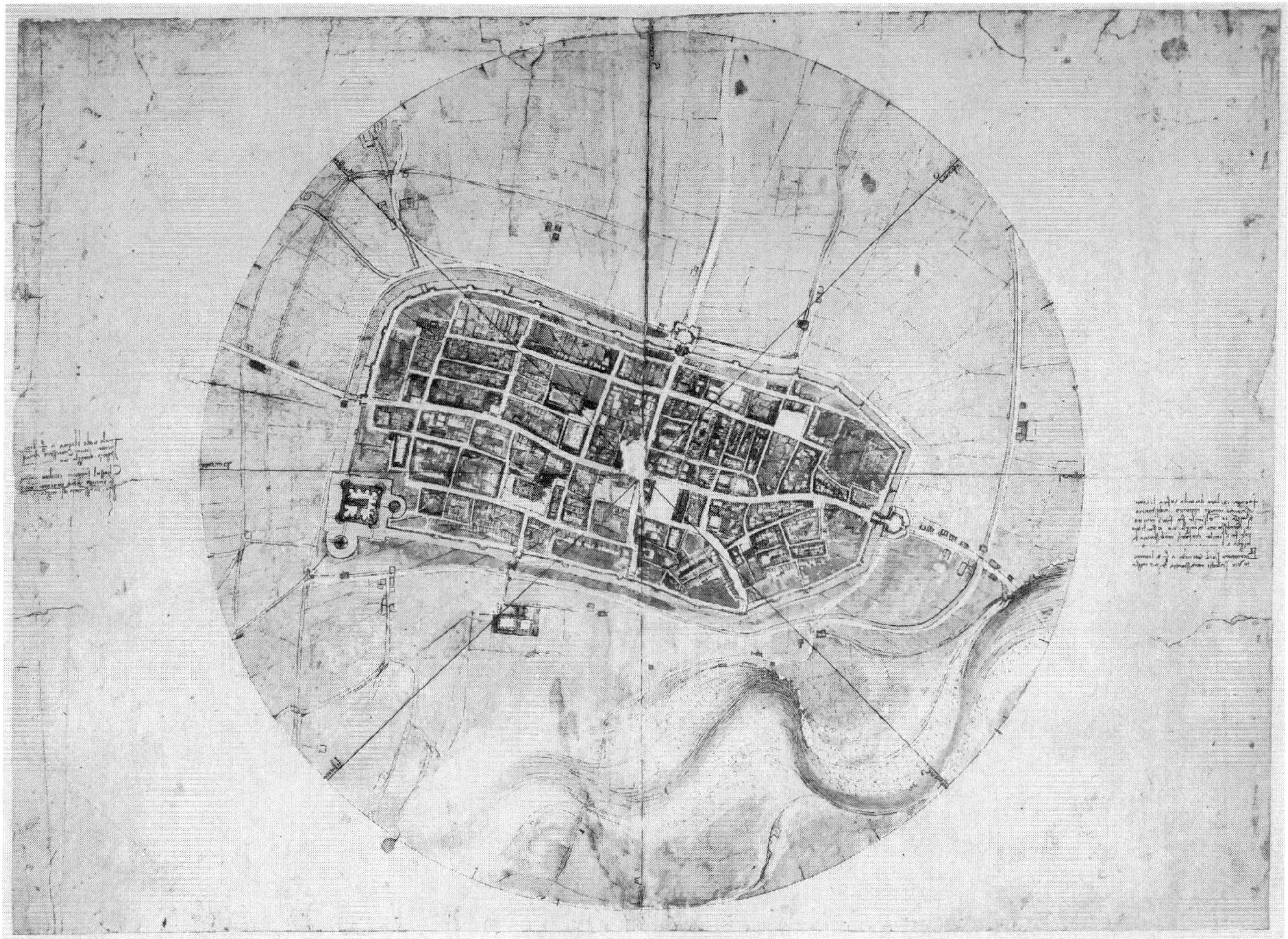

Figure 5.2 Leonardo da Vinci, *Plan of Imola*. 1502. Pen, ink, and watercolor, 17⅛ × 23¾ in. (44 × 60.2 cm). Windsor Castle, Royal Collection, 12284. Photo Credit: Scala/Art Resource.

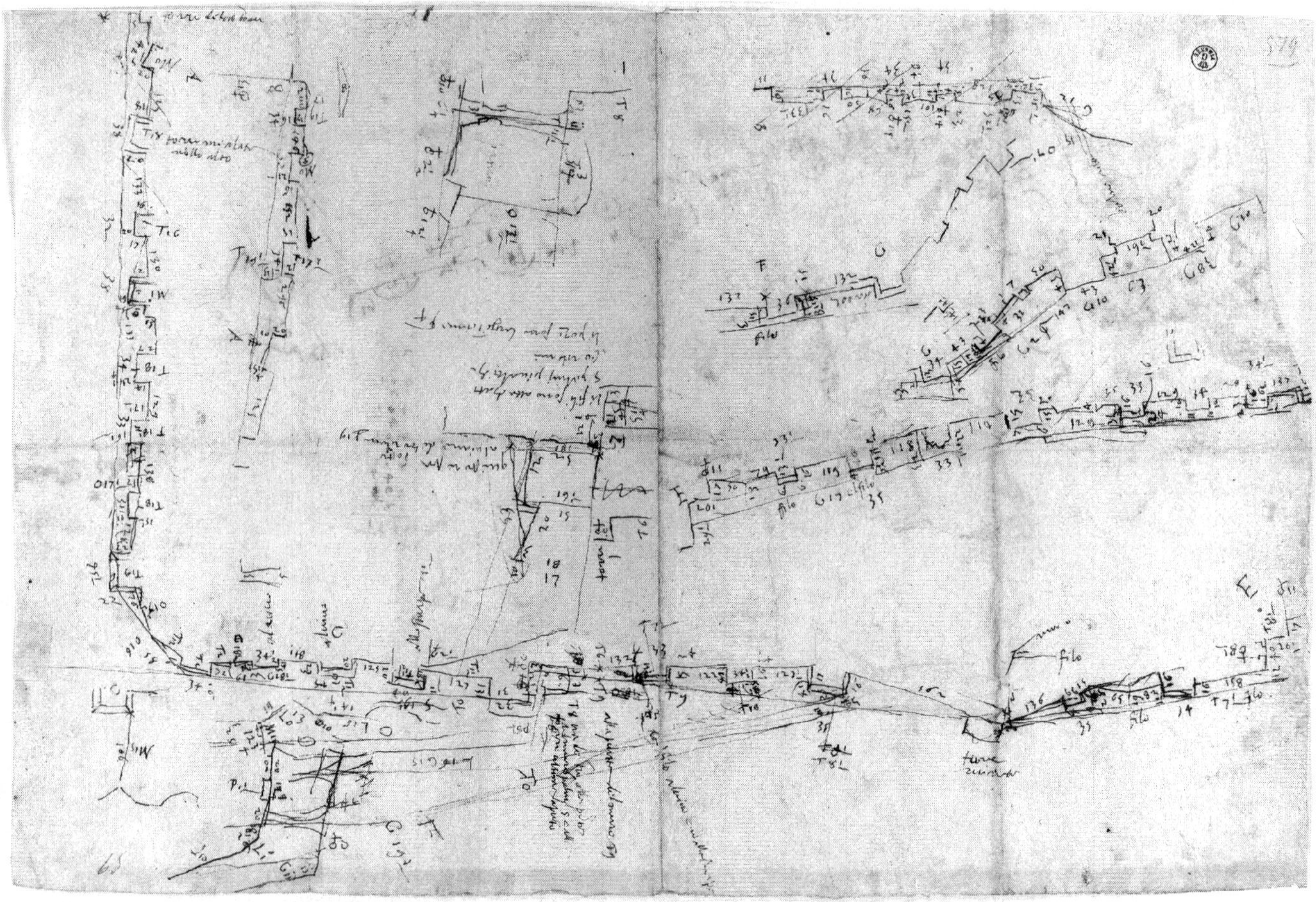

Figure 5.3 Antonio da Sangallo the Younger, survey of the Aurelian Walls from San Saba to the Porta San Sebastiano. 1537. Pen and brown ink. Florence, Galleria degli Uffizi, Gabinetto dei Disegni e delle Stampe, Uffizi U892 A recto. Su concessione del Ministero per i Beni e le Attività Culturali.

Leonardo Bufalini and His Ichnographic Map of Rome

This interest in renewed fortification arose in Rome just as Leonardo Bufalini (d. 1552) arrived in the early 1530s. In his will, Bufalini, who was born in Udine, a city in the Friuli-Veneto northern region of Italy, was called a *faber lignarius*, that is, a carpenter or woodworker. A list of woodworkers in Antonio Blado's print shop where his map was printed named "Leonardo Venetian carver" (*Leonardus Venetus Intagliator*), which probably referred to Bufalini and which suggests that he himself cut the woodblocks for his map. On that map, addressing the reader, Bufalini mentioned that he created it "not only with the geometer's square and compass, but also with the mariner's [i.e., a magnetic] compass." The bottom tier of the map displayed his self-portrait, a surveying instrument, and carpenter's instruments (fig. 5.4).[17]

Sparse but telling documentary sources show that aside from making his map, Bufalini also worked as a military engineer. Some of that evidence comes from a treatise on military engineering by the military engineer and courtier Francesco De Marchi (1504–1576), who was involved in the refortification of Rome in the 1530s and 1540s under Pope Paul III. Pope Paul III's effort involved a massive plan to fortify Rome and envisioned the construction of eighteen bastions around the walls. De Marchi mentioned Bufalini as one of the participants in a debate in front of Paul III on these fortifications. He described Bufalini as "a talented architect, the one who measured all of Rome inside and out and had it printed with hills, theaters, temples, streets, and other things indicated, in which I helped perhaps for six months for my pleasure and to learn more."[18] It seems likely that Bufalini took part in measuring and surveying the walls as part of Pius III's fortification project. Documents show that Bufalini, again described as a woodworker (*fabro lignario*), was part of a team that constructed a bastion at the Vatican.[19]

Bufalini created his map with a woodcut or relief technique. After he designed the map, he carved the blocks, which were then used for printing. Woodblock printing of images was accomplished on the same letterpress and used the same inks as those used for books. Thus was Antonio Blado, whose print shop was on the Campo de' Fiori and who usually printed books, able to print the Bufalini map in 1551.[20]

Although Bufalini's map has often been understood as having been made purely in the interests of military defense, in her definitive study, Jessica Maier has argued persuasively that Bufalini also made it to present Rome as a magnificent city, worthy of the ancient Romans. She points to the fact that his map displays not only contemporary buildings but also ancient structures no longer extant. In his letter to the reader at the lower margin, Bufalini confirmed that his map displayed the city through time from ancient to contemporary:

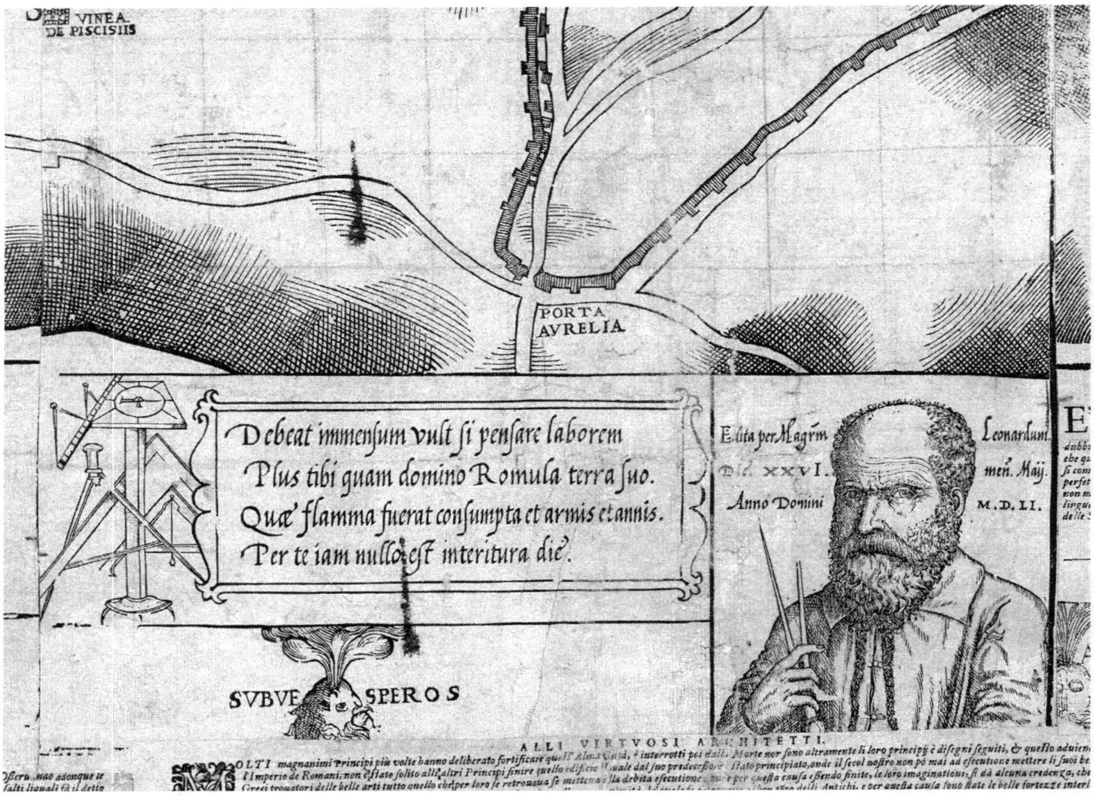

Figure 5.4 Detail of figure 1.6, Bufalini, *Roma*. London, British Library. Maps S.T.R.175. Portrait of Bufalini and his instruments with Latin inscription addressed to Bufalini: "If one wishes to consider the immense effort [of this undertaking], the Romulan land owes more to you [Bufalini] than to its own Lord, which [land] has been consumed by fire and arms and years. Through you now in no age is it about to perish." Copyright © The British Library Board.

> Whoever you are, Leonardo Bufalini from Friuli asks that you not consider [it] worthless, because he believes that he gives to you now the most beautiful of all things, which is to say Rome, and this twin of her; nor indeed did he consider it satisfactory to you to have placed before your eyes the one Rome that is inhabited today unless he also added the ancient one, once mistress of the whole world, as if called back from the grave with very much labor, great expense, and long-enduring images.[21]

Bufalini here plainly stated that he was presenting the city as a whole, timelessly—the ancient city along with the modern. His map revealed contemporary surveying practice in its accurately measured walls. Its inclusion of ancient no longer extant structures required knowledge of such structures through the study of ancient ruins and ancient texts. Bufalini was part of the group of practitioners, antiquarians, and learned humanists who were comparing ancient physical remains and ancient texts, and he was apparently much admired in some of these circles. For example, the renowned humanist Onofrio Panvinio wrote that Bufalini's map was the fruit "of incredible labor and persistent study of twenty years."[22]

The instruments surrounding Bufalini's self-portrait included (at the top left) a *bussola*, the surveying instrument first described by Raphael in his letter to Leo X, with a magnetic compass at the center of a graduated circle divided into degrees and a pivot arm to take sightings. The other instruments in the image are carpenter's tools—a mason's level and geometer's compasses. Bufalini apparently identified as much with his carpentry as with his surveying skills. Between the instruments at the left and his portrait is an inscription in a cartouche: "If one wishes to consider the immense effort, the Romulean land owes more to you [Bufalini] than to its own Lord [the pope, Julius III], which land has been consumed by fire and arms and years. Through you now in no age is it about to perish."[23] In this declaration Rome had been saved from extinction by an image of the city, namely Bufalini's map. Bufalini had thus done more than the popes to save the capital city from the ravages of fire, arms, and time itself.

Guidebooks and Maps: A Complex Relationship

Bufalini's 1551 map seems to have been influenced by an earlier, smaller map that appeared in the 1544 edition of a guidebook by one Bartolomeo Marliani (1488–1560). Marliani was a learned scholar of Greek. He moved to Rome in the 1520s where he worked at translating Greek texts, including *The Clouds* by Aristophanes. He also studied the topography of Rome and wrote his guidebook, the earliest edition of which, *Antiquae Romae topographia libri septem* (The Topography of Ancient Rome in Seven Books) first appeared in 1534 during the rule of Pope Paul III.[24]

Marliani's activities are illuminated by a remark by the German humanist and antiquarian, Georg Fabricius of Chemnitz (1516–1571), who was in Rome at mid-century. In his book on ancient Rome published in 1550, Fabricius wrote that he had studied the works of the ancients as well as more recent books, including those by Marliani, "who very diligently wrote about everything." Fabricius reported that "when we were going to his [Marliani's] house, [he] indicated to us the site of the hills from the Capitoline and guided us to the beginnings of the streets which are in the city." Fabricius provided intriguing information about where the exchange of information occurred: "he was seen at different times at [the shops of] Salamanca and Tramezzino to very kindly hear our opinion."[25] That is, he was in the print shops of Antonio Salamanca and Francesco Tramezzino at the center of Rome—shops that printed and sold antiquarian tracts, maps, and images of the ancient and contemporary city, and seem to have also served as places of communication among individuals fascinated by antiquities. Here we catch a glimpse of

Marliani as a kindly and communicative authority on the topography of the ancient city sharing his knowledge with others.

Marliani's guide to Rome was a small pocketbook-size book, an erudite work primarily based on ancient literary sources—Cicero, Pliny the Elder, Varro, Sallust, and Suetonius, among others. Also using epigraphical sources (inscriptions usually on stone), Marliani attempted to establish when and why ancient Roman structures were founded, where precisely structures mentioned in ancient sources were located, and what events occurred in and around them. Basing his work in part on earlier guidebooks, especially those of Flavio Biondo (1392–1463) and Andrea Fulvio (c. 1470–1527), he proceeded both chronologically and topographically, beginning with the founding of Rome by Romulus and Remus and including a description of the physical layout of the new city.[26]

Marliani's guidebook represented a kind of empirical investigation that combined the observation of structural artifacts with the study of relevant ancient texts. Indeed, his approach was much like that of the members of the Academy of Virtue in Rome in the 1540s. They used an ancient text, Vitruvius's *De architectura* as a tool to investigate Roman ruins—which in turn were used to study Vitruvius's text.[27] Marliani also used ancient histories as he investigated the physical remains of antiquity and tried to understand them in light of ancient texts.

Marliani's guidebooks influenced other books largely based on them—for instance those written by Lucio Fauno, a pseudonym—as has now been firmly established—for the humanist Giovanni Tarcagnota (1499–1566). Tarcagnota (using the name Lucio Fauno) worked extensively as a translator from Latin to Italian, his output including works by the Florentine Neo-Platonist, Marsilio Ficino (1433–1499), and Biondo's Roman topographical writings. Fauno/Tarcagnota also wrote his own guide to Rome in Italian—*Delle antichità della città di Roma*, published by Michele Tramezzino in Venice. [28] Although Fauno/Tarcagnota depended heavily on Marliani in writing his guide, his book was far more accessible than Marliani's. Written in the vernacular, it provided a greatly simplified account.

Marliani's own editions varied notably one from the other. Most significant from the cartographical point of view is the edition of 1544. This edition contained images including three maps of early Rome. One of them is a map on a two-page spread depicting the city surrounded by its walls during the period of the Roman emperors.[29]

This map of Rome during the empire was an ichnographic plan—the first such plan of the city since antiquity—in which the Aurelian Walls with their towers are shown as measured. Marliani explained that they measured the perimeters of the city using ancient feet (*pedes antiqui*) and that the units of measurement on the map were *stadia* and paces (*passi*). A scale on the lower left shows the equivalent on the map of eight *stadia* (fig. 5.5). But, Marliani explained, although they mea-

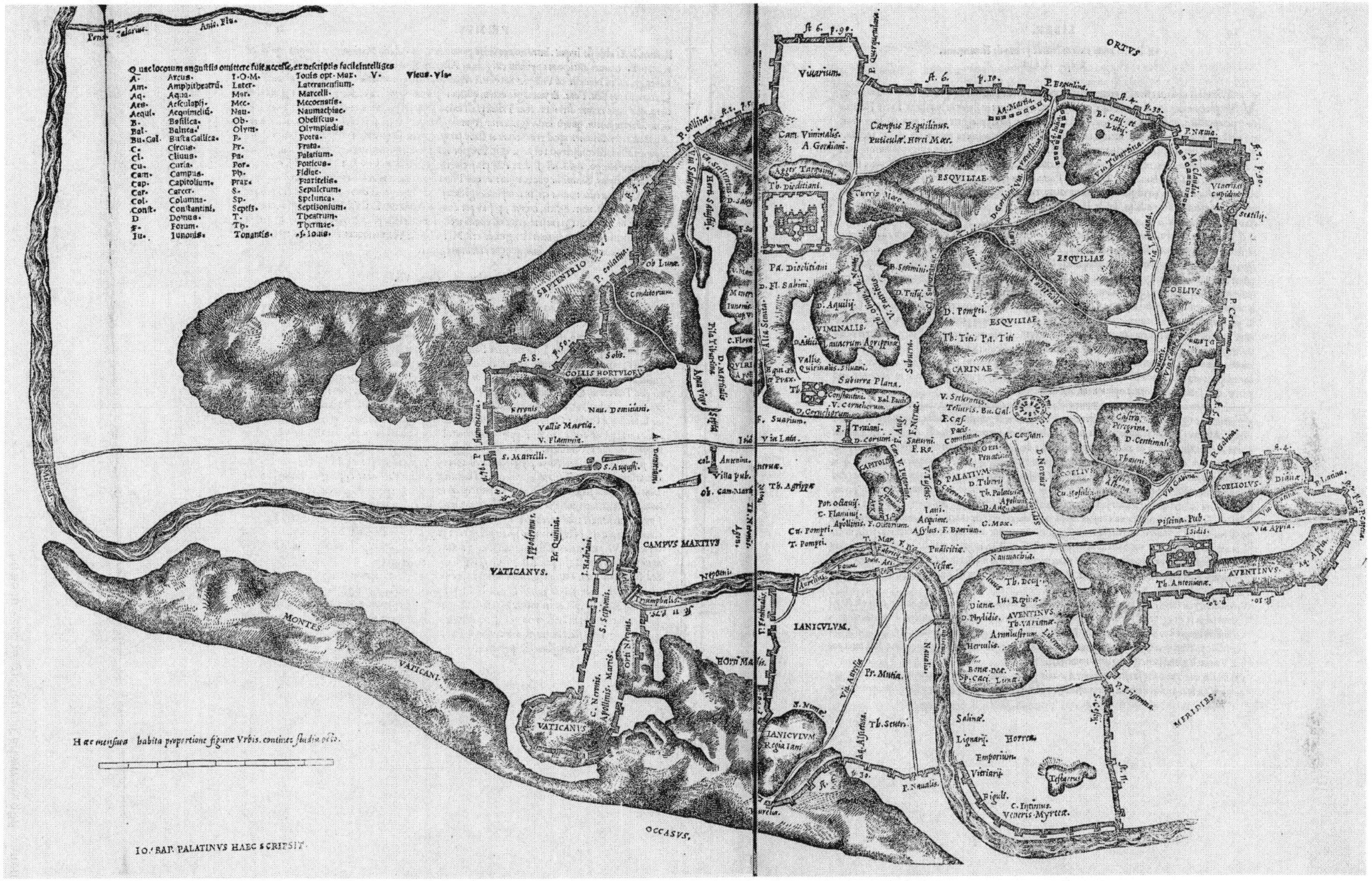

Figure 5.5 Bartolomeo Marliani, *Urbis Romae topographia B. Marliani* (Rome: Valerij, Dorici, et Aloisij fratris, Academiae Romanae Impressorum, 1544), 12–13. Map of Rome during the empire. Courtesy of HathiTrust.

sured the walls using the measure of ancient feet, which they discovered carved in stone in several places, they found another measure high on a column in the Church of the Twelve Holy Apostles with an inscription stating its height as being nine feet. This foot was larger than the one they had used to measure the walls. (Marliani provided two side-by-side scales to show the comparison.) This variation, he observed, may be because the marble was eroded or because measures used in different times were different.[30]

Marliani's map was oriented with the north at the upper left. The elevations of hills and valleys were indicated by shading and crosshatching. Objects within the walls—bridges over the Tiber, streets, some ancient monuments, gates in the walls, prone obelisks, and remains of aqueducts—were all in their proper places and correctly positioned in relation to one another. The labels of these locales and objects were written in various styles of roman and italic lettering, and the inscription on the lower left—"Io. Bap. Palatinus haec scripsit" (Giovanni Battista Palatino wrote [this])—indicated that Giovanni Battista Palatino (ca. 1515–ca. 1575), a calligrapher who was writing a book on forms of lettering, designed the identifying labels for the map.[31]

In this 1544 edition, Marliani explained that if we had wished to erect this figure and the following "according to natural perspective," it would have happened that, with all parts of the hills hidden except for one side, "no one would be able to locate a building on its own site." For that reason, he explained, so that the map is useful to everyone, "we thought it better [. . .] that we have placed the figures [as] flat planes, rather than amuse certain people with a worthless picture." Nevertheless, he assured his readers that he had taken care to shade the sides of the hills according to their altitude so that the valleys were easily distinguished.[32] Marliani thereby articulated two guidelines that would be followed by Bufalini in his map to be published seven years later—the importance of orthographic projection as a tool for being able to locate structures accurately in space, and the use of shading to indicate topographical features such as elevation.

As Ehrle notes in his 1911 foundational study that accompanied a facsimile reprint of the Bufalini map, Marliani's map shared some remarkable similarities with Bufalini's large map first published in 1551. Some hypothesize that Bufalini may have actually been involved in creating the maps for the 1544 edition of Marliani's guidebook. Whether or not Bufalini actually had a hand in Marliani's measured map, it is certain that he was aware of it, especially since the 1548 Italian edition of Marliani's guidebook was published by the same publisher—Antonio Blado—who brought out Bufalini's map a few years later. It is likely that Bufalini was in Blado's shop in 1548 carving the blocks for his own map.[33]

Conflicting Archaeologies

Marliani kindly pointing out urban features and discussing them in printshops presents only one side of Roman antiquarian studies. Roman antiquarians often engaged in bitter disputes. The most remarkable controversy focused on the location of the ancient Roman Forum. (The argument about the ancient Roman Forum was about the precise location where the ancient Romans gathered to conduct their civic and religious activities. It would have been a far smaller area than the present-day tourist site called the Roman Forum.) On one side of the sixteenth-century controversy, Bartolomeo Marliani believed that the ancient meeting place had been located between the Arches of Septimius Severus and Titus. On the other side, the antiquarian fascinated with waters, Pirro Ligorio, and his friend, the learned scholar of ancient Greek, Benedetto Egio (d. 1568/1570), believed that it had been located between the Palatine Hill and the south end of the Capitoline Hill. (In the sixteenth century, well before its modern excavation and archaeological study, there was insufficient archaeological evidence to specify with certainty the precise location of the ancient site.)[34]

Pirro Ligorio, who was presumably trained as a painter, had become an ardent student of Roman antiquities—ruins, ancient inscriptions, and coins. His habit, especially in his later, post-Roman years, of fabricating inscriptions (often giving the place and date of discovery, thereby making them seem to be realistic) marred his subsequent reputation. Yet there is no doubt that he was one of the most knowledgeable antiquarian investigators of his time. As a probable autodidact Latinist with little actual reading skill in Greek, his interactions with learned humanist friends and interlocutors were crucial. Ginette Vagenheim has investigated Ligorio's relationships within an informal academy dedicated to the study of antiquities, the Accademia dello Sdegno (Academy of Indignation), and has emphasized the collaborative nature of his writings. His friends and interlocutors in addition to Egio included the humanist Dominican monk Onofrio Panvinio. Ligorio used the library of the Latin poet, antiquarian, and collector Angelo Colocci and also knew the humanist antiquarian Fulvio Orsini (1529–1600). He extensively communicated with these learned men while composing his encyclopedic writings on antiquities, and he in turn imparted to them his own vast knowledge of material objects and ancient ruins.[35] Ligorio's learned friends undoubtedly assisted him in his bitter dispute about the location of the Roman Forum.

The most important of those friends for the argument about the Forum was Benedetto Egio. Egio was a priest from Spoleto active among those exploring antiquities in the 1530s and 1540s. He was a translator of both Greek and Latin and an expert epigrapher. Egio had many friends among humanists and antiquarians in Rome, including Onofrio Panvinio. In addition, he was a friend and adviser

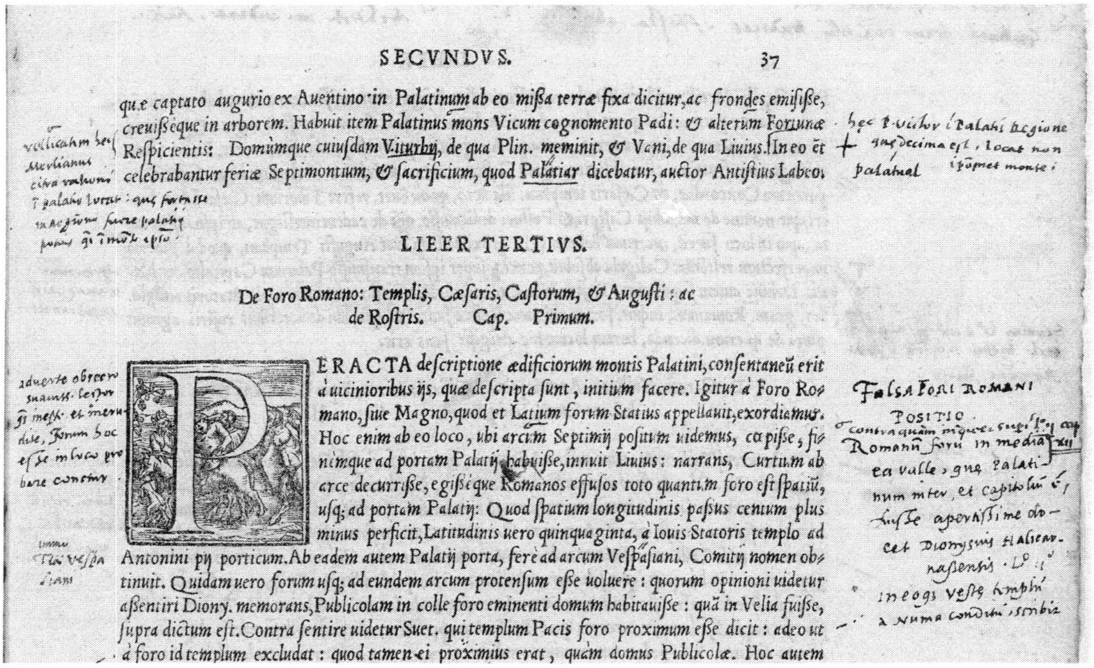

Figure 5.6 Bartolomeo Marliani, *Urbis Romae topographia B. Marliani* (Rome: Valerij, Dorici, et Aloisij fratris, Academiae Romanae Impressorum, 1544). BAV Ross 1204, 37 (detail), where on the right margin is written "Falsa Fori Romani Positio." © 2018 Biblioteca Apostolica Vaticana, by permission of the Biblioteca Apostolica Vaticana with all rights reserved.

to Pirro Ligorio. Egio probably translated Greek texts and inscriptions for him, which they would have discussed together. Although few biographical details of Egio's life are known, his friendship with Ligorio is well documented.[36]

Ligorio and Egio's main antagonist in the Forum argument was Bartolomeo Marliani. The dispute was carried out in printed texts and on maps and in the form of name calling—the names called were not those of the actual antagonists but rather of characters in the ancient comic play by Aristophanes, the *Clouds,* which Marliani had translated. In *The Clouds,* an uneducated peasant named Strepsiades calls Socrates "Socratidion" (Little Socrates), a contemptuous diminutive. These two names, the rude peasant Strepsiades and the disparaged Socratidion, came to be the epithets thrown back and forth between Marliani, who called Pirro Ligorio (and Egio) "Strepsiades," and Pirro Ligorio (and Egio), who called Marliani "Socratidion."

On the margins of a copy of Marliani's 1544 edition of the *Topographia urbis Romae* in the Vatican Library (BAV, Ms. Ross 1204), Egio wrote extensive negative notes (fig. 5.6). The animosity between Egio and Marliani was one between two learned scholars of ancient Greek. In the annotations of his copy of Marliani's *Topographia,* Egio calls the author "Socratidion" three times.[37]

On the basis of paleographic evidence within the marginal notes, Marc Laureys and Anna Schreurs convincingly suggest that Egio created these annotations over time—that he carried Marliani's 1544 guidebook with him as he went out to study ruins and inscriptions, and as he checked on citations in ancient texts. In BAV, Ms. Ross 1204, Egio mentioned his friend Pirro Ligorio five times, including an occasion when the two jointly transcribed an inscription.[38] In his own extensive writings, Ligorio frequently described Egio in laudatory terms—for example, "M. Benedetto Egio da Spoleto, a very erudite man in Greek letters as in Latin"—and acknowledged his help. It is evident from various sources that they often went out together to study and discuss Roman antiquities, ruins, and inscriptions.[39] Egio, a classical scholar with an expert knowledge of both Greek and Latin, and Ligorio, a painter and architect with a comprehensive understanding of Roman ruins and other ancient remains, possessed between the two of them thoroughly complementary skills for investigating Roman antiquities.

Even before the salvos in the Strepsiades-Socratidion dispute, the location of the Roman Forum was a topic of discussion. This is evident from the letter to readers that Giovanni Tarcagnota (alias Lucio Fauno) added to the first (1548) and all subsequent editions of his guidebook (a work, as we have seen, based largely on Marliani's). Because all the editions of his guidebook were published by Michele Tramezzino in Venice, also the publisher of Pirro Ligorio's maps and *his Libro* [. . .] *delle antichità di Roma* in 1553, it is reasonable to think that the print shop itself was a site for the debate. Fauno/Tarcagnota had long been associated with the Tramezzino shop, and it seems likely that his levelheaded discussion of the location of the Forum (in which he sided with Marliani), first published in 1548, helped to ignite the Strepsiades-Socratidion conflagration.[40]

In his guidebook, Fauno/Tarcagnota devoted an appended letter *To the Readers* (*Ai lettori*) to the topic of the location of the ancient Roman Forum. His tone, unlike that of the Strepsiades-Socratidion interchange, was restrained. He described Ligorio's position without naming him and disagreed with it. In an intricate set of arguments using statements in ancient texts—including those of Livy, Tacitus, Plutarch, Pliny, and Suetonius—Fauno/Tarcagnota argued that the Forum extended in a straight line from the Capitoline Hill in the direction of the Arch of Titus (where Marliani also thought it was). Fauno/Tarcagnota maintained his reasonable tone throughout: "This subject of the antiquity of Rome," he wrote, "is for the most part so doubtful and various because none of the ancients have discussed it clearly and because almost all the ancient places are destroyed; [such] that it would be wrong reasoning to say one is able to say directly, 'thus it is,' [or] 'thus it is not,' because for the most part it is necessary to follow conjectures."[41]

Unlike Fauno/Tarcagnota, Ligorio was not one to engage in calm discussion. He was highly critical of Marliani's guidebook, and in his only book published in his lifetime (out of thousands of pages of writing), he harshly criticized what he thought were errors being made about the topography of Rome. His *Libro* [. . .] *delle antichità di Roma* treated ancient circuses, theaters, and amphitheaters and included a second part, *Paradosse,* devoted entirely to criticizing the errors of others (obviously Marliani and his supporters). In his view, the most basic error was the incorrect placement of the ancient Roman Forum. Ligorio suggested that most of the many other errors then circulating about the placement of monuments — he itemized more than eighty-five — derived from this initial mistake concerning the Roman Forum.[42]

Although Ligorio did not mention names, in the *Paradosse* he made clear in marginal notations of a manuscript draft of the work, now housed in the Bibliothèque Nationale in Paris, that he was taking aim at Marliani among others. Here, repeatedly writing "contra il Marliano" or "contra Marliale" in the margins, he deplored Marliani's influence and stubbornness. "Even people," he raged, "who have seen clearly what I have said do not want to retreat from having written false things, and their badly founded opinions appear not only in print but in marble through the counsel of the poorly informed Bartolomeo Marliale, for whom the truth would not persuade — even though Cicero and Romulus, builder of the city, might come, he would not cede a single point to them."[43]

It happens that modern archaeology agrees with Marliani and Flavio Biondo against Ligorio about the location of the Roman Forum. But Ligorio's placement of the Forum between the south end of the Capitoline Hill and the Palatine was not without reason. He based his argument in part on references in ancient texts. He also argued that the area between the two arches was occupied instead by the Via Sacra and that the Forum and the Via Sacra could not be located in the same place.[44]

Ligorio placed particular trust in the late antique Regionary Catalogs. These catalogs comprised lists of buildings arranged under each of the fourteen Augustan regions of the ancient city in which they were presumed to be. In the sixteenth century, the catalogs were known under their supposed authors, Publius Victor and Sextus Rufus. They were greatly augmented and amended in the late fifteenth and sixteenth centuries, beginning especially with the humanist Pomponio Leto and his Roman Academy.[45]

Ligorio's *Libro* [. . .] *delle antichità di Roma,* along with the scathing *Paradosse,* was published in 1553 in Venice by Michele Tramezzino. Michele and his brother, Francesco Tramezzino, had owned a print shop in Rome, but they fled to Venice during the Sack of 1527. Francesco returned in 1528 to set up the shop in Rome anew. It became a gathering place for antiquarians and scholars, one of whom was undoubtedly Pirro Ligorio, who had arrived in Rome in 1534. The *Libro* [. . .] *delle antichità di Roma* included not just the book but five separately published images — graphic reconstructions of the Circus Maximus, the Circus Flaminius and the Castrum Praetorium, and two (very different) engraved plans of Rome, one published in 1552, the second in 1553.[46]

Pirro Ligorio's use of copperplate engraving signaled a new trend in the printing of images, one which would soon dominate the lively print market of Rome. His 1552 Roman map may well have been inspired by — and clearly in certain aspects was based on — the Bufalini map published the year before. Although Ligorio's map was created by copper engraving rather than a woodcut, and although it was much smaller than Bufalini's, the orientation was exactly the same (north to the left), with the Vatican in the lower left. Ligorio's walls were without doubt derived from Bufalini's walls. And both maps showed ancient and contemporary structures. Pirro was explicit in his bifocal display of the ancient and contemporary city — on the upper left of the map is written "Site of the city of Rome with those remains of ancient monuments which now are observed, created by Pirro Ligorio, Neapolitan (fig. 5.7)."[47]

Despite Bufalini's evident influence, Ligorio's own vision of Rome differed in fundamental ways. Ligorio's map was filled with buildings that were constructed with perspective in contrast to Bufalini's orthographically measured plans. Indeed, Ligorio's 1552 map of Rome, a copper engraving incised by "G.L.A." (Georgius Lilius Anglus — George Lily, an Englishman resident in Rome), contained no measurements despite the fact that Ligorio would have been perfectly capable of creating measured drawings. Ligorio was concerned rather to draw images of intact ancient buildings derived from his on-site study of their ruins. Thus, his map was filled with images of ancient buildings — some in ruins and some that he had reconstructed, as well as modern buildings and other structures. He depicted no longer extant ancient structures, such as the Circus of Nero — which he placed with its obelisk behind the Castel Sant' Angelo — intact alongside contemporary structures, such as St. Peter's, including an obelisk standing on its south side. The city walls were depicted with their gates, the major roads into the city were delineated, and the bridges over the rivers were indicated. Some of the structures were labeled — ancient structures with roman letters, contemporary with italic. Ligorio, like Bufalini, indicated hills with crosshatch shading, but such topographical features were far less prominent on his map than on Bufalini's.[48]

Ligorio's second plan of Rome, a copper engraving incised by Giulio de'Musi and published in 1553, came out before his *Libro* [. . .] *delle antichità di Roma* and the *Paradosse*. This is evident from the publisher Tramezzino's note to the reader at the upper-left-hand corner of the plan. We have already published, it declared, the description of the city of Rome, including most of the old and new buildings with images of signs "having been drawn by Pirro Ligorio with industry and recently produced at last with our own copper plates." The cartouche included a startling

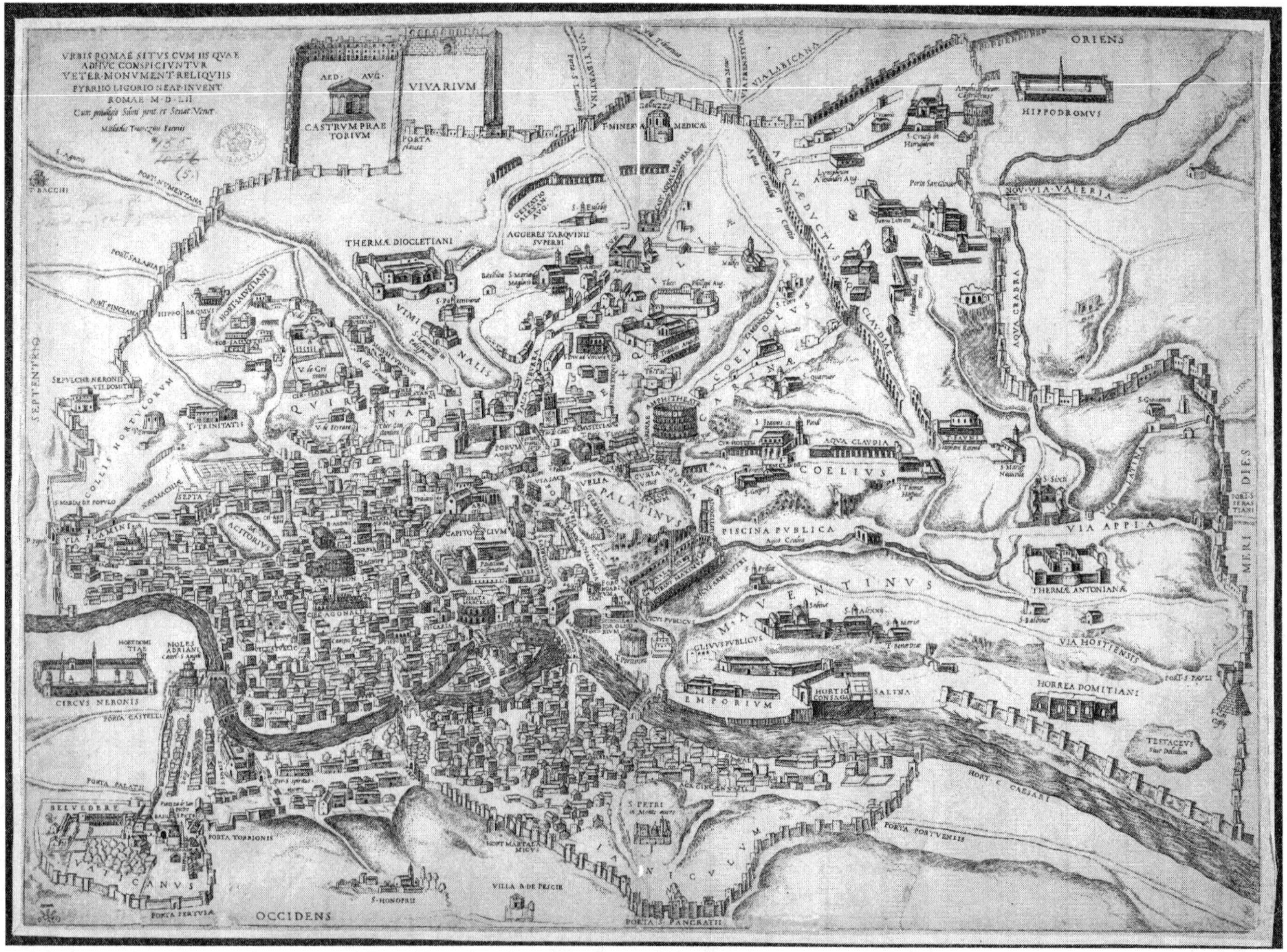

Figure 5.7 Pirro Ligorio, *Urbis Romae situs cum iis quae adhuc conspiciuntur veter. monument. reliquiis Pyrrho Ligorio Neap. invent.* (Venice: Michele Tramezzino, 1552). London: British Library. Maps* 155. (5.). Etching and engraving, 18¼ × 21½ in. (40.2 × 54.5 cm). Copyright © The British Library Board.

attack on incompetent antiquarians: "Accept it [this plan] freely," the address advised, "so that you are no longer deceived along with inept antiquarians, blinder than a serpent's slough, who like Little Socrateses teach their Strepsiades nothing but nonsense. In addition, you will see very clearly that which is made not at all heedlessly, several things in this are named and located against the perverse opinion of earlier antiquarians, from the *Paradosse* of Pirro Ligorio, very soon about to be published (fig. 5.8)."[49]

The 1553 map itself was oriented with the north to the left and St. Peter's and the Belvedere (the only contemporary buildings shown) in the lower left corner. The map was a kind of visual representation directly derived from the Regionary Catalogs, although with material added from other sources. The Aurelian Walls were displayed with a statement at the upper left that "the city at the circumference has 13,550 *passi* [about 20 km]." But, rather than measurement, the focus of the map was to delineate the fourteen regions the emperor Augustus had created in 7 BCE. The catalogs provided headings that named these ancient districts or regions of Rome and then a list of the important structures and other features within those regions. This plan of ancient Rome reproduced the lists taken from the catalogs in all capital letters, and positioned them on the approximate sites where Ligorio thought the structure or feature was. The map provided a very rough indication of elevation by shading and depicted the ancient gates, bridges, and aqueducts. The area between the Capitoline and the Palatine was crowded with both labels of sites, including the Roman Forum, placed where Pirro believed it had been, and a few structures. At the top left was a picture of the wolf suckling the twins Romulus and Remus and farther to the right the profile of a bearded man, the founder of Rome—"Romulus Conditor" (Romulus Founder). At the bottom right of the map was Tramezzino's printer's mark, and above it, the mark of the engraver of the copper plates, Giulio de' Musi.[50]

Marliani promptly and vigorously responded to the attacks published in Ligorio's 1553 map and the *Paradosse*. Without changing the colophon of the 1544 edition of his *Urbis Romae topographia*, he issued two new versions with changes made to the front and back matter.

In the first new version, he added an essay following the title page on the topic of ancient names. He noted that he did this after the discovery of various inscriptions cut in stone—"and especially from those which were recently discovered in the Roman Forum." He was referring to fragments of incised stones that were discovered in the Roman Forum in 1546–1547, known as the Capitoline *Fasti*. They contained a chronological list of Roman magistrates and triumphant generals that had been carved into stone during the early years of Augustus's Principate. The significance of this discovery for enlarging an understanding of ancient Roman history was immediately recognized. Cardinal Alessandro Farnese at first claimed

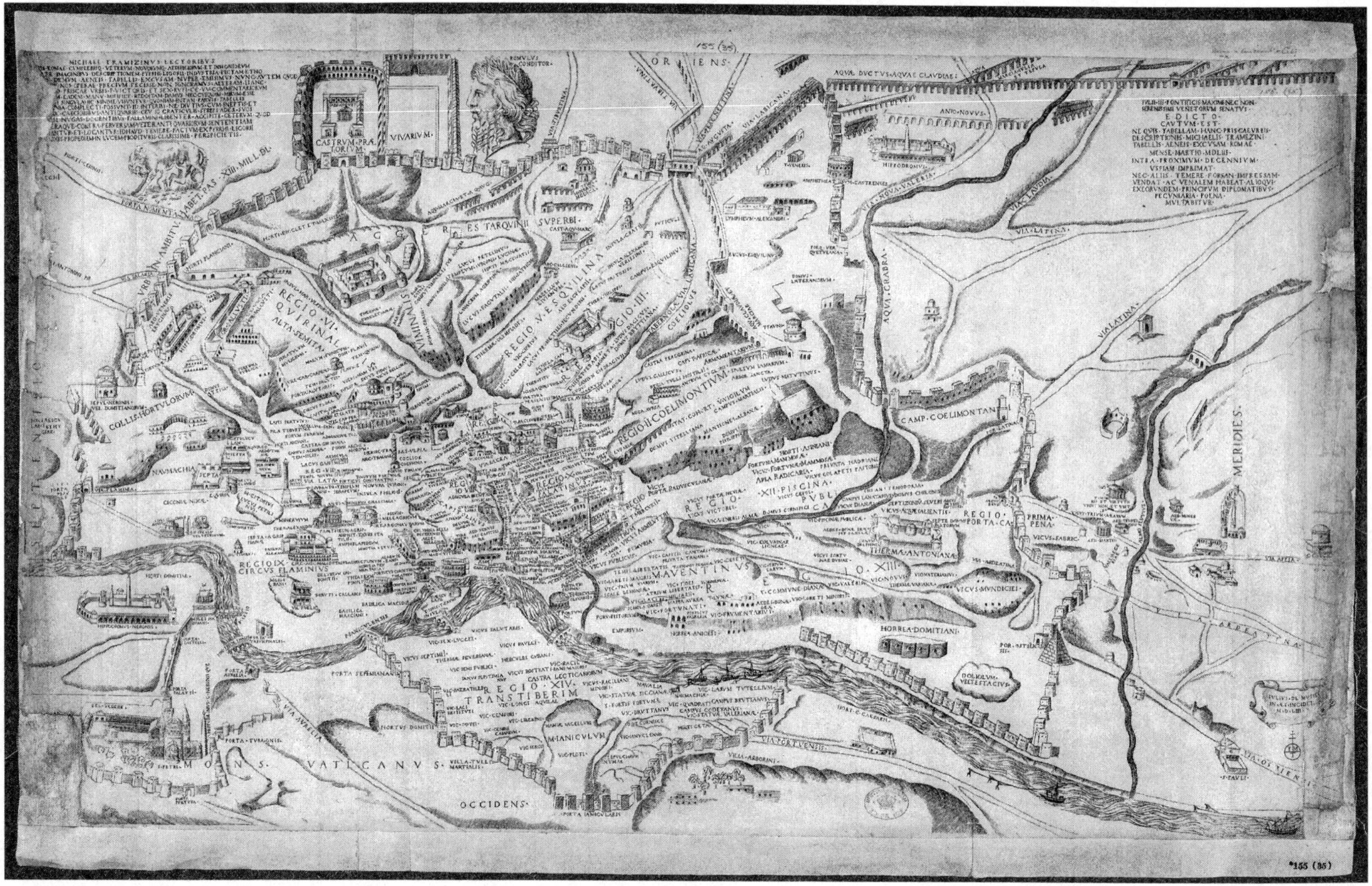

Figure 5.8 Pirro Ligorio, *Urbis Romae* [. . .] (Venice: Michele Tramezzino, 1553). London, British Library. Maps* 155. (35.) Etching and engraving. 16½ × 28 in. (42 × 71 cm). © British Library Board.

Figure 5.9 Capitoline Fasti in the Sala della Lupa of the Palazzo dei Conservatori, Rome, Musei Capitolini. © Roma, Sovrintendenza Capitolina ai Beni Culturali. Photos by author.

the stone fragments and kept them in his palace, but he then gave them to the city, where they were (and are) kept in the Palace of the Conservators (now part of the Capitoline Museum) (fig. 5.9). It was Marliani who published the first edition of the *Fasti* in 1549, and in 1560 (after several other editions had gained prominence), he brought out a commentary.[51]

Marliani also took up the Strepsiades-Socratidion exchange in his address to readers at the end of this new version of the 1544 *Urbis Romae topographia*. He first suggested that the changes and corrections that he would list were due to new knowledge gained from the *Fasti*. Before he began the list of his own corrections, however, he mounted a vigorous defense of his work against his name-calling adversaries.

> I admonish that you do not judge that this [book] was done through error, even as we hear that a certain Strepsiades barks. For he stammers certain dreams concerning his own certain Roman Forum against the opinion of all the most learned men, and to those who do not understand, he arrogantly professes himself teacher. [. . .] And although it is shameful to respond to a worthless rascal

of this kind, nevertheless, so that those too little learned in the topography of the city are not lead into error by the nonsense of this man, I promise very soon to show that the things written by me about the Forum are more true than the Pythian Oracle.[52]

Very soon thereafter a lengthy addition appeared bound into the front of the second new version of Marliani's 1544 guidebook—again without a change in the colophon—that provided a lengthy rebuttal to Ligorio's *Paradosse*. This addition, which thus far has been found in very few copies, was entitled "B. Marliani topographiae urbis Romae haec nuper adiecta" (This newly added to the topography of the City of Rome by B. Marliani). In it Marliani vehemently defended his own work and attacked Ligorio's view on the location of the Roman Forum.[53] On the new title page, Marliani dissented "against the new and stupid opinion of a certain Strepsiades of the *Clouds* of Aristophanes, a very ridiculous argument," and he warned the reader that "certain figures of the city having been printed under the name of the same Strepsiades, Magister, are certainly false," here referring to Ligorio's maps and their indications of the location of the Roman Forum. Throughout the twelve-page "Nuper adiecta," he castigated Strepsiades without ever naming Pirro Ligorio. Strepsiades, he railed, first read his *Urbis Romae topographia* repeatedly, taking many things from it, and then, "after he was saturated with honey, he pollutes the remaining things with dung." One by one, Marliani cited ancient texts that recounted events from which one could infer topographical information. He adamantly insisted that the ancient Forum was where he had placed it.[54]

Near the end of his life, Marliani became a complete recluse. He lived alone in a small dwelling at the Tor Sanguigna close to the church of Sant'Agostino near Piazza Navona. He had a white beard and always dressed in black. A cobbler's apprentice brought him his material needs (presumably food) and handed it in through a hole in the wall. He was apparently wealthy, although how he obtained such wealth is unknown. How he had earned a living is also unknown. He gave many pious donations for orphans and for the dowries of poor maidens either for marriage or to enter a religious house. In the last year of his life, he founded a confraternity, the Company of Saint Apollonia (Compagna di Santa Apollonia), under the auspices of the church of Sant'Agostino and approved by Pope Pius V with a bull of January 16, 1566. The purpose of the confraternity was to aid poor maidens with dowries.[55]

We know these things about Marliani's life because of a legal dispute over his will. The revealing dispute took place between the Company of Saint Apollonia and the executor of the will, one Bartolomeo Ruscone. Ruscone had found a cache of money under a floorboard in Marliani's dwelling and had immediately closed his own shop—he was an embroiderer—and lived thereafter like a gentleman.

Marliani himself had dressed elegantly, although always in black, and was never without his chain of the order of the Cavalieri di San Pietro. He continued to study and measure the city of Rome, and it turns out that he was a collector. According to one witness, five sacks of medals were found in his rooms. He donated his many translations from Greek to the library of Sant'Agostino (the Biblioteca Angelica), where they remain to this day, and he was buried in the church itself.[56]

Ligorio's Continuing Cartographical Activity

Whatever the effect of his onslaughts against Marliani, Ligorio continued his own researches undaunted (until his departure from Rome in 1566) as well as his collaboration with Michele Tramezzino in the publication of maps. Ligorio's efforts to reconstruct Rome itself reached its zenith in his 1561 representation of ancient Rome. The large engraved plan comprising twelve sheets, the whole measuring 50¾ by 57 inches (129 × 145 cm). It was published by the Tramezzino brothers in Rome. It was a bird's-eye view of the city packed with reconstructed ancient buildings—some recognizable monuments, others generic structures such as houses (fig. 5.10).[57]

In his classic study of this large 1561 depiction of ancient Rome, Howard Burns underscores that Ligorio used every possible source available to him—the Regionary Catalogs and other ancient texts, the remains of ancient buildings, available images of the ancient city in ancient paintings and reliefs (such as those on Trajan's column), and most importantly, images on ancient coins, of which Ligorio was an expert. Ligorio created his image not as a contemporary artist would create it but rather as if he himself were an ancient artist creating a representation of the ancient city as it would be represented in antiquity.[58]

As in his other depictions of Rome, the image was oriented with the north at the left. The fourteen ancient regions are noted in large capital letters, although those letters are barely visible among the dense clutter of buildings. The Roman Forum lay predictably between the Capitoline and the Palatine, where Ligorio thought it was. (In the worn state of the first [1561] edition, most of the letters are obscured, but they are plainly visible in the second edition, published by Giovanni Scudellari in 1587.) Yet the topographical feature of the height of the two hills was for the most part obscured by the numerous buildings. The Tiber River was shown with all its bridges intact and with Tiber Island stylized in the shape of a boat centered by an obelisk.[59]

In a cartouche on the upper left of the plan, Ligorio described his sources: "Portrait of Ancient Rome from the vestiges [and] ruins of buildings, from the testimony of ancient authors, the evidence of coins, from inscriptions on bronze, lead, stone, and tiles collected by Pirro Ligorio, Roman Citizen, and restored and

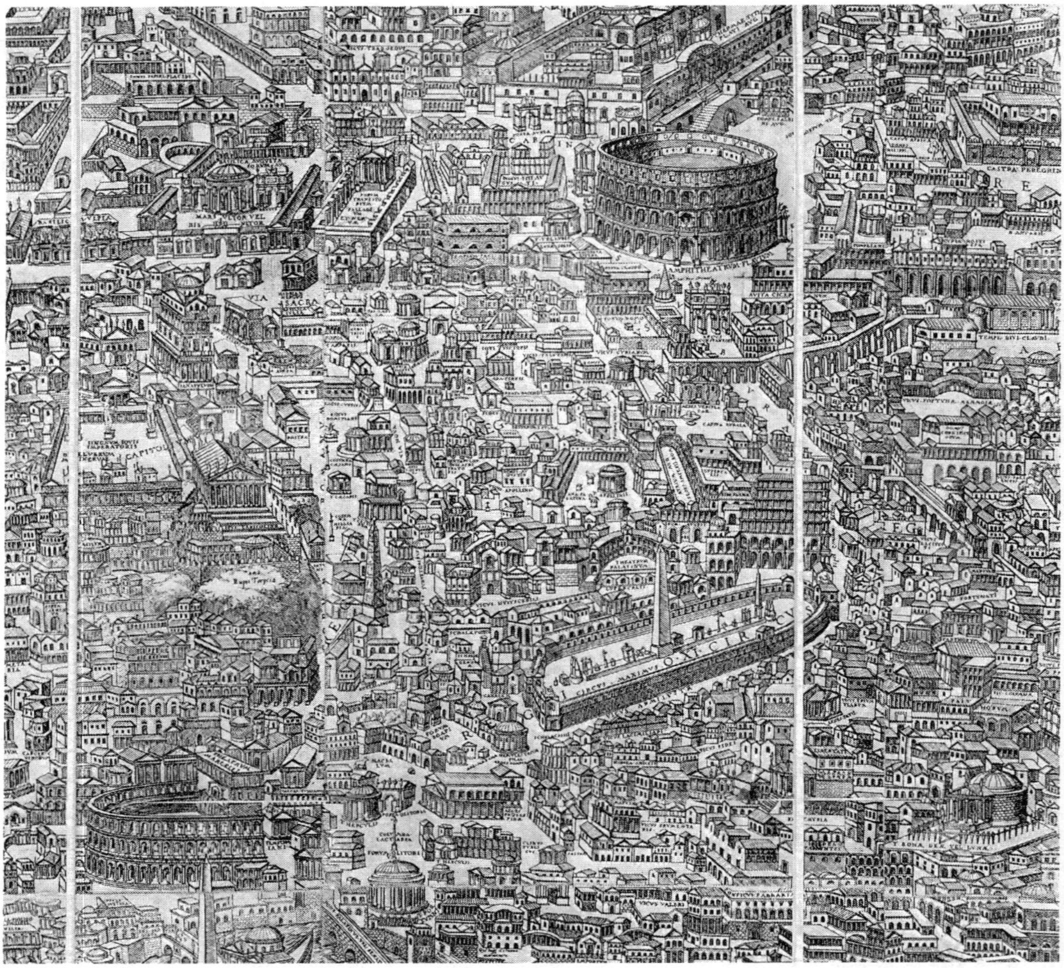

Figure 5.10 Detail of Pirro Ligorio, *Anteiquae urbis imago accuratissime ex veteribus monumenteis formata* (Rome: Michele and Francesco Tramezzino, 1561). London, British Library. Maps C.25.d.9. (1.). Etching and engraving on 12 sheets (detail: part of three sheets with Colosseum on upper right, Circus Maximus below, and the Capitoline Hill at mid-left). Whole 50¾ × 57 in. (129 × 145 cm.). © British Library Board.

described on this plate for the fourteen regions into which the Emperor Caesar Augustus divided the city."[60] This was an exact description of what it was. It was not a map, but a portrait, an effigy, as in this label, or, as in the title across the top, an *imago*. Although its perimeters are taken from Bufalini's map, it is in some sense an answer to Bufalini—not a measured map of Rome but a re-creation (in the strongest sense of the word) of ancient Rome and its hundreds of structures.

6

Maps, Guidebooks, and the World of Print

Print shops dominated the center of Rome, which was second only to Venice as a center of printing in Europe. These shops clustered in the *rione* of Parione between Campo de' Firoi and Piazza Navona. They displayed prints on their walls and sold thousands of single-sheet engravings, including maps and plans of the city. We have seen that they could serve as gathering sites for conversation about Roman topography and antiquities. It is also true that their proprietors were fiercely competitive with one another and participated in rivalries that could escalate into the theft of engraved copper plates and even murder.[1]

By midcentury, a flourishing trade specializing in single-sheet prints from copper engravings, or sometimes etchings, was producing and selling hundreds of images. In the sixteenth century, copperplate engraving represented a new technology that had advantages over the woodblock. Metal plates could be made larger than woodblocks; lines on the plate could be engraved with more precision, allowing greater articulation on the print; and the plates could be changed and modified with relative ease. Engraved copperplates lasted longer than woodblocks and thus had a longer printing life. The copperplates were printed on a roller press. It consisted of a wooden hand-operated machine in which the plate was

Figure 6.1 Detail of Vittorio Zonca, *Novo teatro di machine et edifice per varie et sicure operationi con[n] le loro figure tagliate in rame e la dichiaratione e dimostratione di ciascuno* (Padua: Fran. Bertelli, 1621), 76. Engraving of a roller press for copperplate engraving. Image courtesy History of Science Collections, University of Oklahoma.

placed on a bed and the paper was put on top and printed with one or more rollers turned by a crank or wheel (fig. 6.1).[2]

The creation of maps of Rome, the study of Roman antiquities, and infrastructure projects in the physical city are often treated as separate subjects and studied by specialists working in separate subdisciplines. In the sixteenth century, however, the three activities involved not just products but practices that were extensively intertwined and closely connected.

The variety of views of Rome was impressive. Jean-Marc Besse and Pascal Dubourg Glatigny have created a typology of Roman maps that distinguishes three main categories: maps in books, such as Marliani's 1544 map of Rome under the emperors; freestanding maps sold individually that could be put in atlases, such as Pirro Ligorio's maps of 1552 and 1553; and large wall maps, such as the Bufalini map of 1551 and Pirro Ligorio's 1561 *Anteiquae urbis imago*.[3]

Whatever their form, city maps and images were virtually always collaborative projects—projects that involved not only image making and printing but also a deep engagement with the physical topography of Rome. The study of ruins, surveying, antiquarian study of texts and ancient objects, drawing, copper engraving, and printmaking were among the practices of mapmaking.

The Printers of Rome

The Sack in 1527 devastated printmakers along with everyone else. One example is the shop of the brothers Michele and Francesco Tramezzino (printers of works by Pirro Ligorio and Fauno/Tarcagnota). The Tramezzino brothers were established printers in Rome. Michele was working as an apprentice in the bookshop of the Scoto family in Rome as early as 1522 and at that time wished to establish his own shop. When he managed to do this is unknown, but the brothers fled to Venice in 1527. Michele remained in Venice. Francesco returned to Rome in 1528 to set up his house and shop in Via del Peregrino. It is evident that the brothers continued to work together—perhaps helped by being established in their separate locations, two of the most active printing centers of Europe.[4] Pirro Ligorio would have met Francesco in his Roman shop. As we have seen, most of his maps were published in Venice by Michele, although his large 1561 city view of ancient Rome was printed in Rome by both brothers.

The friendship between the Tramezzino brothers is clear from an accord made between them on January 10, 1562. Michele's lawyer—none other than the Roman jurist Luca Peto—called Francesco to his office in the Rione Sant'Angelo to ratify a document with which Michele, having decided that he was of the age "to live without pain," divided the common goods of the brothers to the explicit advantage of Francesco. (Why Michele did this is unclear, but it may have had to do with the fact that Francesco had a wife and children, whereas Michele did not.) Michele continued his activities of book editing, printing, and selling in Venice until 1574. When he died in 1579, he had produced at least 257 editions, including reprints and maps.[5]

Often Michele Tramezzino's books had received privileges (limited monopolies usually given for ten years) from both the Venetian Senate and the pope—probably, it has been suggested, because of the brothers' friendship with Antonio Blado, the papal printer. Compared with Michele, Francesco worked more as a bookseller than an editor and printer. Some of the books that he did edit or at least sponsor were printed by nearby printers, including Antonio Blado.[6] Francesco's shop on Via del Peregrino served a purpose beyond selling books in that it (as well as the shop of one of Francesco's rivals, Antonio Salamanca) functioned as a gathering place for antiquarians and humanists who were investigating Rome.

Pirro Ligorio's contact with Michele must have been established through Francesco's shop in the center of Rome.[7]

Other printers in Rome included a great surge of immigrants. Most important was Antoine Lafréry (1512–1577), who Italianized his name to Antonio Lafreri after he arrived from the Franche-Comté in eastern France. Lafreri is first in evidence in Rome in 1544, the date of three copperplate engravings that he had signed. He owned a flourishing print-publishing business and was also associated with humanist archaeological investigations. He became the most important publisher of prints in Rome and indeed the most important print publisher in all of Italy and the rest of Europe at that time. One of Lafreri's rivals was the abovementioned Antonio Salamanca (1478–1562), who had arrived from Salamanca, Spain, around 1505. Once competitors, the two joined forces in a partnership that began in 1553. The partnership lasted until Salamanca's death.[8]

It was Lafreri's idea to create special title pages for specific general topics; customers could then buy a title page and select a group of prints—many hanging on display in the shop—to go with it. For example, one rubric, the *Geografia*, focused on maps and topographical views;[9] a second was devoted to devotional prints.[10] A third, the *Speculum Romanae Magnificentiae* (Mirror of Roman magnificence) with its title page engraved by Lafreri's countryman Etienne Dupérac (in Italian, Stefano Duperac, ca. 1535–1604), comprised an album of views of Rome, including statues, buildings, monuments, architectural plans, city views, and maps (fig. 6.2). The extant versions of the *Speculum* each contain a different selection of prints forming a total of more than six hundred images.[11]

Lafreri's production of prints was important for antiquarian investigation and was also fundamentally tied to the market. Eager consumers touring Rome as pilgrims, humanist scholars, artist and architect practitioners and others could acquire a print or two depicting a favorite site—or if they were well heeled, an entire collection, picking and choosing among the examples in a shop. Lafreri's shop was preeminent, but such consumers would have had more than twenty shops to choose from located right in the center of Rome while walking from St. Peter's to the Piazza Navona.[12]

Antonio Lafreri died intestate on July 20, 1577, under suspicious circumstances (murder was suspected). His death ignited a conflict that reveals the fierce rivalry within the Roman print business. A papal investigative commission (the Commissariato della Camera) sequestered Lafreri's goods and then found his reputedly closest relative, his sister's grandson, Stefano Duchetti (d. 1583) and named him heir. After hearing the news in October, Stefano's uncle, Claudio Duchetti (fl. ca. 1565, d. 1585), who had been working in Sicily, hurried back to Rome to claim his share of the inheritance. The rival shop of Lorenzo Vaccari (fl. 1574–1608) and his partner Donato Rasciotti (fl. 1572–1598) also seem to have been in-

Figure 6.2 Stefano (Étienne) Duperac, title page of *Speculum Romanae Magnificentiae* (Rome: Antonio Lafreri, 1573–1577). Engraving, 16 15/16 × 11 5/8 in. (43 × 29.5 cm). The Metropolitan Museum of Art, New York. Open access.

volved. Vaccari questioned Claudio Duchetti's right to the inheritance of Lafreri's shop, whereas Claudio Duchetti accused Rasciotti of stealing a large number of drawings from the shop. The following day, one of the Lafreri shop employees, the Flemish printer Egidio Rouer, known in Rome as Egidio Fiammingo (active in Rome 1570s), resigned and went to work for Vaccari. The day after that, a promising young engraver from the Lafreri shop, Gerolamo da Modena, disappeared. His body was found eleven days later in the Tiber River near the Ponte Sisto.[13]

Giovanni Antonio Dosio, Skilled Artisan, and Bernardo Gamucci, Noble Antiquarian

Murderous rivalry characterized some aspects of Roman printmaking, but cooperative partnerships also flourished. An example of the latter can be seen in the work of two men from the Tuscan city of San Gimignano. One was the artisan-trained Giovanni Antonio Dosio (1533–1611), and the second was Bernardo Gamucci (dates unknown), the author of an important guidebook. Dosio was trained in Florence as a goldsmith and sculptor and arrived in Rome at the young age of fifteen. He worked on architectural and sculptural projects while also creating numerous drawings of Roman monuments and cityscapes. He came to be deeply involved in Roman antiquarian culture, aided by his friend, the humanist, translator, and poet Annibale Caro (1507–1566). Dosio and Caro both participated in the vibrant antiquarian circle around the noble cardinal Alessandro Farnese.[14]

Dosio created a map of Rome published in 1561 (fig. 6.3). His view shows the city from the viewpoint of the Pariolian Hills (Monti Parioli, hills to the northeast of Rome outside of the Aurelian Walls) with the city and its walls embedded in an enormous landscape. The map is oriented with the north on the right. At the bottom is a key of letters and numbers with labels including eighteen gates, five bridges, aqueducts, baths, circuses, arches, columns, and churches (called *Templa*), all of which show great attention to detail.[15] In this 1561 plan, Dosio ignored the dispute over the location of the Roman Forum. Although he labeled such items as the Arch of Titus, the Palatine Hill, the Capitoline Hill, and the Forum Boarium with names or indicated them with letters or numbers, he did not mark the Roman Forum itself. (His labeled drawings of the Roman Forum in the Uffizi show that he clearly believed the Roman Forum to be where Marliani thought it was—extending from the Arch of Septimius Severus.)[16] Yet the prospective audience for the map undoubtedly was not only the contentious group of antiquarians who lived in Rome but awed visitors eager to be informed. In the immense expanse of landscape into which the dense center of the city is set, the visual emphasis on the city walls, their gates, and the visibility of the roads leading into the city from all

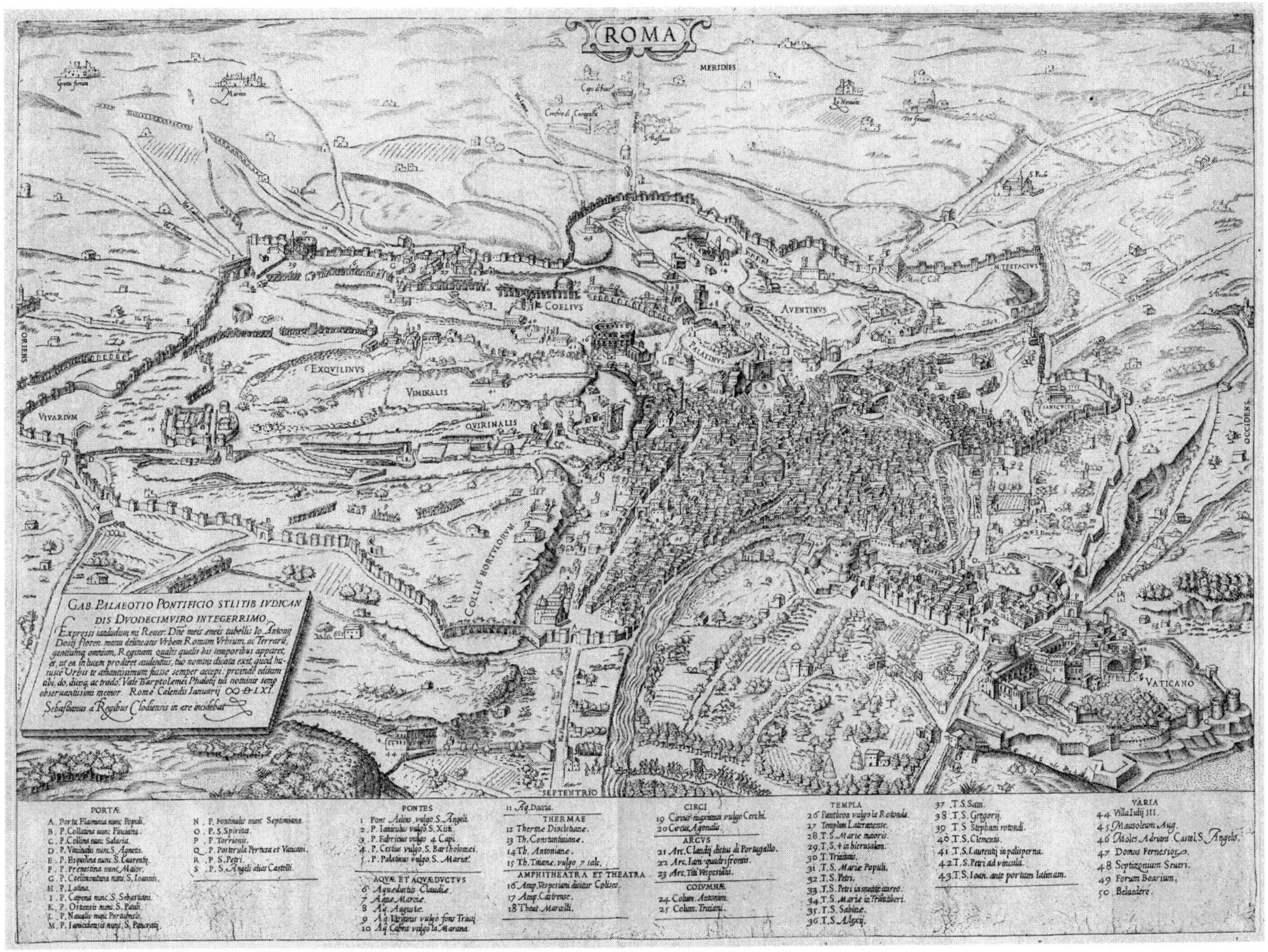

Figure 6.3 Giovanni Antonio Dosio (designer) and Sebastiano del Re (engraver), *Roma* (Rome: Bartolomeo Faletti, 1561), engraving. 16½ × 21½ in. (42 × 55 cm). Photo courtesy of the Newberry Library, Chicago, Novacco 4F 250.

Figure 6.4 Giovanni Antonio Dosio, Arch of Septimius Severus, Florence, Galleria degli Uffizi, Gabinetto dei Disegni e delle Stampe, Uffizi 2567A. Pen and ink on paper. Su concessione del Ministero per i Beni e le Attività Culturali e del Turismo.

directions provides an evocative scene that seems to express the viewpoint of visitors entering the city from afar for the first time.

Dosio spent more than twenty years in Rome between the time that he arrived and about 1576, and he drew numerous views of architectural ruins and Roman sites (fig. 6.4). He published his own book of engravings of Roman ruins and antiquities in 1569. His drawings were transformed into engravings by Giovanni Battista de' Cavalieri (ca. 1525–1601). Dosio's book contains fifty full-page engravings of ancient Roman buildings and gates, each containing a brief identifying and descriptive phrase. Unlike Pirro Ligorio, who wanted to reconstruct the ancient city as he believed it had been, Dosio wished to record the monuments of the ancient city "as they are seen today" (ut hodie cernuntur). As Carolyn Valone in her foundational study of Dosio's Roman years put it, "classicists and historians alike have come to appreciate the veracity and clarity of Dosio's record of Rome as it existed in his time."[17]

Dosio also worked as a sculptor and a military architect. One of his patrons was Torquato Conti (1519–1571), who was from a noble Roman family. Conti had close ties to humanists and antiquarians in Rome in part through his family connections with the Farnese family and in part through his long friendship with Annibale Caro (who was also Dosio's friend and adviser). The Conti family was established in the area of the Roman Forum where today the fourteenth-century Conti tower still stands. Conti employed Dosio to carry out excavations near the church of Santi Cosma e Damiano, abutting the Roman Forum (as recognized today), leasing the land from the canons of the church.[18]

This is where in 1562 Dosio discovered numerous fragments of the *Forma Urbis Romae*, incised stone pieces of the ichnographic map of ancient Rome created in the third century CE. As is known today, what he discovered were pieces of an enormous map about sixty by forty-three feet (18.3 × 13 m) that depicted the ground plan of every structure of the city (fig. 6.5). Incised on 150 slabs of marble, it hung on the wall of the *aula* or grand room of (what is today recognized as) the Temple of Peace. It was drawn to the scale of 1:240 and must have been the result of land surveys. Although it was little understood when it was discovered in the sixteenth century (and is still today the focus of intense research and controversy), the discovery created great excitement—an actual surveyed record of the ancient city was a precious finding, particularly at a time when antiquarians were obsessed with discovering where structures mentioned in ancient texts were actually located and how to identify ruins and fragments on the ground. Torquato gave the stone fragments to Cardinal Alessandro Farnese, who entrusted them to the humanist antiquarians Onofrio Panvinio and Fulvio Orsini (1529–1600).[19]

That Dosio's San Gimignano compatriot, the nobleman Bernardo Gamucci, admired him is clear from his description of the discovery of the *Forma Urbis Romae*—made, Gamucci wrote, by "Giovan Antonio Dosi da San Gimignano, a

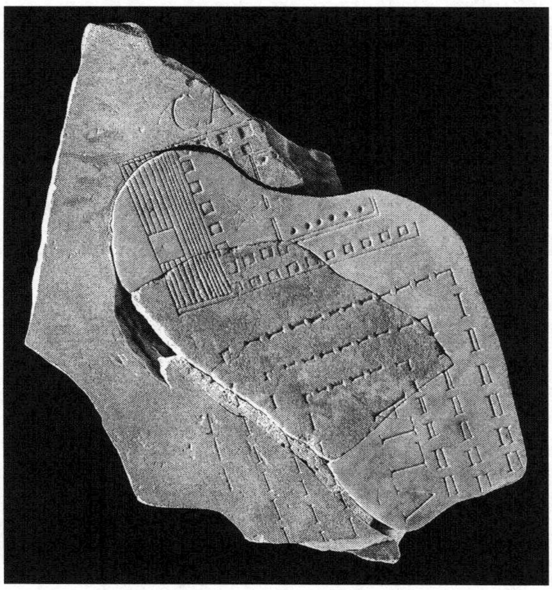

Figure 6.5 Forma Urbis Romae, Fr. 18bc, Forum Romanum. Section of the Roman Forum including the Temple of Castor (*aedes Castoris*), the Basilica Julia (*basilica Iulia*), and the Etruscan Street (*vicus Tuscus*). Roma Antiquarium Comunale © Roma, Sovrintendenza Capitolina ai Beni Culturali.

young virtuoso, architect, and antiquarian of not small promise." Gamucci used Dosio's drawings of Roman monuments and sites in his popular guidebook to Roman antiquities, *Libri quattro delle antichità della città di Roma* (Four books on the antiquities of Rome), first published in 1565. The 1565 edition of this guidebook included a foldout map of Rome with letters and numbers keyed to sites and structures mentioned in the text. Little is known about Gamucci himself, an antiquarian from a prominent family in San Gimignano who clearly had studied and measured ancient buildings and ruins.[20]

In the guidebook, Gamucci treated contemporary buildings as well as ancient ones. He ended with fulsome praise for the work of more recent times, such as renovations and construction on the Vatican and St. Peter's, works by Bramante (particularly the Tempietto in San Pietro in Montorio), Antonio da Sangallo, Raphael, and "the divine" Michelangelo. Gamucci especially praised the urban projects of Pius IV, such as his new streets and new fortifications.[21] For Gamucci, the antiquities of Rome and the marvels of the contemporary city were closely bound together; indeed, contemporary Rome had emerged out of the ancient city.

Gamucci's approach to controversial topics was cautiously empirical. Was a gate mentioned by Macrobius, the Porta Ianualis, one of the four gates of the city of Romulus, or rather, as some argue, a gate of the founder's royal palace? "I do not wish," Gamucci wrote, "to either deny or affirm what it is, because my intention in all this work is none other than to show those things true that with the authority of writers or with the certainty of the buildings can still be verified in our times, submitting in this always to the judgment of those who are more understanding than I."[22]

The guidebook often provided measurements for buildings being discussed—for example, the Basilica of Maxentius, for which he gave measurements for various parts that were shown on the accompanying plan.[23] For Trajan's Column, he provided the height, the diameter, the number of steps inside, and the number of windows; he gave the dimensions for the Circus Maximus and the dimensions of some of the obelisks found in the city; and with assistance from his brother, Raffaele Gamucci, "Geometer mathematician of our times in Rome not inferior to any other," he provided some dimensions of the Pantheon.[24]

In his foldout map of the city, Gamucci labeled the Roman Forum in its Marliani location between the Arches of Septimius Severus and Titus, where "Forum Rom[anum]" is plainly written. Gamucci considered the topic of the Forum "a truly worthy subject and [one] described by many ancient and modern writers with authority and care." But he had had difficulty because the authors writing before him had had "so many controversies" on the subject. He wished nevertheless to find the truth from "histories, and the rules and orders that one draws from architecture," but even more from what "was discovered every day in digging among the ruins" and from inscriptions, statues, and buildings.[25] Thus, Gamucci insisted that from empirical methods the truth about the location of the Forum could be known with certainty. In these hands-on practices, he was surely greatly helped by Giovanni Antonio Dosio.

Stefano Duperac, Engraver and Printer, and Onofrio Panvinio, Learned Humanist and Antiquarian

Like Dosio, the Parisian Stefano Duperac was trained as an artisan and subsequently acquired antiquarian learning, thereby rising on the social scale. Duperac worked in the Rione Parione and designed the cover for the *Speculum* in Lafreri's shop. He had close ties to antiquarians and carried out his own investigations of Roman antiquities, creating notable maps of Rome, large and small, ancient and modern.[26]

Duperac was a painter and engraver who, after a fruitful stay in Venice, worked in Rome between 1559 and 1578. While in Rome he engraved images of ancient monuments and other sites in the city, many of them published by Antonio Lafreri.[27] Early on, between 1561 and 1563, he designed and painted the thirteen landscapes in the frieze of Pius IV's loggia in the Vatican. His landscapes have been described as "vast naturalistic views," and they certainly influenced his maps. During his work on the loggia he would have met Pirro Ligorio, who was then architect of the Vatican Palace and creator of the complex iconography of the loggia decoration. Both Duperac and Ligorio, who were probably friends or at least acquaintances, traveled in the circle of antiquarian scholars anchored by the antiquarian Fulvio Orsini and the humanist Augustinian monk Onofrio Panvinio.[28]

Emmanuel Lurin has demonstrated Duperac's extensive collaboration with Panvinio, showing that the humanist monk tutored the engraver in antiquarian studies while he in turn provided engravings for Panvinio's books. Panvinio was from an impoverished family of the lesser nobility of Verona and entered the order of the Augustinian Hermits at the age of eleven after the death of his father. His skill as a Latinist and historical chronicler quickly became known by his order, especially after his publication at the age of nineteen of its history. When sent to Rome in 1549 to continue his studies, he became enamored of Roman history, topography, and antiquities, and he devoted the rest of his short life to their study, also maintaining his interest in ecclesiastical history. He received permission to live outside of the monastery and became a member of the "family" of Cardinal Alessandro Farnese sometime in the early 1550s, serving as his librarian and historical adviser. He was a prolific writer. Some of his books appeared in his lifetime, others posthumously, while yet others remained in manuscript. Panvinio prepared an edition of the Capitoline *Fasti* with commentary that was published in 1557 without his permission and to his outrage by the painter and collector Jacopo Strada (1507–1588). Panvinio remedied Strada's (in his opinion faulty) version with his own edition in 1558. Both versions appeared before the commentary on the *Fasti* published in 1560 by the guidebook author, who happened to have been Panvinio's fellow Augustinian, Bartolomeo Marliani.[29]

Panvinio's writings included a history of the popes and the *XXVII Pontificum Maximorum elogia et imagines accuratissime ad vivum aeneis typeis delineatae* (Elegies and images of twenty-seven pontiffs delineated with copper engravings very accurately to life). (See Figures 0.1 and 0.2.) His death at the age of thirty-eight prevented the publication of his aspiring masterwork on Roman antiquities in one hundred books, the first of which would concern Roman topography. These books are almost entirely extant in several manuscripts in the Vatican Library, especially in the so-called Codex Orsini (Vat. lat. 3439).[30]

The extensive collaboration between Panvinio and Duperac—both of whom were part of the larger circle that included Benedetto Egio and Pirro Ligorio—was based on reciprocal interests and needs. Panvinio had developed an extensive knowledge of ancient Roman and ecclesiastical history, knowledge that he shared with Duperac. At the same time, Panvinio thoroughly understood the cognitive and discursive possibilities of images. He believed that within historical studies there was a complementarity between painting and writing. Although nature refuses permanence to humans themselves, painting (e.g., through the portrait), he thought, allows us to attain it. He gave an important place in the Codex Orsini to the description of places, objects, and rites, and it is filled with images created by artists he had employed, including Duperac. Images of ancient places and antiquarian objects extended one's knowledge of them and were crucially important. The Codex also contains copies of some of Pirro Ligorio's drawings. Duperac's skill at image making was crucial to Panvinio's enterprise. Duperac used his skill as a painter and an engraver and the antiquarian learning that he acquired from Panvinio to ascend the social and professional ladder. He ended up in the exalted position of architect to King Henry IV of France.[31]

Although Panvinio was a humanist and textual scholar, he also made a map (undoubtedly with the help of Duperac). Panvinio's map of Rome, *Anteiquae urbis imago* (Image of the ancient city), was engraved by Duperac and published in 1565. The small map was printed with a privilege of Pius IV and the notice that it was *delineata* (designed) by Onofrio Panvinio. The map included a legend on the lower right with a list of eighty-nine toponyms that correspond to numbers on the map.[32]

Duperac's work on Panvinio's map seems to have inspired him to strike out on his own. By himself, he created three maps of Rome, two of the ancient city and one of the modern. The first, *Specimen, seu perfecta urbis antiquae imago* (Form or perfect image of the ancient city) is a small image of the city depicting only its ancient features. It was published by Lafreri in 1573 and is found in many collections of the *Speculum*. It was in fact an almost faithful copy of the map that Duperac had helped to design and had engraved for Panvinio in 1565. He showed some ancient structures in elevation, and sometimes he made them whole again—for example, the Colosseum was completely intact as was the Circus Maximus, which featured a standing obelisk. Duperac showed himself to be firmly on the side of his friend Pirro Ligorio in his placement of the Roman Forum in Augustan Region VIII between the Capitoline and Palatine Hills. In an inscription addressed to *studioso Lectori* (the studious reader), Duperac, "architect[us]," emphasized that his image of the ancient city was created with information by ancient writers and from remains that have survived to the present, which have been "very accurately drawn."[33]

Seemingly having gained confidence, Duperac next created a large plan of the ancient city, an etching that was first published in 1574, perhaps in anticipation of the jubilee that Pope Gregory XIII planned for the following year (fig. 6.6). It was printed by his fellow countryman and Lafreri's rival, Lorenzo Vaccari (as we know from the ten-year privilege from Pope Gregory XIII, dated March 15, 1574, and attached to a fragment of two sheets of the map in the British Library). Although the original copper plates still exist, the earliest extant print of the entire map, which is held at the British Library in addition to the fragment mentioned above, is a single example printed by Francesco Villamena (ca. 1565–1624), probably in the 1590s. Duperac has made the image of Rome strikingly different from the smaller image of the year before. The new version is made up of eight separate sheets that combine to make a large map 41¼ by 62½ inches (105 × 159 cm) titled *Urbis Romae sciographia ex antiquis monumentis accuratiss. delineata* (Scenographic view of the city of Rome very accurately drawn from ancient monuments). The map was oriented in the traditional (sixteenth-century) way, with the north to the left. Duperac depicted the city surrounded by a beautiful, almost undulating landscape dotted with trees, bushes, and shrubs. He emphasized the roads leading to Rome from all directions and the clear delineation of the continuation of these roads within the city walls. He filled the city itself with precisely rendered, intact ancient buildings, many labeled with italic letters. He made the Aurelian Walls prominent and depicted just outside of the walls, the *pomerium*, the ancient sacred boundary of the city. On the lower right, Duperac placed a plaque containing a dedication to King Charles IX of France—thereby revealing his ambition to return to his home country.[34]

In his dedication to the French king, Duperac emphasized his own extensive investigations of Roman ruins. He reported that he had spent fifteen years "with great labor" on the study of these ruins "with very diligent investigation, very

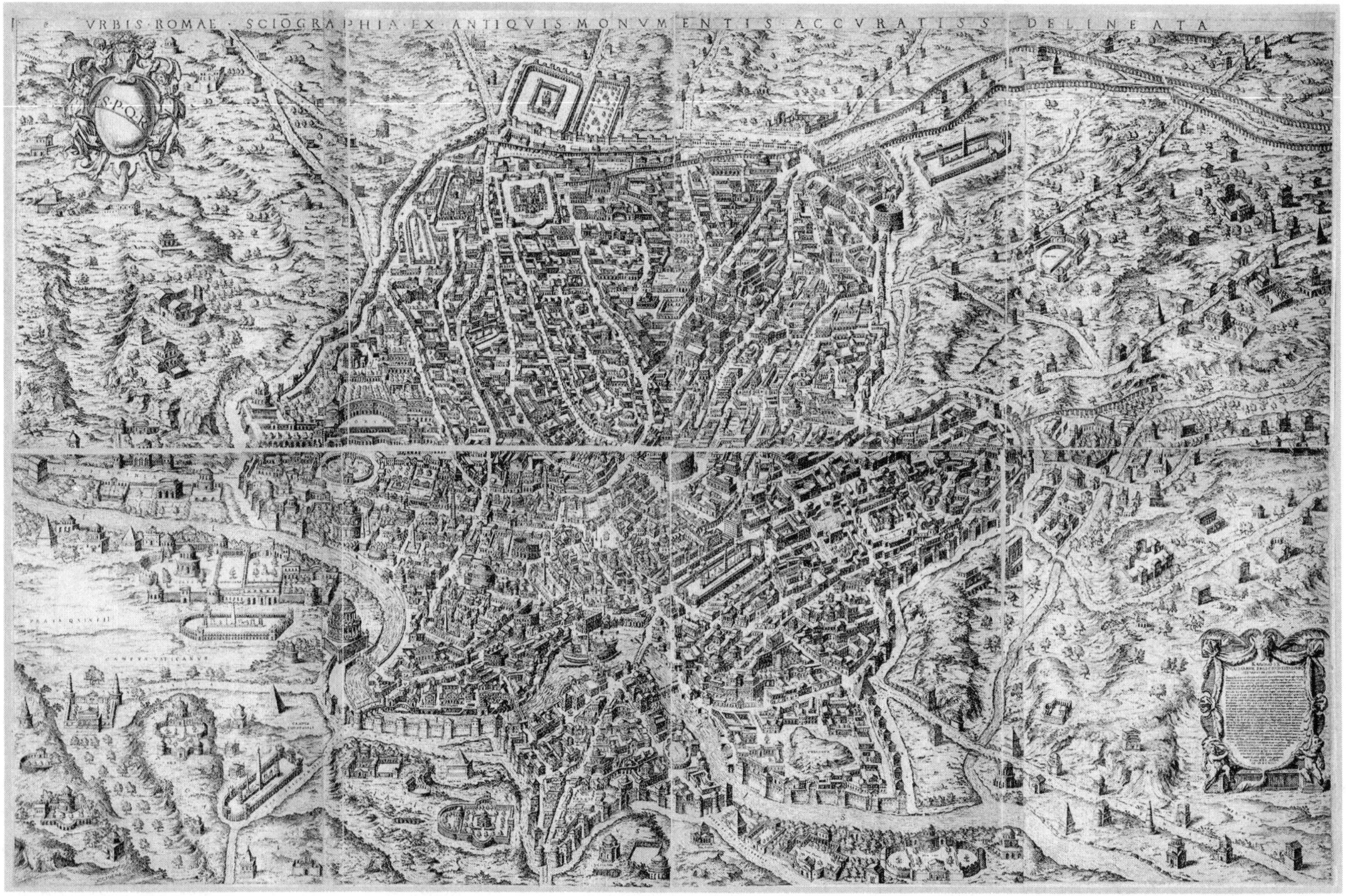

Figure 6.6 Stefano (Étienne) Duperac, *Urbis Romae sciographia ex antiquis monumentis accuratiss. delineata* (Rome, 1574). Etching on eight sheets. (Original sheets joined digitally by Bob Korn.) London: British Museum.

acute observation, [and] very accurate description." He had also studied inscriptions, broken stones and ruins, and remains of every kind of ancient thing and had described them very diligently, "noting constantly in what place each thing was discovered and excavated." He had, he noted, consulted relevant writings of the ancients, "having called in the judgment of very erudite men." Finally, with the permission of Cardinal Alessandro Farnese, he had carefully inspected the marble fragments of the ichnographic map discovered near the church of Santi Cosma e Damiano (the *Forma Urbis Romae* discovered by Giovanni Dosio in 1562). He noted that despite the difficulties of arranging the pieces correctly, "it was of great help to me" in making the map.[35] Duperac thus presented himself as an antiquarian who had worked with other antiquarians and scholars as well as artifacts and who had made an extensively researched and highly accurate representation of ancient Rome.

From this magnificent image of ancient Rome, Duperac turned to the creation of a plan of the modern city. The first edition of Duperac's panoramic view of contemporary Rome, *Nova urbis Romae descriptio* (New description of the city of Rome) was an etching published by Antonio Lafreri or his shop in 1577 (fig. 6.7). (Although Lafreri was named on the lower left of the map, the date, December 1, 1577, would have been more than four months after Lafreri's suspicious death on July 20.) The spectacular view of the contemporary city, composed on four sheets and dedicated to the French king Henry III, measured about 31¼ by 39¾ inches (79.4 × 100.7 cm). It showed a bird's-eye view of the city nestled in a sweeping landscape of the surrounding countryside with roads leading to and through the Bufalini-inspired walls, which stand out prominently. Duperac situated St. Peter's and the Vatican in the upper right rather than the lower left, very unusual for this period; the north was on the right. The clearly delineated buildings were for the most part modern and shown in elevation. Many buildings, churches, ancient monuments, but also houses, were displayed individually as they appeared rather than in stereotypical, identical forms. There was a notable precision of detail. By far the greatest number of buildings precisely drawn and labeled were churches and basilicas, which are labeled as Temples.[36] Duperac's final image of Rome, with its many roads leading through the appealing, rolling landscape into the city and its numerous churches, was surely aimed at pilgrims and other visitors coming from afar—and perhaps it also expressed his own longing to go home.

Duperac used the technique of etching (as opposed to engraving) in his 1577 map. In this technique, a wax coating is applied to a prepared copper plate, and an etching needle is used to scratch through the wax to make the image. An acid applied to the wax cuts away into the metal, creating the lines on the plate that will fill with ink. Duperac's specialty of etching was relatively rare in the Rome of the 1570s. It allowed him to work more rapidly than by engraving, and with fluidity

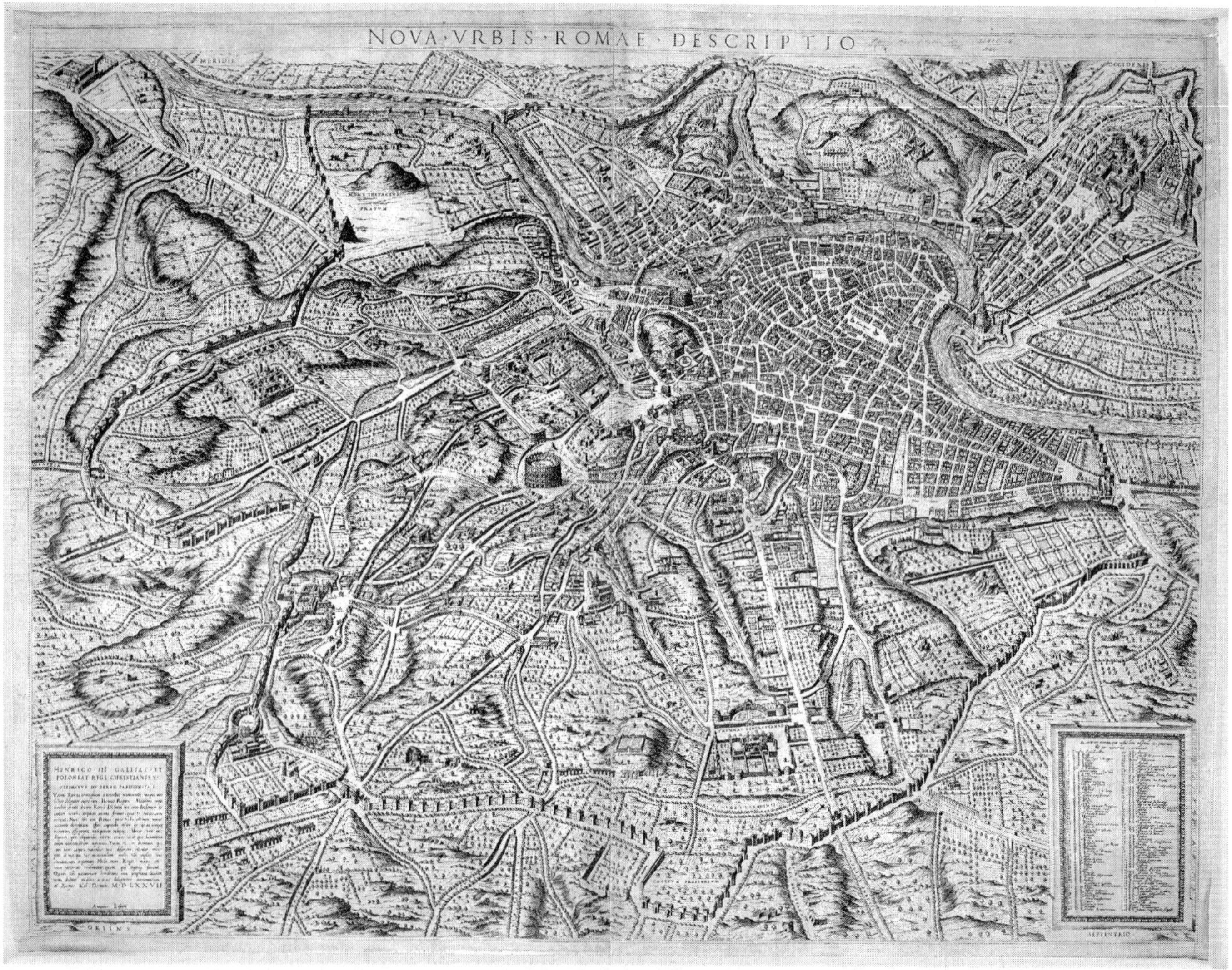

Figure 6.7 Stefano (Étienne) Duperac, *Nova urbis Romae descriptio* (Rome: Antonio Lafreri, 1577). London: British Library. Maps* 23805 (8.). Etching 31¼ × 39¾ in. (79.4 × 100.7 cm). © The British Library Board.

Figure 6.8 Stefano (Étienne) Duperac, *Roman Forum with Arch of Septimius Severus*. Engraving/etching. From *I vestigi dell'antichità di Roma* (Rome: Lorenzo della Vaccheria, 1575), No. 3. Rijksmuseum, Amsterdam.

and a refinement of line, giving him the ability to produce interesting effects of volume, material solidity, and chiaroscuro. Duperac's 1577 plan was tied to his collection of thirty-five views of ancient monuments, *I vestigi dell'antichità di Roma* (Vestiges of the antiquities of Rome), first published in 1575 (fig. 6.8). The creation of both plan and collection of views was guided by the careful analytical description of urban space that was also an evocative living portrait of the city.[37]

Duperac returned to Paris in 1578 to become a painter, architect, and garden designer (influenced by Pirro Ligorio) for French noble and royal families. His daughter Artemis married a French nobleman (and secret protestant) Jean Bourdin, pointing to his upward social mobility. Duperac was a brilliant engraver and etcher of maps and other images, but his higher status seems to have ensured that he never touched a burin again.[38]

Mario Cartaro: Engraver, Printer, Surveyor, Cartographer

Mario Cartaro (ca. 1540–1620) was an engraver who was also a printer and proprietor of his own print shop. Like Duperac, he investigated Roman antiquities and created maps of Rome. From the nearby city of Viterbo, Cartaro arrived in Rome around 1560 accompanied by his younger brother Cristofano. He was initially among the suspects in the group around Lafreri's rival, Lorenzo Vaccari, in the murder of the young engraver Gerolamo da Modena. Although Gerolamo's murderer was never identified, Cartaro must have eventually evaded suspicion—the papal commission handling the case gave him the task of dividing Lafreri's estate and its large number of prints and plates. Cartaro himself was a prolific printmaker. Over a hundred of his prints are known. He seems to have run a flourishing business and hired other printers and engravers, including his brother Cristofano. His cartographical works include matching terrestrial and celestial globes. He remained in Rome until 1590 and then went to Naples, where he created the drawings for printed maps of the Kingdom of Naples with the help of a mathematician, Nicola Antonio Stigliola (1547–1623), and where, in addition, he became employed as royal architect and engineer.[39]

Cartaro created three maps of Rome. In 1575—clearly for the occasion of the jubilee—he incised a small map of the city, *Urbis Romae descriptio* (Description of the city of Rome), with a disproportionately large Vatican Hill and St. Peter's in the lower left and the Roman Forum placed between the Arches of Septimius Severus and Titus, where Marliani had placed it.[40]

A year later Cartaro created a large (35⅞ × 44½ in./91 × 113 cm) image of modern Rome: *Novissimae urbis Romae accuratissima Descriptio* (Very accurate description of the very new city of Rome). In his dedication in the upper-left corner to the Senate and the People of Rome, Cartaro declared that he himself both designed the image and engraved the plates: "the site of the city of Rome [. . .] having been drawn diligently by himself and represented accurately with copper plates," and that he had included the ancient buildings that remained as well as modern ones. Below the dedication is a scale: *mille palmi Romani* (a thousand Roman palms), which equals about 733 feet (223.4 m or, the equivalent of 3 cm on the scale that appeared below). The plan was an ichnographic map based on that of Bufalini with the addition of realistically drawn buildings. Cartaro placed the city within a vast panorama of vineyards and surrounding countryside. He made the Vatican disproportionately huge. A cartouche on the upper right declared "Roma Renasces" (i.e., *renascens*: Rome being born again) and displayed a picture of the wolf suckling the twins Romulus and Remus and beneath that, a privilege of ten years. Of the 126 numbered rubrics, the first sixty-five—more than half—identified churches. On the map itself, Cartaro drew the ancient monuments diligently and with accuracy notwithstanding that he also reconstructed as intact many ancient buildings that existed only in a ruined state during his lifetime (fig. 6.9).[41]

In 1579, Cartaro created his large map of ancient Rome. Again it is an ichnographic plan following Bufalini, which also includes realistic-looking buildings, and it is oriented in the usual way, with the north to the left and Vatican Hill at the lower left. This plan of ancient Rome is crowded with intact ancient buildings inside of and (in contrast to Cartaro's 1576 plan of contemporary Rome) right up

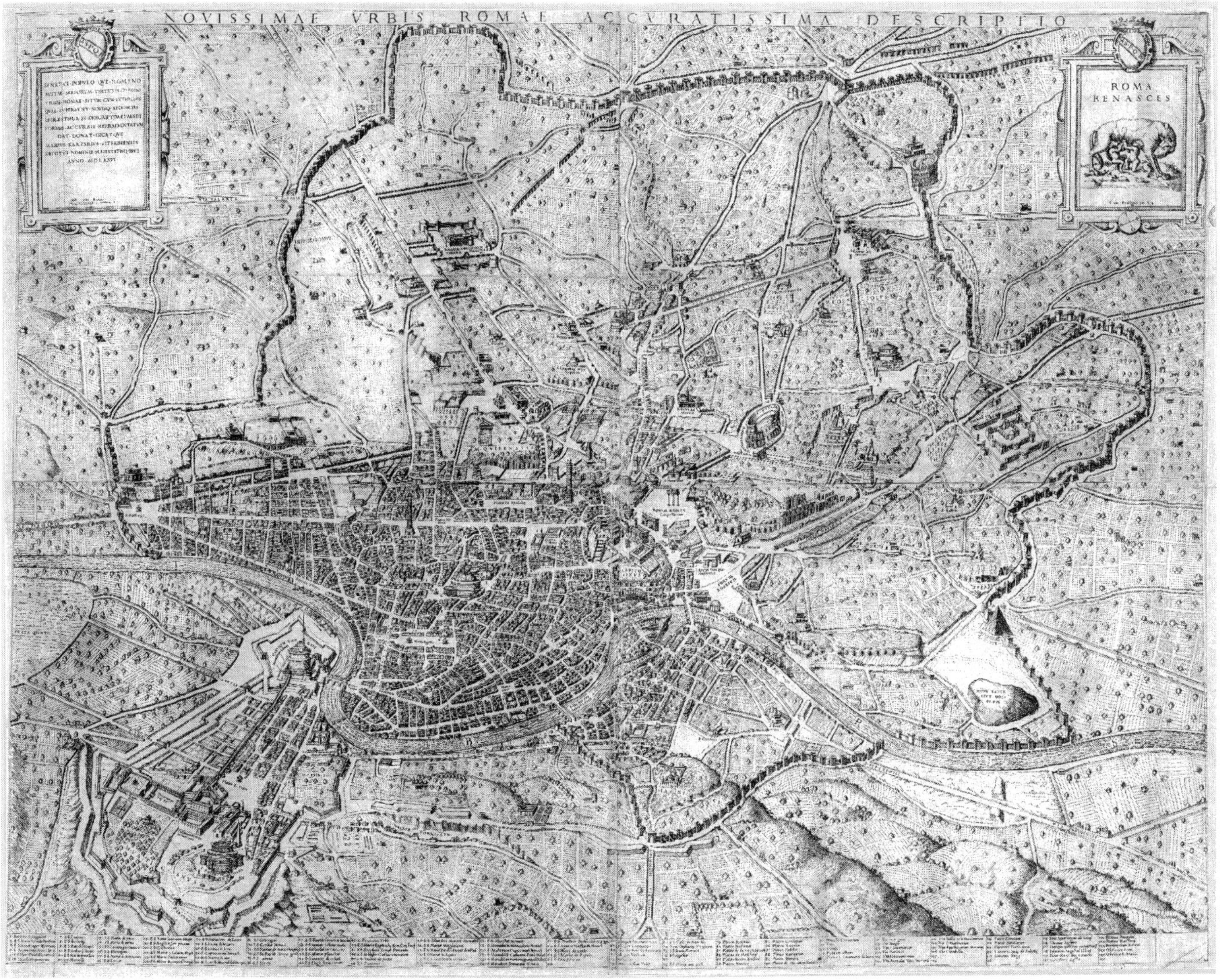

Figure 6.9 Mario Cartaro, *Novissimae urbis Romae accuratissima descriptio* (Rome, 1576). Etching, 35⅞ × 44½ in. (91 × 113 cm). Rome. Biblioteca dell'Istituto di Archeologia e Storia dell'Arte. Roma X.648, inv. 47738.

to the Aurelian Walls with only a small amount of relatively empty countryside outside of the walls. Major roads into the city are particularly visible. Sixty-eight numbers indicate buildings named in the legend on the lower edge of the map—often the sites of modern churches not actually depicted. As Cartaro explained, the numbers were there to show the location of the modern buildings when their ancient counterparts had diverse names (fig. 6.10).[42]

On a cartouche in the upper-right corner, Cartaro addressed the "Studious Reader." He noted that since the boundaries of the ancient city were uncertain, he had enclosed it within the extant walls, although he believed it had been larger before the destructions of the barbarians. He emphasized that "we will have situated ancient buildings in their original places with very true intervals as much as it was possible to gather from the extant vestiges today." He continued, "you will have therefore Rome both old and new [i.e., the large plan of 1576], with two plates fashioned by us of the same magnitude." With the same proportions being observed in both, one could compare the two cities, ancient and modern. Finally, Cartaro emphasized the care with which the plan had been made, "having been supported with the study of very many learned men, [and] having been strengthened with the diligent inspection of all ancient fragments that are extant with the use of mathematical instruments," and not only buildings on their sites but also clearings, forums, piazze, streets, and alleys.[43] He thereby claimed both learning and careful observation as the basis for his maps.

Mapping Sistine Rome

The intense activity of reconstruction that Pope Sixtus V pushed forward during his five-year papacy between 1585 and 1590, led to cartographical representations that emphasized this energetic pope's urban accomplishments. The most important and spectacular of these was the 1593 map of the city by Antonio Tempesta (1555–1630). The map was published by Nichlaus Van Aelst (ca. 1550–1613), a prolific Flemish printer who ran a print shop near Piazza Navona. Tempesta himself was a versatile and highly productive painter, draftsman, and printmaker. He has been called "one of the most prolific engravers of all time." Born in Florence, he trained in the Florentine Accademia del Disegno but spent most of his working life in Rome. He painted numerous frescos in the Vatican Palace and in other palaces such as in the Villa Farnese in Caprarola about 25 miles (ca. 40 km) from Rome and Santo Stefano Rotondo in Rome. Beyond fresco painting, he worked as a printmaker, creating over one thousand prints between 1589 and 1625 of birds, animals, hunting scenes, mythical scenes, landscapes, religious scenes, and much more.[44]

Tempesta's unique map, *Recens prout hodie iacet almae urbis Romae cum omni-*

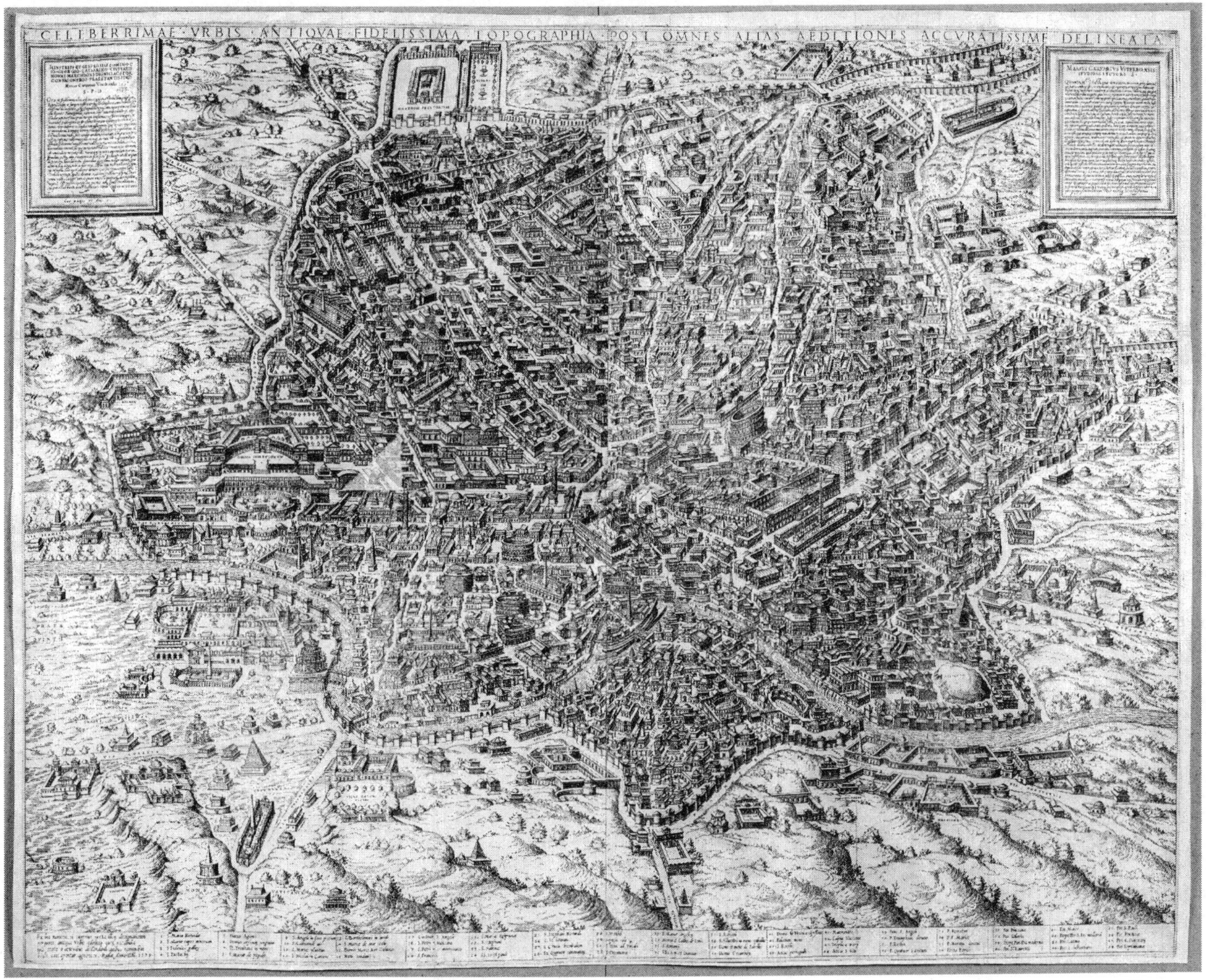

Figure 6.10 Mario Cartaro, *Celeberrimae urbis antiquae fidelissima topographia post omnes alias aeditiones accuratissime delineata* (Rome: Mario Cartaro, 1579). London: British Library. Maps* 155. (10.). Etching, 35⅞ × 44½ in. (91 × 113 cm). Copyright © The British Library Board.

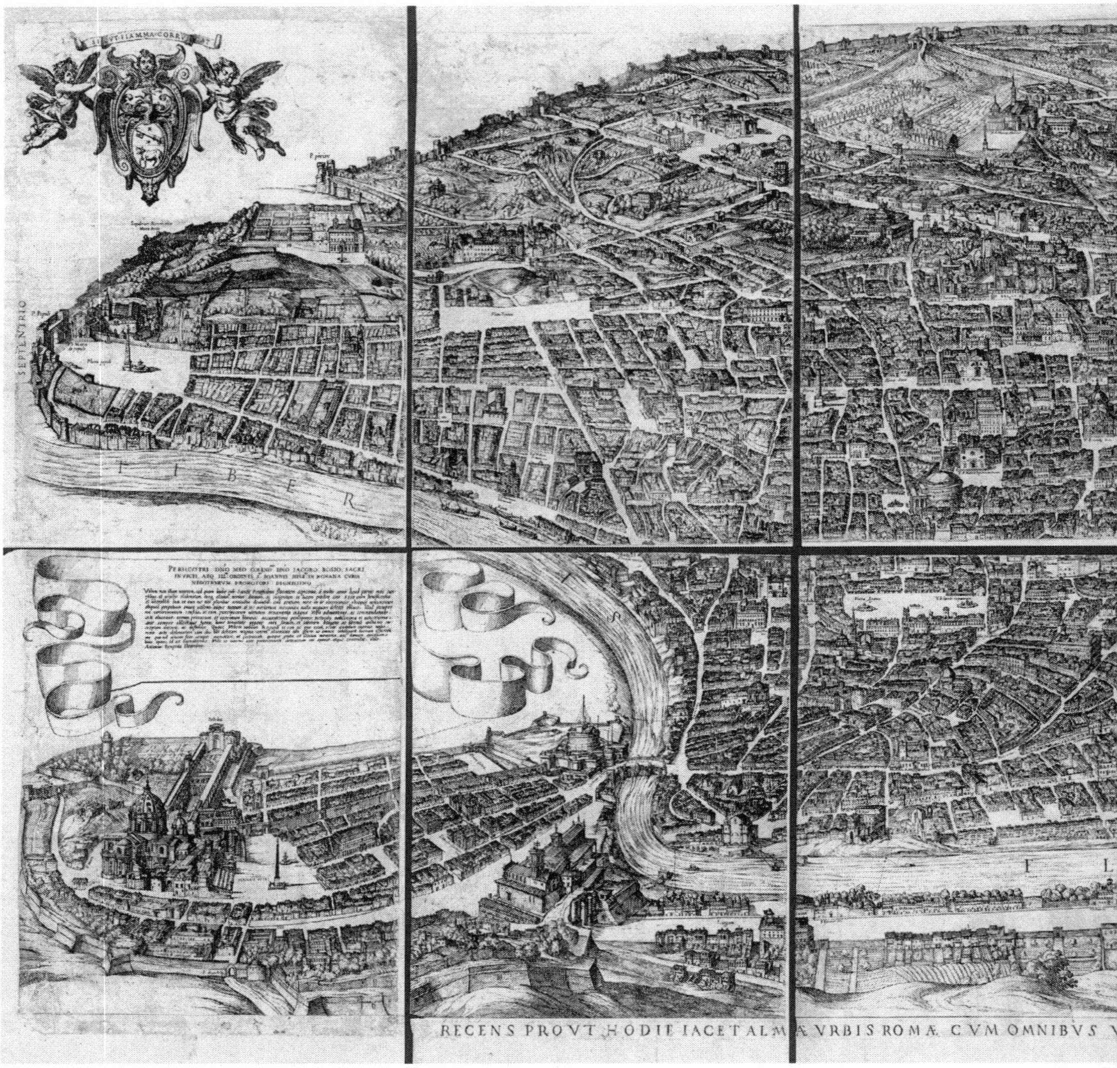

bus viis aedificiisque prospectus accuratissime delineatus (A recent view of the bountiful city of Rome as it lies today with all the streets and buildings very accurately drawn) was first published in 1593 (fig. 6.11) and was reprinted a number of times, including a replica in 1606. Tempesta conceived and began the work during the reign of Sixtus V. As his fluid lines reveal, he used the technique of etching, dividing the map into twelve sections. He announced on the lower right corner of the map that he alone had carried out the entire work—he had "designed, drawn, and incised" it (*invenit, delineavit et incidit*). He presented a vivid representation of Sistine Rome with emphasis on Sixtus V's urban innovations. The map, based in part on the Cartaro's 1576 image of modern Rome, presented a large (40¼ × 96 in./103.5 × 244 cm) image of the city seen from the viewpoint high above the Janiculum Hill and with the north to the left and the Vatican in the lower left corner.[45]

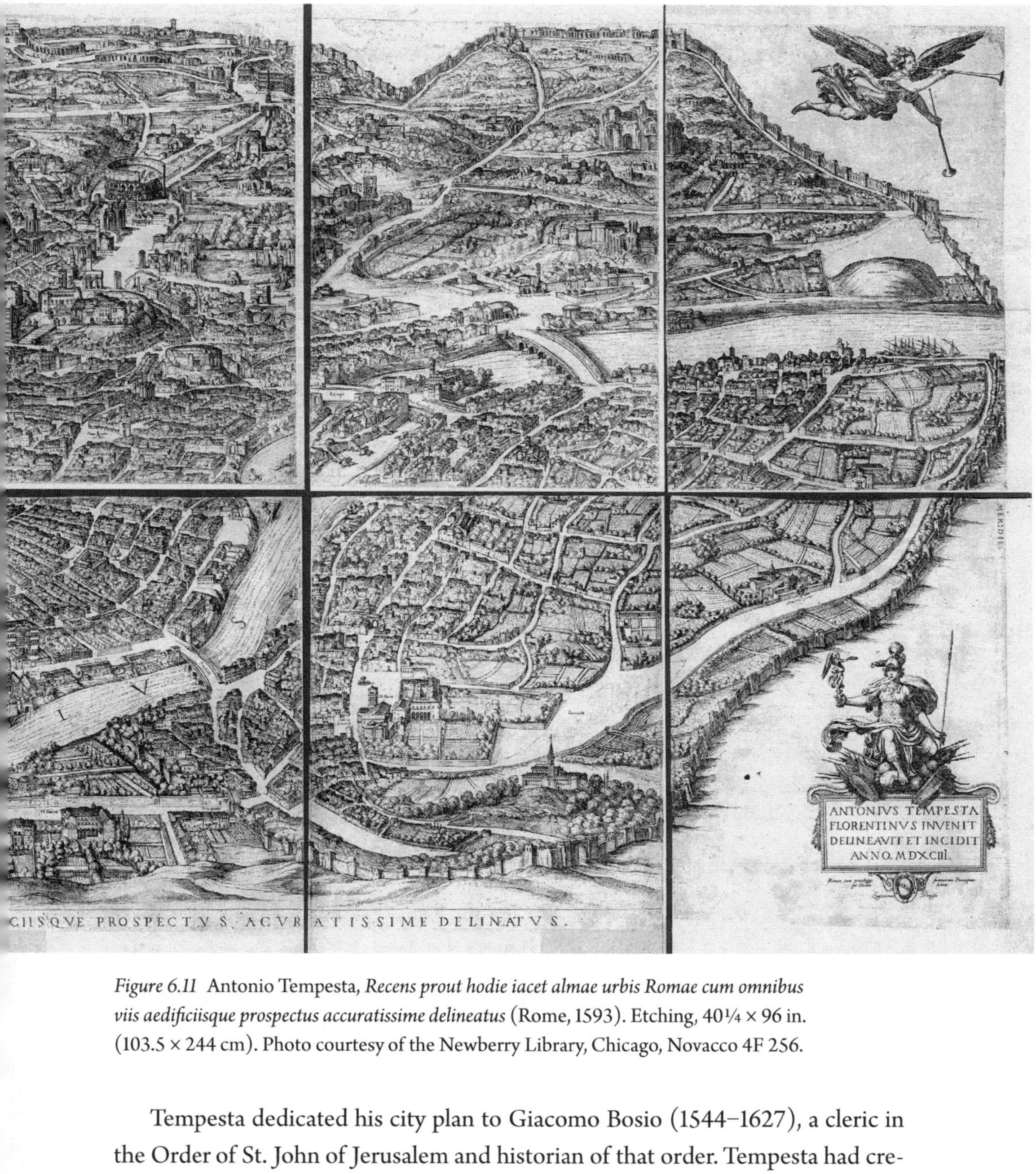

Figure 6.11 Antonio Tempesta, *Recens prout hodie iacet almae urbis Romae cum omnibus viis aedificiisque prospectus accuratissime delineatus* (Rome, 1593). Etching, 40¼ × 96 in. (103.5 × 244 cm). Photo courtesy of the Newberry Library, Chicago, Novacco 4F 256.

Tempesta dedicated his city plan to Giacomo Bosio (1544–1627), a cleric in the Order of St. John of Jerusalem and historian of that order. Tempesta had created a design for the frontispiece of Bosio's three-volume *History of the Knights of Malta,* and he may also have worked for Giacomo's brother Antonio Bosio (1575–1629), one of the founding figures of Christian archaeology, who had worked closely with Onofrio Panvinio. Tempesta's map features Giacomo's coat of arms in the upper-left corner. In his dedication in the ribbon-ornamented cartouche on the left, Tempesta notes that "we look not at that ancient city but as [it is] today, flourishing under the holy popes," and he allowed it to be published only now, after many years "with hardly a little of my own costs and vigilance."[46]

Although dedicated to a patron, the map presented the city shaped by Sixtus and his immediate predecessors. Tempesta's map displayed the dynamic economic life of the city—the river mills, the ferries, the commercial boats, the wide roads for carriages, and the Ripa, the main commercial docking point for the city—and new, wide streets, those created by Pius IV, Gregory XIII, and Sixtus V. But unlike earlier maps, such as those of Duperac and Cartaro, Tempesta's streets came to an abrupt end at the Aurelian Walls. Most of the walls extended to the edge of the sheets, but where they did not, they were bordered on the outside by white space. Having ended the image of the city at the walls, he displayed not the various approaching streets that pilgrims would use to arrive at the city but instead the new streets created by the recent popes. He trumpeted the urban accomplishments of papal rule, especially the rule of Sixtus V.[47]

◆ ◆

Intrinsic to the massive production of Roman city views and maps were the ongoing attempts to transform the physical city and its infrastructures, whether successful or not. Mapmaking is often studied as a practice separate from bridge repair, aqueduct construction, waste disposal, sewers, flood control, and the construction and paving of new streets. But reenvisioning Rome on paper and remaking Rome as a physical entity went hand in hand. The new, straight streets on maps reflected new, straight streets on the ground. Moreover, the city renewed, although not so evident in the maps and city views, included a vision of a reformed population. It is to the physical streets of Rome and the people on those streets that we now turn.

7

Reforming the Streets

A famous visitor on horseback, the French essayist Michel de Montaigne (1533–1592), entered Rome for the first time on November 30, 1580. He was accompanied by his entourage carrying baggage, including many books. The group entered through the Porta del Popolo, the sumptuous new gate at the northern entrance to the city that Pope Pius IV had ordered reconstructed nineteen years earlier. Customs officials at the gate inspected the bags at length and confiscated the books to be scrutinized for heretical material, only returning them some weeks later. Montaigne was not a pilgrim per se but a pious tourist who must have felt a special affinity to Rome because of his interest in and deep knowledge of classical texts intensified no doubt because his native language (or one of them) was actually Latin, thanks to an upbringing guided by his father's pedagogical views. Montaigne was curious to see the world and especially the city that once had been the capital of the ancient Roman Empire and now was the capital city of the Catholic Church.[1]

His impressions of Rome, recorded in his travel journal partly by an unidentified secretary and partly by himself, provide a glimpse of the city and its streets at this particular moment in time. In the secretary's words, the cosmopolitan atmosphere was evidenced by Montaigne's annoyance that so many greeted him in French (making it difficult, one presumes, for him to practice his Italian). He marveled at "the sight of so great a court so thronged with prelates and churchmen," and Rome "seemed to him more populous in rich men, and coaches, and horses, by far, than any other [city] that he had ever seen." The appearance of the streets, "and especially in the multitude of the people," reminded him of Paris. Further, Rome was populated along the Tiber on both sides, whereas the hills, where he "took a thousand walks," were occupied by only a few churches and the palaces and gardens of the cardinals.[2]

Montaigne was acutely aware of ancient Roman ruins. From the Arch of Septimius Severus, he judged that the ancient street level had been "more than two pikes length" (about 40 ft.) below. Obliquely bearing witness to muddy and unpaved streets, he observed that "almost everywhere, you walk on the top of old walls which the rain and the coach ruts uncover." He commented on the great amount of theft (the well-off should "give their purse in keeping to the bankers of the city") and on the dangers of the streets at night. He remarked on the appearance of Roman women, opining that they were not as beautiful as they were reputed to be and observing further that "the most singular beauty was found among those who put it on sale."[3]

Montaigne described his attendance at Christmas mass and his audience with Pope Gregory XIII. He observed that the pope was "very lavish in public buildings and the improvement of city's streets." He witnessed a papal procession that occurred on January 3, 1581, passing under the window of his rented room (on Via di Monte Brianzo, across the street from Santa Lucia della Tinta) on the left bank of the Tiber not far from Ponte Sant'Angelo. First, two hundred members of the papal court mounted on horses went by, then the pope himself with a red hat and velvet hood on a white steed, giving a benediction every fifteen steps. There followed another horse, a mule, a "handsome white charger," and a litter with two robe bearers.[4]

Montaigne also described three public executions—one "a famous robber and bandit captain," Catena. The condemned man was transported on a cart attended by a procession of masked men with two monks on the cart with him preaching. He was then "strangled" (i.e., hanged) on gallows. After he died, the body was cut into four quarters, an action that brought cries from the crowd. Then Jesuits and others, climbing to higher vantage points, shouted to the people to take this as an example. On the same day Montaigne saw the execution of two brothers who had committed murder. The brothers were first torn "with red-hot pincers." Then the executioners cut off their fists and put just-killed capons on the wounds. Then they killed the brothers on a scaffold—first they clubbed them with a large wooden mace, then they cut their throats. Montaigne reported (with evident equanimity) that some say they were executed in this manner because this is how

Figure 7.1 Annibale Carracci, study for an execution. Royal Collection Trust RCIN 901955. Brown ink on paper, 7½ × 11½ in. (190 × 292 mm). Windsor Castle Royal Library, 1599. Royal Collection Trust / © Her Majesty Queen Elizabeth II 2017.

they killed their master; others that this was the routine method of execution in Rome (fig. 7.1).[5]

◆ ◆

Processions through the Streets of Rome

Montaigne's travel journal provides a vivid glimpse of the socially, legally, and physically complex system of the Roman streets, which threaded through neighborhoods and served many functions. An important one involved ritual enactments of piety, power, honor, and authority. The basic form of most public ritual in Rome was the procession—either as an introduction to other events, such as a mass or dedication, or as an event in itself. Carnival, Easter, and saints' days could be opened or marked by processions. Other, more secular processions frequently took place—the entrance of princes, ambassadors, and other important visitors into the city; the marriages and funerals of nobles.[6] Perhaps most spectacular was the triumphal procession in honor of Marcantonio Colonna on December 4, 1571, after he led the papal fleet in the defeat of the Turks in the naval Battle of Lepanto on October 7, 1571.[7] The crowning of a new pope occasioned a particular kind of great procession through the city called the *possesso*.

The Possesso

By the elaborate, highly orchestrated *possesso* through the streets of Rome, the popes signaled their supremacy over the city and over the urban population. A few days or weeks after each papal coronation, papal dominance of the physical city was ritually reenacted. The new pope processed through the streets with a huge entourage from the Vatican to the Basilica of St. John Lateran, the archbasilica of the city and official seat of the pope in his capacity as the Bishop of Rome. There, further ceremonies ensued. In 1471, Pope Sixtus IV, at his own inaugural procession, changed the name of the ceremonial procession from *processo* (procession) to *possesso* (possession). Thus did the pope *possess* St. John Lateran, the seat of the bishopric of Rome, as he also took symbolic possession the city itself. In the fifteenth century, the great procession often encountered threats and attacks from the Roman population. Even in the sixteenth century, the solemn parade was sometimes met not with adulation but with signs urging good government.[8]

The immense procession was organized in strict hierarchical order, and disputes over precedence erupted frequently. The parade included representatives from numerous Roman groups—the Roman barons, hundreds of clergy, abbots, cardinals, the heads of the thirteen *rioni* (the *caporioni*) each carrying the flag of his district, the three conservators of the communal government, the marshals (*marescialli*), and finally, the new pope himself riding on a white horse near the end. The route of the procession was bedecked with tapestries and enhanced by temporary triumphal arches. These displays and indeed the specific route varied from one pope to the next. The pope would stop along the way to perform traditional ritual acts. In one, he threw coins to the poor. In another, he received a token of ritual subjugation from the Jews at the Arch of Titus—built in the first century CE to celebrate the victory of the emperor Titus over the city of Jerusalem. This ceremony culminated when the pope threw the Torah to the ground.[9]

The *possesso* changed in many particulars over the sixteenth century. Increasingly, it demonstrated the discontinuity between the new pope and the preceding one, while it also sought to display the power of the popes over the Capitoline government and the city itself.[10]

One changing feature involved the practice of throwing coins to the poor. The master of ceremonies for Pius IV's coronation on January 6, 1560, reported that before the *possesso*, the pope came to the loggia to throw the money. In the frenzy to grab the coins, the crowd became violent: "infinite men fell one over the other," and "about twenty-six men suffocated."[11] Pius IV's successor, Pius V, considering the deaths that had occurred in 1560, dispensed with throwing money to the poor. "He did not throw money," an *avviso* reports, but instead, "made all the poor assemble in the Campo Santo and then go out through a single gate and to each had three giuli given." He also sent 25 scudi each to poor monks and the poor of the monasteries.[12] Gregory XIII also refrained from throwing money, instead giving 15,000 scudi to various charitable institutions, while Sixtus V gave money to hospitals and houses of religion.[13]

Although the papal routes all led from St. Peter's to Saint John Lateran, there were significant deviations. Pius V chose a novel route by which he displayed his opposition to antiquarian interests and pagan influences. He avoided processing by way of the ancient Via Sacra through the Forum, the traditional route of the *possesso*, and avoided going through the Colosseum as well.[14] Sixtus V, in his *possesso* of May 5, 1585, emphasizing his power over the city, took the traditional route but saw to it that papal symbols dominated the Capitoline Hill. Famously accompanying the other ambassadors were a group of ambassadors from Japan, underscoring the universality of the church.[15]

Visiting the Seven Churches of Rome

From the mid-sixteenth century, a new practice, visiting the seven churches of Rome in one day, emerged as a novel form of procession. To visit the churches and basilicas of Rome had always been considered a devotional practice and was the goal of the many pilgrims who flocked to Rome. Filippo Neri, founder of the Oratorians, began the practice of visiting all seven major churches in a single day—St. John Lateran, St. Peter's, San Paolo Fuori le Mura, Santa Maria Maggiore, San Lorenzo Fuori le Mura, San Sebastiano, and Santa Croce in Jerusalem. As Martine Boiteux has described it, the route was actually a course at the margins of Rome, leaving and reentering the gates several times (fig. 7.2). Neri began the practice by himself. In the early 1550s, he was joined by a small band of twenty-five to thirty of his proselytes. The practice became increasingly popular and was adopted by Pius IV and then with great enthusiasm by both Gregory XIII and Sixtus V.[16]

In his book *Roma Santa*, the pious English priest Gregory Martin lavished much attention on the many churches of Rome, including the major destinations of pious visits. He particularly detailed lists of relics housed in each of them, pointing to the focal point of devotion on arrival at each of the seven churches.[17] From time to time, the designated churches were changed—most importantly, Sixtus V substituted Santa Maria del Popolo for San Sebastiano. It was closer and more convenient and would soon (as a result of the decision to make it one of the seven churches) benefit from the aura of a newly erected obelisk at Piazza del Popolo.[18]

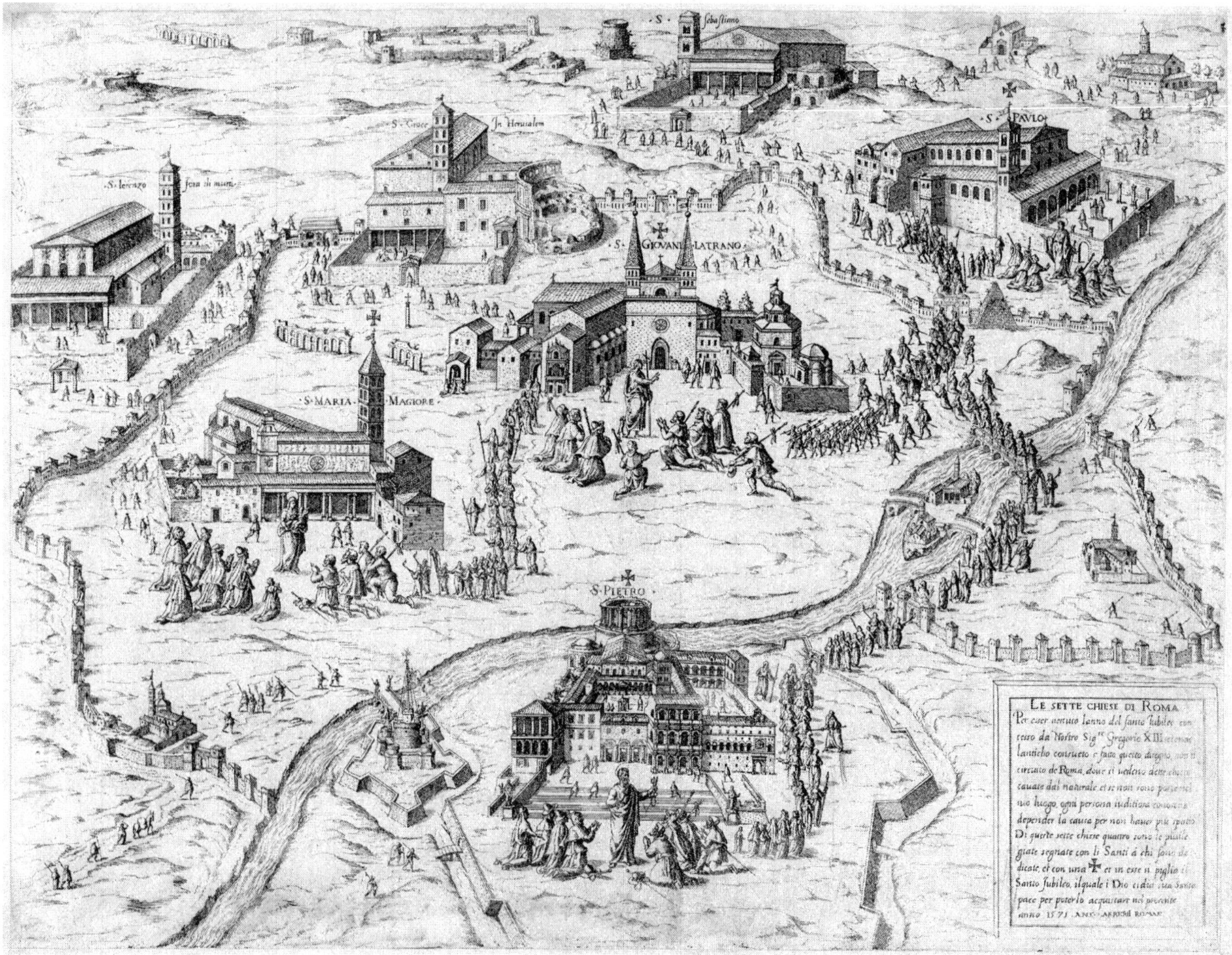

Figure 7.2 Giovanni Ambrogio Brambilla (attributed), *The Seven Churches of Rome*. In *Speculum Romanae Magnificentiae* (Rome: Antonio Lafreri, 1575). Etching and engraving, 15⅝ × 20¹⁄₁₆ in. (39.7 × 50.9 cm). The Metropolitan Museum of Art, New York. Open access.

In an important way, the visit to the seven churches of Rome was a reenactment of the *possesso* on a regular basis—an intermittent assertion that the streets were controlled by the papacy. Led by the Oratorians and often including the pope himself, the visit became a perfect modality for exerting Christian and papal authority over the streets and people of Rome.

Renovation of the Streets

The *possesso*, visits to the seven churches, and other theatrical cavalcades through the streets of Rome enacted the pope's and the church's authority over the city. The popes also tried to shape the streets in more permanent ways. The popes who ruled between 1560 and 1590 (especially Pius IV, Gregory XIII, and Sixtus V) undertook the construction of new, wide, and straight streets. Such streets facilitated processions and the passage of pilgrims, allowed horse-drawn coaches (the new rage of elite Romans) to pass by one another, and functioned as urban symbols of papal power.

The papal practice of creating new streets and widening and straightening old ones was firmly established in the fifteenth century. Sixtus IV ordered the construction of the Ponte Sisto in time for the jubilee of 1475 and had the medieval street Via Santa on the right bank of the river cleared of obstacles and repaved. He thereby created the northern segment of Via Lungara from the Ponte Sisto to the Vatican.[19] At the turn of the sixteenth century, Pope Alexander VI (ruled 1493–1503) ordered a wide, straight street to be constructed in the Borgo called the Via Alessandrina (fig. 7.3). It was the first street in which the straightness was to be accentuated by buildings all of the same height on either side, creating a uniform avenue. The pope issued the bull *Super aedificiis in via Alexandrina construendis* that required all owners of houses or properties on the street to build (or augment) their houses to the height of seven *canne* (ca. 17 yd. or 15.6 m).[20]

The most important new street of the early sixteenth century was the Via Giulia. Julius II (ruled 1503–1513) created this monumental avenue on the left bank of the Tiber. The new street began across the river from Santo Spirito and ended at Ponte Sisto. Continuing the work of Sixtus IV, Julius also had the Via della Lungara widened and straightened. Thus, two streets joined by the Ponte Sisto facilitated the passage of pilgrims to St. Peter's.[21] Julius's successor, Leo X, turned his attention to the north, creating Via di Ripetta (formerly Leonina). The new street formed one arm of what would become the great trident of streets leading from Piazza del Popolo. The center street was Via del Corso. The third arm, now Via del Babuino, was constructed under the Farnese pope, Paul III (fig. 7.4).[22]

After the traumas of the Sack in 1527 and the flood of 1530, most of Paul III's efforts regarding urban topography were directed toward receiving the Holy Ro-

Figure 7.3 Giovanni Antonio Dosio, *Via Alessandrina seen from Piazza S. Pietro*, 1560. Florence, Galleria degli Uffizi, Gabinetto dei Disegni e delle Stampe, Uffizi 2580A. Pen and brown ink on paper. Su concessione del Ministero per i Beni e le Attività Culturali e del Turismo.

man Emperor, Charles V, in 1536. Charles and his huge entourage would come by the Via Appia in imitation of an ancient triumph. The emperor entered through the Porta San Sebastiano and then processed through widened and straightened streets, which perhaps became models for the future. A new street was built between the Arch of Septimius Severus and the Arch of Titus—a project that required the demolition of numerous structures including houses.[23]

After Paul III, the major contributor to Roman urban street design was Pius IV. This pope erected more new city gates than almost any pope before or since. Among them was the famous Porta Pia, designed by Michelangelo (fig. 7.5). Pius IV's new street, the Via Pia (now XX Settembre) extended on the Quirinale toward the Aurelian walls and through the elegant new gate. This was a street for aristocratic Rome. Lined by gardens and vineyards and eventually palaces, it provided a suitable avenue for the entrance and exit of noble retinues and luxurious carriages.[24]

Pius had other new streets constructed as well. The Via Tabernola, now known as Via Merulana, joined two basilicas—Saint John Lateran and Santa Maria Maggiore—in a straight line. The initiation of this new street project has often been (incorrectly) attributed to Gregory XIII, who took it over when he became pope and completed it in time for the jubilee of 1575. For the first time, the eastern part

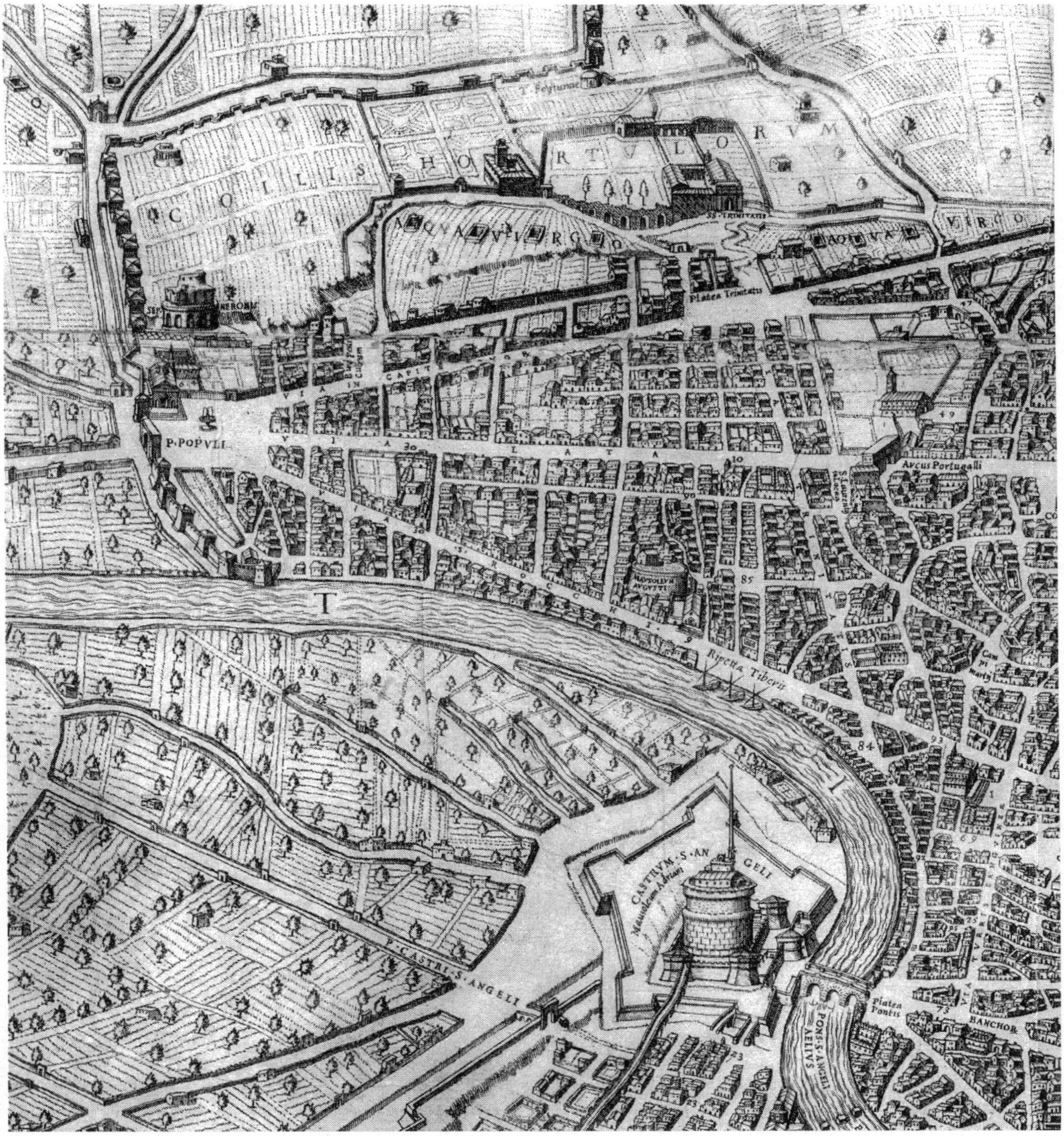

Figure 7.4 Detail of fig. 6.9. Mario Cartaro, *Novissimae urbis Romae accuratissima descriptio* (Rome, 1576). Etching. Trident of streets from Piazza del Popolo. Piazza del Popolo (north) at left. From top as named on map: Via in Capite (now Via del Babuino), Via Lata (now Via del Corso), and Via S. Rocchi (now Via di Ripetta).

of the city was developed, including the Qurinal, Viminal, and Esquiline Hills.[25] Pius IV also had constructed the Via Angelica leading out to the north from his newly developed area of the Borgo, the Borgo Pia.[26]

By the 1560s a major new form of transportation — the coach — had begun to transform the streets of Rome, making wide, straight streets ever more desirable and the usual narrow and congested Roman streets ever more inadequate. The

Figure 7.5 Porta Pia, Rome. Designed by Michelangelo for Pope Pius IV. Photograph May 2017 © Bob Korn with kind permission.

coach, or *cocchio,* was imported from Hungary in the first half of the sixteenth century and had by the 1560s become an important new status symbol. An extant accounting list from 1594 reveals 883 coaches in the city. At that time six hundred families were able to afford coaches, and some wealthy aristocrats owned three or four.[27]

Horse-pulled coaches and buffalo-pulled carts crowded the streets of Rome and damaged street pavements. They also became an important source of revenue. An edict dated August 1568 ordered that all owners of coaches, carts with iron-rimmed wheels, carriages, and animals that pull these conveyances (and wear iron "horseshoes") were to be taxed "in order to restore in part the great damage that they make continuously on the brick paving and the streets of Rome." A series of penalties for failure to pay were enacted, including whipping and imprisonment for coachmen who failed to comply.[28]

Under Gregory XIII legislation was enacted that required coaches rented out for hire (*ad vecturam*) to be licensed, and it levied a tax of 2 scudi per month on them. The tax was used by the Congregation on Streets, Bridges, and Fountains to pay the street taxes of impoverished petitioners who could not afford the *gettiti* required for cleaning and paving. Other regulations forbade coaches to carry men and women together unless they were related and forbade the transport of prostitutes.[29]

Pope Sixtus V, one of the great renovators of streets in the late sixteenth century, created long, straight streets, most importantly Via Felice (now Via Sistina, Via delle Quattro Fontane, and Via de Pretis). Whether he and his advisers had in mind a unified street plan for the city as a whole is debated.[30] Equally controversial are Sixtus's reasons for his initiatives on streets. Did he mainly want to facilitate the progress of pilgrims from one pilgrimage church to another? Was secular urban development foremost in his mind? Or was the prominence of his own papacy and the ascendancy of his own family of paramount importance?[31] All of these motivations could have coexisted and probably did.

Whether or not Sixtus V had in mind an urban master plan, it is clear that he wanted to develop the hills of Rome and to make the basilica of Santa Maria Maggiore a central focus. Santa Maria Maggiore stood adjacent to the majestic Villa Montalto being developed by Sixtus and his sister Camilla Peretti. It was in Santa Maria Maggiore where Sixtus created his own mausoleum as well as the mausoleum for Pius V, the pope who had made him cardinal. The new street Via Felice connected Trinità dei Monti on the Pincian Hill (the church at the summit of today's eighteenth-century Spanish Steps) with Santa Maria Maggiore on the Esquiline and then extended to Santa Croce in Jerusalem. The never-completed Via Felice was to have continued from Trinità dei Monti to the Piazza del Popolo. Sixtus provided dramatic markers by erecting ancient Egyptian obelisks at the ends of some of these new streets.[32]

In the papal bull *Decet Romanum Pontificem* of September 13, 1587, Sixtus V encouraged building construction on the new streets by providing special dispensations for those who lived or built there. Sixtus V wanted to revitalize the basilicas and facilitate thoroughfares to them. Any who built on the new streets Via Felice or on Via Pia were given special immunities. Except for committing treason, they could not have their properties confiscated. If a person lived in a building on any of these streets for two years, he would gain the same rights as Roman citizens. If such residents had incurred debts outside the state, they must not be troubled (i.e., they would be immune from such debts). Artisans setting up shops in buildings on the new streets would have none of the usual taxes imposed by the guilds and would be excused from guard duty during plagues or wars.[33]

An illustration of the streets of Rome that Sixtus V planned or in some cases built can be found in a book praising the works of the pope by Giovanni Francesco Bordini (ca. 1536–1609). Bordini was an ordained priest and a Roman-born follower of the founder of the Oratorians, Filippo Neri. In 1588 he published a small book of poems praising the works of Sixtus, which included fifteen engravings displaying the pope's accomplished or planned works. One of the images was a plan of the new streets supposedly envisioned by Sixtus—a system of straight wide streets that connected central nodes to one another—the Lateran, Trajan's Column, the Porta San Lorenzo, the churches of Santa Croce in Jerusalem, and Trinità dei Monti. It was a polycentric system (fig. 7.6). The accompanying poem was titled "Concerning the very wide streets that Sixtus V Pontus Maximus has opened and paved in the form of a star from the Esquiline Hill to diverse places." The illustration prominently displayed the basilica of Santa Maria Maggiore. A wide street runs from Santa Croce in Jerusalem to Santa Maria Maggiore and then on to Santissima Trinità dei Monti. Another street—which was never built—ran from the Trinità to Santa Maria del Popolo. The poem went on to describe Santa Maria Maggiore as the center of a star from which to visit the seven churches.[34]

Paving the Streets

The Sistine vision of new straight streets with obelisks standing at crucial intersections as focal points may well have been marred by an urban reality—muddy and debris-encumbered streets that were either unpaved or had degraded paving. Images of Rome found in diverse sources from the early 1570s regularly displayed central locations of the city (such as the Piazza on which stood the Column of Marcus Aurelius) as unpaved (fig. 7.7).[35] Paving required maintenance. It is not surprising that the popes most focused on urban streets were also anxious to get those streets paved and to maintain that paving.[36]

Uncertainty and disagreement existed concerning just what paving materials to use—whether brick (*mattone*) or stone (*selice* or *selce*). The latter, popularly called *sampietrini,* were cobblestones made from basalt that were beveled before being hammered into place in the street—a paving material still in use today in historic parts of Rome. In the sixteenth century, stonecutters hewed the stones to make irregular polygonal shapes and square tops that workers pounded into the street one next to another to form the pavement surface (fig. 7.8).[37]

There was doubt about whether stones should be used at all for pavement. Basalt, the volcanic stone that is used for the *sampietrini* did not absorb moisture, which often appeared on the stone-paved surfaces in the morning. Some feared that the use of stone would create excess humidity and bring about bad air or miasma, thought to cause disease. In his *Inter multiplices curas* published on August 23, 1565, Pius IV ordered that the streets be paved with baked bricks, because, it explained, Rome was subject to great humidity, and stone-paved streets make the air of the city more humid than it would be if the streets were paved in brick. The bull forbade construction workers—under penalty of exile, fines, and corporeal punishment at the discretion of the masters of the streets—to pave with

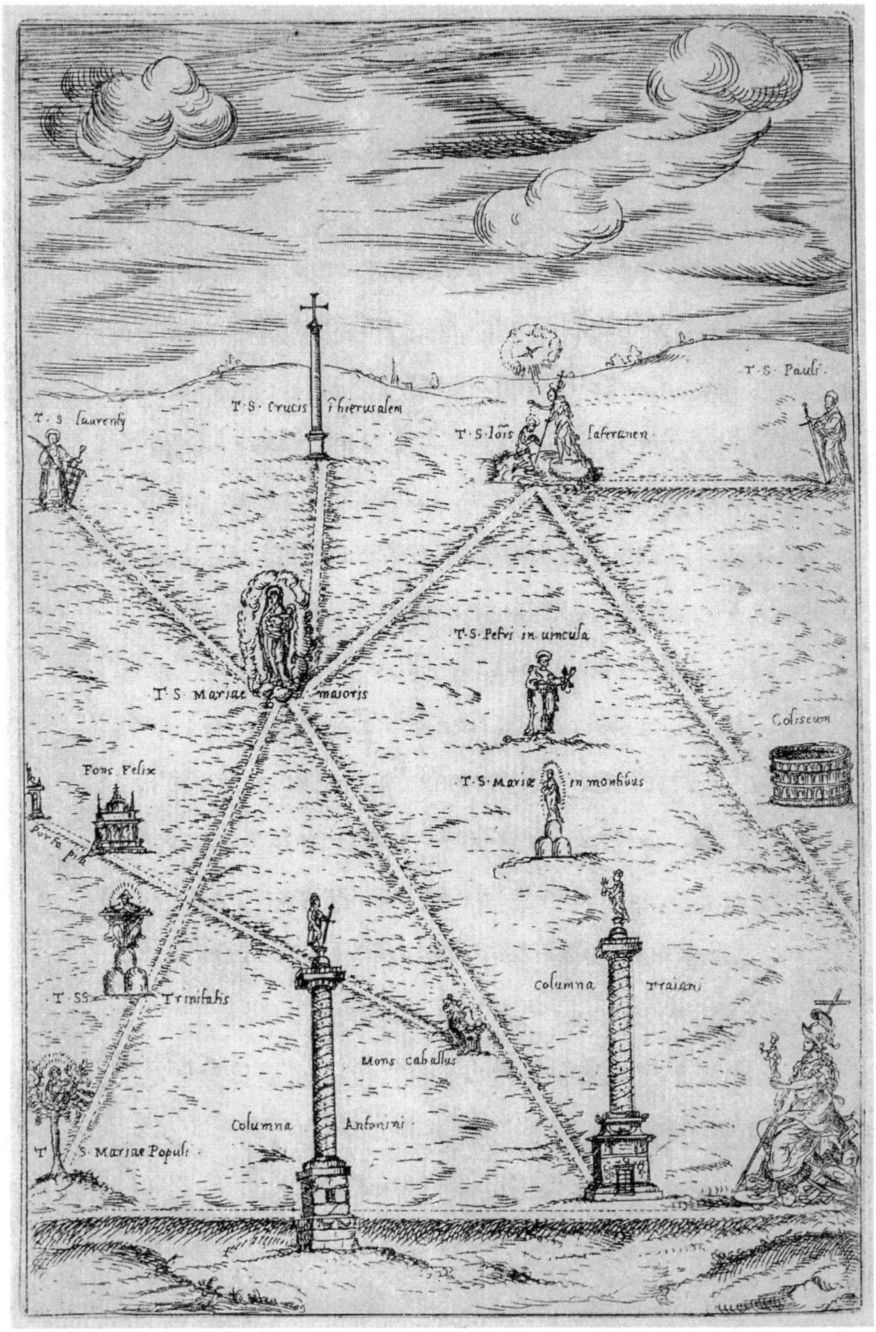

Figure 7.6 A plan for streets in Sistine Rome. Giovanni Francesco Bordini, *De rebus praeclare gestis a Sixto V. Pon. Max.* (Rome: J. Tornerius, 1588). Mispaginated; sheet between 46 and 51. Courtesy The Library of Congress.

Figure 7.7 Stefano (Étienne) Duperac, *Column of Marcus Aurelius*. Engraving/etching on paper. From *I vestigi dell'antichità di Roma* (Rome: Lorenzo della Vaccheria, 1575), No. 34. Rijksmuseum, Amsterdam.

Figure 7.8 Sampietrini stones used for street paving. Piazza Monte di Pietà, Rome. Photograph May 2017 © Bob Korn with kind permission.

stone. Owners of property, under penalty of 200 ducats, were also prohibited from paving the streets in front of their houses with stone. Instead, the street had to be paved with brick and then had to have borders (called *guidae*), which could be made with stone—whether peperino, travertine, or basalt.[38]

Pius IV died four months later in December 1565. The extent to which the new law was put into effect is unclear.

Four years into the rule of the succeeding pope, Pius V, a flurry of paving activity took place. This is evident especially from a surviving notarial book that Ottavio Gracchi created in 1569–1570 in his capacity as notary for the magistracy of the Presidenza delle Strade. The book is filled with licenses given to a variety of masons (*muratori*) to pave specified stretches of streets and piazze. This register reveals that Pius V for the most part carried out his predecessor Pius IV's desire to pave the Roman streets with brick. Of the fifty-one licenses recorded, most were issued to pave an area of street or square with brick—"License for paving with brick" (*Licentia faciendi ammattonatum*). Only five record licenses to pave with stone (*Licentia faciendi siliciatum*), one records a license for paving with both materials, and two grant licenses for paving with stone or gravel (*seleciatum seu imbrecciatum*).[39]

The most sustained discussion on street paving at this time can be found in a small manuscript tract written by Guidobaldo Foglietta, a little-known inventor from the Marches. Foglietta begins his "Discourse on Brick or Stone Paving in Rome" by pointing to the durable streets of ancient Rome. He thought that roads should still be made in the ancient Roman way, but he admitted that the expense would be too great. He provided therefore his own ideas about how the streets could be paved in a more durable way.[40]

Two things had to be considered—the agents impacting on the pavement and the pavement's resistance. The agents Foglietta had in mind were the wheels of carts and carriages. He observed that the narrow wheels of carriages damaged the street and suggested making the rims of the wheels a third wider. Foglietta also suggested making the iron rims out of one piece, as in barrel hoops, thereby using fewer nails. He explained that the heads of the nails in the rim wear unevenly and were almost like teeth. Each head acted "like the point of a scalpel that cuts and separates the stones violently."[41]

With regard to materials for paving, Foglietta suggested that, leaving aside the large stones used in antiquity (too expensive), "excluding bricks as very fragile material," and also excluding travertine (for its "very frangible porosity"), there were two alternatives. First, one could use small paving stones (*selici* or *selce*), that is, basalt *sampietrini*. Or, one could make a kind of concrete with a mixture of river stone, lime, and sand (the *modo di astraco*), which should be constructed two palms high. The *astraco* would be a material that resists more than brick and could be repaired easily.[42]

But, Foglietta remarked, when he was in Rome, he noticed that the masons (*muratori*) did not take care to make good foundations for the streets—which were often dug up in any case to make aqueducts and drains. The Roman streets had to sustain "the violent motion of carriages [loaded] with travertine, columns, and other heavy materials, and they should have strong foundations." The earth should be pounded and well tamped before construction begins. They should also be built so that the water runs off of them. The surfaces should be level, and "the material of good composition." And, insisted Foglietta, "much diligence must be used, such as is not practiced up until today."[43]

Foglietta's tract on street paving paid tribute to Pope Sixtus V. And Sixtus V indeed initiated a major campaign of paving the Roman streets—both with brick and with stone. Especially in 1586, the Congregation on Streets, Bridges, and Fountains took up paving the streets with bricks on all but one occasion, and that one with both brick and stone.[44] A notarial book in the Archivio Storico Capitolino contains contracts given by the masters of the streets to various masons (*muratori*) and records fourteen contracts given in 1586 to pave some streets and squares with stone and eight contracts to pave with brick.[45] Another document, this one in the Archivio Segreto Vaticano, lists 121 streets, *rione* by *rione*, that had been paved with stone (*selciato*).[46]

Whether to pave with brick or stone continued to be an issue. On January 27, 1587, an *avviso* noted that "it is discussed to pave Rome with bricks, and to lift the stone paving because of the thousand bad effects that this work [i.e., stone] causes."[47] This report was borne out by the agendas of the last meetings of the congregation before its dissolution. On December 14, 1587, the quality of bricks for street paving was discussed. Giacomo Della Porta was instructed to visit the furnaces of the city to ensure the good quality of the bricks, verify the foreseeable annual production, and establish conditions for a contract. In the next and final meeting of the congregation, on February 3, 1588, it was decided to fix the price of bricks and give the work to the lowest bidder ("al miglior offerente").[48] Whether or to what extent this decision was carried out is unknown.

There was great concern for the condition of the Roman streets on the part of many, but it seems unlikely that this preoccupation translated into permanently paved and repaired streets, in part for structural reasons. Rome had no permanent department of streets (or public-works department) with a permanent budget and permanent employees charged with ongoing paving and street maintenance. The Congregation on Streets, Bridges, and Fountains met only about once per month, and they spent inordinate amounts of time on individual requests from poor widows and others requesting exemption from the taxes required for cleaning and paving. Another common agenda item involved requests by workmen for their pay for work completed. Even with Sixtus V's subsidy (with money that came from taxing the Roman people), much was lacking from the point of view of efficiency: ongoing or permanent funding, ongoing or permanent schedules of maintenance, and permanent employees.

People on the Streets

The streets of Rome were filled with a cacophonous and colorful variety of people—sumptuously dressed nobles and ragged beggars, "honorable" women suitably covered and veiled, both men and women dressed in high luxury, wealthy courtesans and impoverished prostitutes, monks, priests, cardinals, pilgrims from afar, Roman citizens, Jews, Roma people known as gypsies (*zingari*), soldiers, and thieves. The streets were also trampled by animals and rolled over by carriages and carts. Roman streets were frequent scenes of violence, and they had permeable boundaries. The buildings that lined them had doors used by people continually going in and out and also windows from which women, especially, observed and also interjected themselves into the activities below.[49]

The popes struggled to influence and control the city's heterogeneous and growing population. For the church, the status, appearance, and behavior of the people on the streets were as important for the magnificence of the city as were splendid palaces, majestic churches, and straight, wide, paved, clean streets. Yet papal attempts to control the Roman people, although wide ranging, were sporadic and not entirely successful.

The Jews

Even before the conclusion of the Council of Trent, Paul IV Carafa "reformed" the people on the streets by forcing the Jewish population into a single area of the city. The papal bull *Cum nimis absurdum* that created what would become known as the ghetto on July 14, 1555, represented a radical departure from earlier policies. Previously, the Jews were more or less tolerated with the thought that they eventually would convert sometime in the distant future at the approach of the Last Judgment. In contrast Paul IV, influenced by eschatological thinking and the conviction that the Last Judgment was imminent, thought that the Jews must be converted at once and that creating hardships for them (as well as preaching to them) would achieve that end.[50]

The harsh *Cum nimis absurdum* ordered that Jews in all papal territories must live separately from Christians. In Rome they were forced to live in an enclosed space with gates locked at night. Friendships with Christians were prohibited. They could not own real property and were required to sell their houses to Christians within a designated period of time. Jews had to be identified publically—both men and women were required to wear distinctive markings. Jews were prohibited from having Christian servants or wet nurses and prohibited from dining with Christians. The only commerce allowed to them was in the purchase and sale of secondhand clothing (*arte cenciariae*). They were required to observe all statutes of the city, which favored Christians over Jews. Other clauses regulated Jewish record keeping and activities having to do with loans to Christians.[51]

The bull was immediately put into effect. Eleven days after it was issued, Jewish men were forced to wear yellow berets, Jewish women yellow veils. Two months later, the construction of the wall surrounding the ghetto—located in the Rione Sant'Angelo directly across from Tiber Island on the left bank—was completed. Many Jews lived there already, but others living in Trastevere, near Campo de' Fiori, and elsewhere were forced to move. Within six months, Jewish real property was liquidated throughout the city—at a total value said to be 500,000 scudi. The Jews now were constrained to live in crowded conditions in the unhealthiest part of the city and were usually among the first to experience the floods of the Tiber River.[52] Paul IV's successors never rescinded the bull, although some, such as Pius IV and Sixtus V, relaxed enforcement and mitigated some clauses. Pius V strictly enforced them.[53] At the same time, great efforts were made to convert the Jews. For those that wanted to convert, special provisions were made. Most important were catechumen houses in which recent converts were required to live and where they received training and support and were forbidden contact with nonconverted family members.[54]

Prostitutes

Another group subject to attempts at papal control and repression were Roman prostitutes. They increasingly became a focus of conflict. Prostitutes had long been a fixture the city, which had a highly disproportionate male population. At the beginning of the century, enjoying the company of a courtesan was very much part of the social status and image of elite Roman males. Elizabeth Cohen, focusing on ordinary, often impoverished prostitutes, has observed how integrated they were into neighborhoods and dwellings—they were sisters, daughters, aunts, and mothers—often living side by side with people working in other trades.[55]

For sixteenth-century Rome, scholars have traditionally made a distinction between the wealthy courtesans (*cortigiane*), companions of nobles and other wealthy men, and the impoverished *meretrici*, women who were prostitutes (or went in and out of prostitution) as a result of economic necessity. Both Elizabeth Cohen and Tessa Storey have emphasized that this binary view is overly simplistic. For all kinds of prostitutes, official intolerance arose in the sixteenth century. Attempts to circumscribe or banish prostitutes became particularly intense in Rome after the reforms of the Council of Trent, promulgated in 1564. But prostitution was not understood in the same way by everyone, nor was it always regarded from a negative point of view.[56]

Pius V was particularly incensed by the Roman prostitutes and wanted to restrict them to one part of the city or even put them into a ghetto of their own. The Capitoline Council resisted his efforts. On May 20, 1566, at a Public Council meeting, Roman patricians were put on notice that the pope "desired immensely" that the council provide "in every way [...] a place for the prostitutes and dishonest women, because he does not wish at all that they stay and live among honest women." The council members postponed the issue to the Secret Council.[57] On May 25 an *avviso* noted that the pope on Sunday at a Public Council meeting had informed the council members of impending reforms, including the stipulation that courtesans "be in one or two parts of the city" such as "in Trastevere or near the [Piazza] del Popolo."[58] In the Secret Council meeting on May 30, the minutes recorded that "regarding the shutting in of the prostitutes," there was "a little difficulty concerning the resolution," and that it would be better to take it up at another council meeting.[59] Thus began a long and ultimately successful effort of the *Popolo Romano* to deflect, stonewall, and actively resist Pius V's desire to ghettoize Roman prostitutes.

While the council members were procrastinating, the pope took matters into his own hands. On June 29, he ordered the police (*sbirri*) to expel all prostitutes from the Borgo, leaving the conservators of the Capitoline government to find a place to enclose them.[60] On July 22, an *avviso* to the Duke of Mantua reported that the pope ordered twenty-four courtesans "among the most famous" to either repent and convert (meaning to join the Convertite convent, Santa Maria Maddalena della Convertite, located on Via del Corso), get married, or leave Rome within six days and the Ecclesiastical State within twelve. The convent, which was close to Ortaccio (the area of Rome near the Porto di Ripetta and the Mausoleum of Augustus), was dedicated to accepting reformed prostitutes. The pope ordered all remaining prostitutes to move to Trastevere.[61]

The people of Trastevere objected. As reported in an *avviso* of July 27, 1566, in a remarkable show of solidarity, four hundred residents along with the *caporione* of the district went to Cardinal Morone, who also lived in the district and was the titular cardinal of Santa Maria in Trastevere. They begged protection and support against the pope's idea. The Capitoline Council also finally mustered a response to the pope's ongoing pressure. Pleading with him not to go forward with the Trastevere plan, they proposed that instead the prostitutes could be ordered to stay off the principal streets, to stay away from churches, monasteries, and the homes of noble women, and to remain on "other less public streets of Rome." The pope replied that some of them could live along the Via Giulia as long as the rest lived in Trastevere.[62]

The councilors could not accept the pope's compromise and sent another delegation, this one forty strong. The pope was furious. He raged, "All this morning you have rung your bell in congregations to do what? In order to preserve the infamy [. . .] that the most beautiful streets in Holy Rome are lived in by prostitutes, where the blood of so many holy martyrs has been spilled, where there are so many relics, so many devotions, where this Apostolic See is and such a great religion."[63]

Despite the pope's fury, the Capitoline Council found itself still unable to accept Pius V's plan. They sent a letter urging tolerance of the prostitutes and continued to resist the pope's efforts.[64] Pius V continued to exert ongoing pressure on Roman prostitutes—the well-off courtesans and the poor *meretrici* (and those in between). His efforts, which continued throughout his reign, included public whippings,[65] regular expulsions of prostitutes from the city,[66] and restrictions to certain areas of the city.[67]

Pius V succeeded in enclosing the prostitutes in a walled area of Ortaccio, but the ghetto seems to have been short lived. Perhaps the walls built were not particularly permanent or sturdy. Eventually, the never fully successful effort to restrict prostitutes to one area of Rome settled on the Campo Marzio—actually a very large district, large enough to accommodate them and many others as well. Even then, many prostitutes apparently lived in other parts of the city.[68] Pius's policies against Rome's prostitutes were continued by his successors,[69] but papal ambition and urban realities often diverged.

Beggars and Vagrants

Roman streets were also filled with beggars and vagrants. The English priest Gregory Martin vividly described them: "Dayes and nights they wander up and downe. needy, and naked, and harbourlesse, shewing their diseases, calling upon their Creatour for helpe, using one an others limmes, such as them selves lacke, divising and singing spiritual songs to move mens mindes to compassion, asking a piece of bread, or a very little meate, or some patched and foule garment to cover their unsemelinesse or to solace their sores."[70]

Throughout sixteenth-century Europe, attitudes toward the poor and changes in policies and institutions of poor relief developed in response to their numbers, which were growing, especially in urban areas and even more especially in Rome. In Rome and elsewhere, hospitals and other institutions for housing the poor were established. Certain categories of poor people (the "worthy poor" such as widows and orphans) were cared for—and increasingly, controlled—while other impoverished people (the "unworthy poor") were often expelled.[71]

A trend toward punitively disciplining the poor and the expulsion of the "unworthy" poor from the city became the norm in Rome as in other cities. In the late sixteenth century, begging came to be condemned despite the ambiguous legacy represented by the mendicant orders that saw asking for alms as a pious act. In Rome, the expulsion of vagrants and beggars began around midcentury and thereafter became regular if intermittent events. Under Pius IV on September 24, 1564, the Governor of Rome, Alessandro Pallantieri, issued an edict ordering that within three days "all vagabonds whatsoever who are without craft or means" must clear out of Rome under penalty of going to the galleys (being sentenced to being a rower in a galley). The same would happen to "all beggars who are healthy and strong," if within the same time period they do not find themselves some work.[72]

Pope Pius V continued Pius IV's policies. On April 1, 1566, his bull *Cum primum* ordered prelates to take action against beggars and vagabonds who were wandering through the churches during mass and holy rites. Those in charge must keep "paupers and also beggars or those seeking alms" outside of the church doors during those times or themselves face severe penalties.[73]

One solution for dealing with the ever-growing influx of impoverished people that occupied the streets of Rome was to force them into "hospitals," multifaceted institutions that might host pilgrims or care for the sick, including "incurables," or house orphans and other children and poor people.[74] In 1581 Gregory XIII, following a Bolognese precedent, decided to force all of Rome's beggars into a hospital at San Sisto outside of the walls along the ancient Via Appia. This institution was opened on February 27, 1581, with a grand procession of 850 beggars who were marched in and placed under the care of the confraternity Trinità dei

Pellegrini. The contemporary Sienese writer Camillo Fanucci provided a vivid account. First, beggars both male and female were required to register at the Ospedale della Santissima Trinità, giving their names, number of family members, a list of their possessions, and an account of the infirmities that had led to their begging. At the same time every faithful Christian in Rome was urged to donate clothing, mattresses, and other useful goods for the care of the mendicants. Finally, anyone caught begging after a certain date would be imprisoned.[75]

The great procession occurred on February 27. First, all the beggars gathered at the church of the Santissima Trinità dei Pellegrini and heard mass. Men of the confraternity led the huge procession holding a standard and accompanied by men holding large lanterns. Next came prelates and lords, all dressed in "rough sackcloth holding simple staffs." The procession included men carrying torches, monks wearing sackcloth, and choruses singing hymns and psalms. Following were the 850 beggars, the blind being led, and the infirm and lame being pulled in carts. At the end walked the officials of the confraternity. Fanucci remarked that "the spectacle was truly moving, marvelous, and perhaps a similar thing is never to be seen again." The procession ascended and then descended the Capitoline Hill "with greater triumph than the ancient Romans ever made." Finally they arrived at San Sisto, where they were received with great charity. But, added Fanucci, after a certain time, because of the bad air there, or because the hospital was so far from the center, or because the confraternity did not have the resources to support the venture, the hospital was abandoned.[76]

Sixtus V's ideas for ameliorating poverty and clearing vagrants from the streets were far reaching and included establishing centers for textile production—silk and wool—that would employ the impoverished. The most striking plan, one that failed because of the pope's death, was to convert the Colosseum into a wool manufacturing plant along with a residence for those who worked there.[77]

More traditional was Sixtus V's establishment of a hospital for the poor (this time in the city center rather than outside the walls) with his bull *Quamvis infirma et varia* of May 11, 1587. The prologue of the bull emphasizes that the poor should be taken care of in their local communities and that "it should not happen that the poor themselves wander through cities not belonging to them." Yet into "this dear city, nourisher of piety and charity," were received into the arms of the Holy Church from all parts of the world "a great number of poor, pushed by hunger, cold, nudity, and lack of everything, [and] afflicted by many diseases and misfortunes."[78]

The bull presented a vivid, highly negative picture of the beggars of Rome. The hospital established by Sixtus (Ospedale dei Mendicanti) on the left bank at Ponte Sisto would prevent beggars from "spreading out and wandering through the streets and piazze of the whole city [and] from wearying themselves searching for food." It would ensure that "they not fill with groans and shouting, not only public places and the entryways of private houses, but also places of worship," and that they would not interfere with churchgoers attending mass. The new hospital would prevent them from "creating havoc with demands, cries, and clamor in the churches, in which speaking quietly is required." Finally, and most important, they would "not [. . .] wander through the city in uncertain places, not knowing their pastor, without cognition of any divine rule or good behavior, just as brute animals deprived of reason," caring only for their bellies and not their spiritual health (fig. 7.9).[79]

The bull instructed that each beggar must be inspected for bodily health and age so that the indolent are discovered and "their laziness and worthlessness removed from the street" (presumably by expulsion from the city). The newly constructed hospital was made autonomous, giving supervisors exemption from the jurisdiction of other urban or ecclesiastical entities. Indigent pilgrims could also stay there, take three meals, and then be on their way.[80] The bull addressed not only Christian charity; it also emphasized the necessity of getting unseemly, loud, and clamoring beggars off the streets of Rome.

Writing near the end of the century, Camillo Fanucci reported on Sixtus V's hospital at Ponte Sisto. The holy work had not been accomplished. Throughout Rome one could see "poor beggars, and in such great numbers that one is not able to stand or go through the streets if one is not continually surrounded by them." In the hospital itself, there were very few beggars, not more than 150 persons but usually fewer, including prelates and servants.[81]

It is likely that those detained in hospitals were as unfortunate as those expelled from the city. In his fundamental study, Luigi Fiorani pointed to the appalling sanitary conditions of the hospitals made worse by (and certainly contributing to) epidemics such as that of 1589. One contemporary witness (Sanzio Cicatelli, a cleric dedicated to helping the poor), while describing an epidemic of 1591, noted of San Sisto (the hospital on the Via Appia founded by Gregory XIII) that around fifteen hundred sick, poor people were there. Between January and February, five monks died "one can say like holy martyrs with extraordinary torment [. . .] since they died having been pierced by infinite bites of lice, it not being possible to defend themselves against them." Cicatelli noted that the beggars hated the hospitals into which they had been incarcerated and that many fled to gain their freedom.[82]

Violence

The streets of Rome were also spaces of violence—of murders, brutal assaults, thefts, robberies, and other crimes. In the sixteenth century, barber-surgeons and other medical personnel, whose tasks included treating wounds, were required to

Figure 7.9 Fresco depicting the Ponte Sisto and Ospedale dei Mendicanti. Vatican City: Biblioteca Apostolica Vaticana. Notice the mill under the arch closest to the left bank. Date of photograph ca. 1920–1930. Credit: Alinari / Art Resource, NY.

report all suspicious wounds, giving the name and address of the victim, the nature of the wound, and if possible the name of the assailant. These records show that Romans of every age and social level (but male more than female) were ready at the slightest provocation to fly into a rage, hit, and throw stones. Violence was inherent within a patriarchal society dominated by male honor codes and hierarchical status differences, a society in which violent confrontation was often seen as preferable to judicial proceedings.[83]

In addition to everyday violence between individuals and families in Rome, both the city and surrounding countryside were plagued with banditry—organized groups of outlaws that perpetrated murder, theft, and kidnapping. Banditry had complex causes, only in part related to an intractable feudal aristocracy. Many bandits had been exiled from Rome or other cities and had had their property confiscated. For mutual protection they traveled in small groups and lived off of robbery and other violent crimes. Some intermittently hired themselves out as soldiers. Because bandits could leave the city and often received protection from feudal lords in the Roman Campagna, they were partially protected and were often unmanageable.[84]

In the early post-Tridentine decades the popes issued a steady stream of bulls and edicts in an attempt to suppress violent crimes. In addition, they decreed punishments to those who helped or protected bandits and perpetrators of violence, whether this protection took place in the city or (as was more usual) outside. Such protectors could receive heavy fines and be excommunicated; their houses could be destroyed, they could be removed from any offices they held, and they could be exiled. The decrees often prohibited arrangements whereby murderers could be essentially pardoned by various officials or even by family members of the victim. All this was to little avail, as attested by the regular reissuing of similar or identical regulations.[85]

The decrees of the Council of Trent, as well as papal bulls, also prohibited dueling, a common means of defending honor among noble and elite men. Before the council concluded, Pius IV issued a bull on November 13, 1560 (*Ea quae a praedecessoribus*), which prohibited dueling and delineated punishments—including excommunication, confiscation of property, and imprisonment—for those who attended duels as spectators, looked after or cared for fighters, or supported in any way this form of elite fighting. Emphasizing the Council of Trent prohibition, Gregory XIII also prohibited dueling and severely condemned spectators and those assisting the fighters.[86]

The popes also prohibited the carrying of weapons and arms without a license—the types often specified—through the city and papal territories. Pius IV issued a bull, *Cum vices eius*, on March 6, 1562, that forbade carrying small firearms (*archibussetti*) less than two palms (17 in.) in length. "We have become aware," the bull announced, "that a great number of those [murders] are committed and perpetrated every day with those small firearms commonly called *archibussetti* (that can be carried hidden and seem to be very able, almost unavoidably, to cause death)." Despite severe prohibitions and penalties, these murders had not been prevented "because of the great number of killers and their audacity." The new bull prohibited carrying, either openly or secretly, buying or selling, receiving as a tax or a repossession, or for any other reason, a weapon shorter than two palms. Violators would be charged with treason and insurrection.[87]

Pius V, citing his predecessor's prohibition against short firearms, expanded the law to include daggers, knives (except for bread knives), and other kinds of weapons shorter than three palms (about 26 in.). Further, officials were forbidden to issue licenses for such weapons. All previous licenses were annulled.[88] Despite these and other laws, it was difficult for the police (*sbirri*) to distinguish between licensed and unlicensed arms (because of fake and "lost" licenses). Another difficulty was that the *sbirri* were often from a lower social strata than the perpetrators (or the protectors of the perpetrators), making it difficult for them to effectively enforce the prohibitions.[89]

The outpouring of propaganda extolling Sixtus V during and after his lifetime hailed him as a harsh enforcer of justice who ended Roman banditry and brought peace to the streets of Rome. Traditional historiography in the twentieth century continued this view. Certainly Sixtus's harsh pursuit of bandits, his many public executions (ninety-five in 1585 and eighty-eight in 1586), and his orders to display the heads of the executed on Ponte Sant'Angelo contributed to the image of Sixtus as the herald of justice and order. The reality was quite different. Sixtus's efforts had only a superficial influence on the disorders of the city and the papal state. Violence and banditry continued into the 1590s and well beyond.[90]

What Sixtus V's many public executions underscored is that violence in the streets of Rome included the violence of the state. And while Sixtus's executions may have been unusual in their number, they did not represent a novel practice. Montaigne's witnessing of three executions in the previous pontificate is a grim reminder that they were regular events carried out for public moral instruction.

8

Engineering Spectacle and Urban Reality

On Wednesday, September 10, 1586, a huge crowd gathered at St. Peter's Square. In order not to lose their places, many stood into the evening without food. Some built platforms, charged customers for a space, and "earned a great deal of money" by doing so. That morning, two masses had been celebrated. Eight hundred workers had been compelled to confess their sins and take communion. Now these workers, many leading horses, took their places around forty capstans, each to be turned by three or four horses, a total of 140. The capstans were attached to ropes that led to a maze of pulleys positioned on an immense structure called the *castello*. Within the *castello* a 361-ton obelisk, protected by sheathing, lay on a constructed ramp. This was the day that the Vatican Obelisk—which had already been lowered from its previous position in a dark corner at the side of St. Peter's and rolled on its platform to the center of St. Peter's Square—would be raised above its new pedestal (fig. 8.1).[1]

The architect, Domenico Fontana, stood on a specially constructed command tower to direct operations. With the complete silence of the crowd, at the blare of a trumpet, the horses turned the capstans, stopping when a bell sounded, starting again with the sound of the trumpet. Such coordination was crucial to achieving equal weight distribution on all the ropes, essential to the success of the project. By sundown, after fifty-two

Figure 8.1 Lowering the Vatican Obelisk. Domenico Fontana, *Della trasportatione dell'Obelisco Vaticano* [. . .] (Rome: Domenico Basa, 1590), fol. 18r. Courtesy Rosenwald Collection, Library of Congress.

pulls and fifty-two pauses, the obelisk was hanging vertically over its new pedestal, ready for its final lowering the next day. Mortars were discharged at nearby Castel Sant'Angelo, and "the entire city very much rejoiced." A great crowd of drummers and trumpeters flocked to the house of Fontana (in the Borgo nearby) playing to loud applause.[2] This great display was both a triumph of engineering and a dramatic public spectacle. And the crowd was correct in taking it to be the achievement not only of the pope but of the architect Domenico Fontana.[3]

How to Move Obelisks: A Long Debate

It was an accomplishment preceded by decades of uncertainty and debate about whether and how to move the Vatican Obelisk (or *guglia*, as it was called) to a more visible and pleasing site at the front of St. Peter's. Originally, in 41 CE the emperor Caligula had had it transported from Egypt across the Mediterranean and erected in his circus in the Ager Vaticanus, or Vatican region. The obelisk was one of about fifty that the ancient Romans had brought to their capital city during the Roman imperial age. Often they erected the great monoliths in circuses. In addition to possessing them as trophies of the Roman conquest of Egypt, they assimilated them to their own religious and cultural practices. Centuries later, the unmoved Vatican Obelisk was still standing where it had been since 41 CE, but the great basilica of St. Peter's had been constructed next to it, leaving it in a dark corner shadowed by part of the basilica called the Old Sacristy.[4]

In the medieval centuries, still unmoved, the obelisk came to occupy hallowed ground as its location came to be considered the site of the crucifixion of St. Peter and the killings of other Christian martyrs. Its sacred location undoubtedly explains why it still remained standing in the late sixteenth century; other Egyptian obelisks in the city had fallen or had been pulled down (presumably as protection against the pagan demons that resided within) and had often broken. The two obelisks in the marshy Circus Maximus had even sunk underground. As for the standing Vatican Obelisk, it was much lamented that it stood in an obscure, inaccessible location (fig. 8.2).[5]

Moving the monolith presented great difficulties. Earlier in the sixteenth century, Pope Paul III had pushed especially hard to have it transported to the front of St. Peter's. His efforts are known through a later treatise on obelisks (1589) by Michele Mercati (1541–1593). Mercati was a physician and learned investigator of natural history who served as the head of the papal botanical garden in the 1580s. In his treatise on the obelisks he revealed that Pope Paul III had urged the renowned sculptor and painter Michelangelo to undertake the task. From the famed artist's surviving associates, Mercati had learned that Michelangelo (who

Figure 8.2 Unknown artist, Dutch. View of St. Peter's Basilica in Rome during construction ca. 1580–1582. Brown pen on ribbed handmade paper, 9 × 7½ in. (22.9 × 19 cm). Graphische Samlung, Städel Museum Frankfurt am Main. Photo © Städel Museum—ARTOTHEK.

was a competent engineer), when asked why he didn't want to undertake the task, replied "And if it breaks?"[6]

Michelangelo had appreciated the real dangers involved. The Vatican monolith weighed over three hundred tons. More than two thousand years before, it had been expertly hewn from the solid granite bedrock of the Aswan quarry in Upper Egypt, ordered by an unknown pharaoh. Later the Romans transported it to Rome. But specific details about its Egyptian origins and later Roman transport and erection were (and are) unknown, especially because its surfaces did not contain hieroglyphs. Cut from a single stone, obelisks were not only hugely heavy but also fragile.[7] Dropping one from a lifting apparatus could well cause it to break. If such a catastrophe befell the Vatican Obelisk, so close to the center of papal power, it would have challenged the power and reputation of the pope himself to say nothing of the engineer in charge.

In the early 1570s, in preparation for the planned jubilee of 1575, Pope Gregory XIII had once again raised the issue of moving the obelisk. According to an anonymous report in July 1574, Gregory resolved to move the *guglia* to St. Peter's Square "for the greater commodity of viewing for the people who come to Rome in the Holy Year." The pope estimated the cost of moving it the relatively short distance from its old location to the new as 30,000 scudi.[8]

Intriguing evidence of interest in the project can be found in a letter discovered by Lothar Sickel in the state archives of Parma. The letter, dated February 1, 1581, was written by Camillo Paleotti (1520–1594) and addressed to one Antonio Anselmi. It described a proposal (and model) by "a young Sicilian" who had just arrived in Rome and "who displays, despite being without letters, very great judgment about machines." The unidentified young man offered to transport the *guglia* to St. Peter's Square and had constructed a model that Paleotti had seen (probably in late 1580). The model showed "not only the great ingenuity of this artisan but also his good opinion to he who observes it that the thing could succeed." First, the unknown Sicilian calculated the weight of the obelisk to be 1,800,000 *libre* (about 673 tons, an overestimation by almost double its actual 361 tons). He planned to support it not straight up but prone (*distesa*), and for this he had built two *castelli*, one to lift it upward, and another to lower it, and finally a wagon (*carro*) the same length as the obelisk (to transport it to its new location). Paleotti added that a description of the gear and other equipment to make it work would take "too long to write about." But many cardinals and others had seen the model and had judged it and thought that the thing would succeed. Paleotti continued that the young Sicilian had invented another machine that would allow the transport of large blocks of travertine (*li Tivertini*) to a building site in one day instead of the eight days that it ordinarily took. He himself had not seen this model because the Sicilian had left it with the pope but had promised to show it to him later.[9] The young Sicilian whose plan Paleotti discussed has yet to be identified.

Another proposal came from the old Roman hand Camillo Agrippa (1520–1595). Agrippa was a practitioner as well as an author, someone without a university education who laid claim to natural philosophical and cosmological knowledge, a person from a nonelite background who both figuratively in his writings (which often included dialogues with Agrippa's contemporaries as interlocutors) and actually in his life hobnobbed with both learned and noble men.[10]

His reputation was in part based on his invention of a water-lifting mechanism powered by a water wheel, which famously brought water from the newly repaired Acqua Vergine up to the surface to water the gardens of the palace on the Pincian Hill owned by Cardinal Montepulciano (and later sold to Ferdinand de' Medici—now known as the Villa Medici). Agrippa actually built two hydraulic devices, one in 1574 for Cardinal Montepulciano and the second for the Medici in 1577–1578, the latter for a fountain standing on top of an artificial hill called Parnassus constructed in the garden.[11]

Agrippa's written works included a small booklet published in 1583 that explained in step-by-step detail his proposal for moving the Vatican Obelisk. In this tract, he recalled that when he had arrived in Rome almost fifty years before, in October of 1535, many people were discussing such a project, including the well-known architect Antonio Sangallo, the famous Michelangelo, and "infinite others." For the next thirty years, he reported, he thought about the task, while he also found himself "occupied with inventions not less useful than honorable to the public."[12]

In his tract detailing his plan for moving the Vatican Obelisk, Agrippa described his (ultimately unsuccessful) audience with Gregory XIII and the pope's adviser, Michele Mercati. Mercati, in his own treatise on obelisks, recorded that Agrippa had made a proportionate model of his machine and that in 1581 the pope had asked him to demonstrate it. This event took place in the Metallotheca Vaticana, the room in the Vatican palace that Mercati had designed as a display of natural history focusing on minerals and fossils. Mercati was intrigued by the physical properties of the stone obelisks. Agrippa, during his audience with Mercati and Gregory XIII, demonstrated his model and put forth many reasons why the work could be carried out securely and easily. Unconvinced, the pope concluded that the architects of his time could not move such a heavy weight. Agrippa continued to press his case by writing his treatise and dedicating it to the pope's son, Giacomo Boncompagni, who was captain general of the church. All to no avail.[13]

Agrippa's booklet revealed the lively discussions provoked by this difficult undertaking. Recalling the prominent architects who had previously worked on the problem, Agrippa detailed his own plan and provided an illustration of the

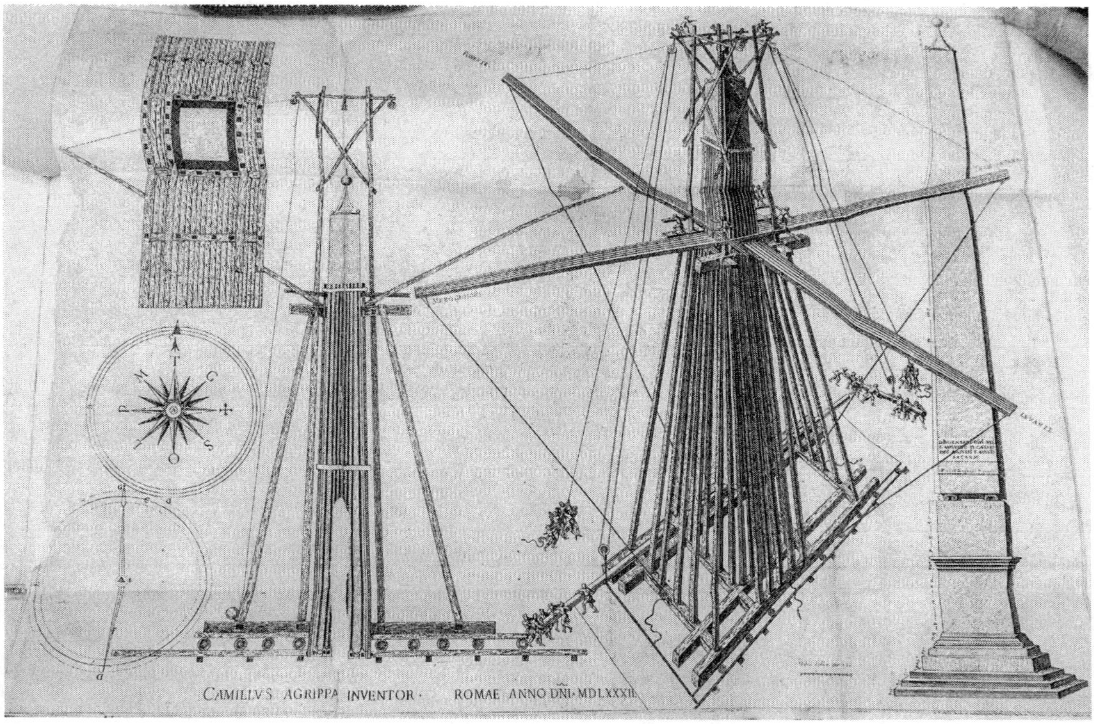

Figure 8.3 Camillo Agrippa, *Trattato* [. . .] *Di trasportar la guglia* [. . .] (Rome: Francesco Zanetti, 1583), foldout image after p. 48. Rome: Biblioteca di Archeologia e Storia dell'Arte, Roma X 245 1.

apparatus that he had designed for the task (fig. 8.3). He proposed moving the obelisk in an upright position instead of first laying it on its side, thereby avoiding the dangerous operation of raising and then lowering it. Through a system of levers and pulleys, the obelisk would be raised, moved to its new location while suspended vertically in the tower, and then slowly lowered by levers onto its new base. The model as he drew it was not functional, since the timbers were unsupported.[14] Nevertheless, several features of his design, including elements of the *castello*, a structure that served as both support and crane, reappeared in the machine that Domenico Fontana would use to move the obelisk a few years later.

Just as important, Agrippa combined a discussion of engineering problems with a philosophical dialogue. The interlocutors in his booklet were two brothers, Fabrizio and Agapito Fossani, Roman magistrates from Rione Colonna. The dialogue, in which Agrippa featured himself as an interlocutor, discussed variations of the center of gravity with respect to variations in the size or magnitude of the machine. Elio Nenci has pointed out that the discussion proceeded on an abstract level and seemed to lose track of real experience or the actual project. Credulity is stretched when the topic shifts from moving the Vatican Obelisk to moving the enormous Trajan's Column, one of the great columns at the center of Rome.[15] Whatever the practical result, the intricate configuration of the two concerns—

practical engineering and philosophical mechanics with the added interest in measurement and practical mathematics—is notable.

Competing Proposals

Soon after the death of Gregory XIII in 1585, Felice Peretti was elected Sixtus V. Without delay Sixtus took up the project of moving the Vatican Obelisk. He ordered that a congregation be assembled to decide how the massive project might be carried out.[16] It would emerge that his own architect, Domenico Fontana, would carry out the task—but the process of choosing the architect and deciding on engineering procedures was complicated.

Fontana arrived in Rome from his home base in Melide on Lake Lugano in 1563. He is first recorded as a stone mason, including a contract for stucco work. He worked with and eventually headed a large workshop composed of family members and compatriots. In 1577–1578 he built a small palace for Cardinal Peretti in Piazza Pasquino. In 1578 Peretti awarded him a contract for his great palace, Villa Montalto on the Esquiline Hill. When Gregory XIII stripped Cardinal Peretti of his benefices, leaving him without the necessary funds, Fontana continued the work, lending his patron the needed money. He was able to provide such support because he and his brothers owned land in Lugano as well as in Rome from which they derived income. From 1581 to 1586, Fontana also served as overseer for the construction of the facade of San Luigi dei Francesi designed by Giocomo Della Porta. He was richly rewarded when his patron was elected Pope Sixtus V and then appointed him papal architect.[17]

The ultimately successful Fontana wrote a highly detailed but also self-serving account of the competition in his spectacularly illustrated, *Della trasportatione dell'Obelisco Vaticano,* first published in 1590. In it, he described and illustrated the famous obelisk transport as well as his other Roman projects for Sixtus. The treatise included remarkably detailed copper-engraved illustrations of the transport of the Vatican Obelisk. These images were designed by Fontana and engraved by Natale Bonifacio (1538–1592), who was from the town of Šibenic in Croatia. The engravings depict the models proposed, the machine actually used, and in successive images they show the entire process of the move in all its detail.[18]

Fontana related that on August 24, 1585, a meeting or *congregazione* of lords and prelates—ordered by Sixtus V—met to discuss how the monolith might be moved most safely and who the "artificer" should be, that is, the person who was most skilled "through intelligence and experience in similar matters." An *avviso* reported on this date that a wooden obelisk was built on the site in front of St. Peter's where the original was to be moved. Fontana reported that the congregation was convened in the nearby palace of Cardinal Pierdonato Cesi to decide on the best plan. In addition to Cardinal Cesi, three other cardinals attended, as did the treasurer general, the president of the streets, the masters of the streets, the senator, the three conservators, and the *caporioni*, among others.[19]

The *congregazione* called on "men of letters, mathematicians, architects, engineers, and other valiant men [. . .] so that each one might give his opinion concerning the execution of such a great undertaking." To facilitate the best solution, the members agreed to convene again twenty-eight days later "to give time to many valiant foreign men who gathered in Rome from various places to show the force of their ingenuity concerning a thing so much desired by Our Lord." In fact, many had already heard of the pope's interest and had already arrived in Rome. Fontana reported that the second gathering on September 18 drew five hundred men from various places—Milan, Venice, Florence, Lucca, Como, Sicily, Rhodes, and Greece. Each carried "his invention" in one form or another. Some brought drawings, some models, some written descriptions. Others explained their ideas verbally to the *congregazione*. From this collection of what were apparently hundreds of proposals, Fontana provided illustrations of only eight (fig. 8.4).[20]

Fontana's illustration of the eight models included Agrippa's in left center and his own in the upper-left corner being lifted to the heavens by winged cupids. We recognize Agrippa's model from the illustration in his own treatise and Fontana's from other illustrations in *Della trasportatione*.[21] The creators of the other models are unnamed. Fontana's account of the September 18 meeting is corroborated only partially by the minutes of a meeting of the Congregation on Streets, Bridges, and Fountains devoted to proposals for moving the Vatican Obelisk.

In the minutes of the congregation held on September 18, 1585, there is no mention of what the actual plans were, but the names of the final bidders and the sum of money for which they proposed to undertake the task are provided. Antonio Ilarione Ruspoli, a Florentine banker, offered to move the obelisk for 9,000 scudi and promised to provide 10,000 scudi security. Domenico Fontana offered to do it for 16,000 scudi. Giacomo Della Porta, at this time architect both of St. Peters and the Roman Commune, did not ask for a contract but asked to be director of the workers who would be paid by the Camera Apostolica. The Sicilian Giacomo Del Duca, who had been a favored student of Michelangelo, made a bid to carry out the task for 7,000 scudi including the expenses of demolishing some houses (undoubtedly those next to the Old Sacristy). Giovanni Fontana, the brother of Domenico, wanted to do it for 13,000 scudi; the Florentine Francesco Tribaldesi (the candidate of Cardinal Medici and therefore also of the Medici Duke in Florence) for 14,000 scudi. Finally, the famed Florentine sculptor and architect Bartolomeo Ammannati, like Giacomo Della Porta, asked to direct the work "for a small sum," the costs to be carried by the Camera Apostolica.[22]

In his own account, Fontana only briefly discussed the proposals illustrated,

Figure 8.4 Eight proposals for moving the Vatican Obelisk. Domenico Fontana, *Della trasportatione dell'Obelisco Vaticano* [. . .] (Rome: Domenico Basa, 1590), fol. 8v. Courtesy Rosenwald Collection, Library of Congress.

except for his own. It is not certain that the models in the illustration were proposed by the seven bidders (minus Agrippa) listed in the archival document, but they may well have been. For the presentation of his own proposal, Fontana reported that he constructed a small wooden model of the *castello*. He made a small obelisk out of lead to go inside the model and fashioned all the working parts in proportion one to the other — including cords, pulleys, and other elements. His account of the presentation reveals a public and highly rationalized process of demonstration and persuasion. All the lords of the congregation were there, he says, as well as all the five hundred masters of the arts. In the presence of this company, "I lifted that *guglia*, and lowered it in a very orderly way, explaining with words the reason and purpose of everything and the principles of each of these movements as well as their following results." A document noted by Cesare D'Onofrio shows that the Camera Apostolica actually paid for Fontana's model, which was constructed by one Colantonio Liante.[23]

Fontana's report of the congregation's decision — when compared with the extant record of the minutes of that meeting — seems to have been disingenuous. "Now having considered with great subtlety," he wrote, "and pondered the arguments, designs, and constructions of each of us, and disputed very much, at last [the congregation] came to this conclusion that the way discovered by me to move and transport the *guglia* was the most easy, secure, and better understood to bring a successful end than all the others that were proffered there, and by common consent of the whole congregation it was chosen and approved to be used in the transport of the *guglia*, leaving all the others behind." However, Fontana continued, because the lords greatly desired that the thing turn out well, they began to fear his age (he was 42 at the time) as too young and with not enough experience to raise great weights. They turned to the Florentine sculptor and architect Bartolomeo Ammannati. (Fontana wrote that Ammannati was 65; he was actually 75) and Giacomo Della Porta (who would have been about 55). Fontana reported that both men were chosen. His model was chosen as the best and "assigned to two valiant architects" to be adopted for the great task.[24]

The minutes of the meeting of the congregation mentioned nothing about this concern with age, which on the face of it seems unlikely given that Fontana was already past forty and had moreover supervised major construction projects, which would have necessitated transporting heavy stones. The minutes did report that the proposals of Antonio Ilarione Ruspoli, Domenico Fontana, Giacomo Della Porta, Giacomo Del Duca, Giovanni Fontana, and Francesco Tribaldesi were rejected and that Bartolomeo Ammannati was chosen as the one to carry out the task. The minutes did not mention which model would be used for the transport, that is, they neither confirmed nor denied that Fontana's model was to be used. It is notable that three members vetoed the decision to give Bartolomeo Ammannati the contract. The naysayers were Cardinal Francesco Sforza di Santa Fiora; the treasurer general, Benedetto Giustiniani; and the prior of the *caporioni*, Gaspare Sanguigni.[25] This was powerful opposition to Ammannati (and likely powerful support for the Fontana workshop). Nicola Navone especially has shown that the Sforza di Santa Fiora family, including Cardinal Francesco Sforza, had supported Giovanni and Domenico Fontana, especially in their early years in Rome in the 1560s and 1570s.[26]

Fontana reported that seven days after the meeting of the congregation he began thinking about the obelisk again. He thought that his plan would succeed but worried that it would not be fully understood by the two architects assigned to carry out the task. (Here it appears that Giacomo Della Porta had been given the task along with Ammannati, despite his rejection as recorded in the minutes of the congregation.) What if they did not carry out his plan correctly? Fontana thought. Even he himself might not follow someone else's invention as well as one of his own devising. He persuaded Sixtus V to his point of view. "Our Lord ordered that I alone must begin the work and execute my intentions."[27]

Although Domenico Fontana ultimately succeeded in wresting the contract for moving the Vatican Obelisk from Ammannati and Giacomo Della Porta, the many discussions about such a project and the various proposals that were made point to a lively tradition of interchange about urban engineering in Rome. That the culture of interchange focused on issues of urban engineering is even more in evidence after the election of Sixtus V, when the congregations turned their attention to the project. After all, Domenico Fontana had long worked as the loyal and trusted architect/engineer of Sixtus and the Peretti family, especially at Montalto. One at least imagines that the pope could have simply given him the contract to carry out the task without further ado—especially given the project's proximity to St. Peter's. Instead, a congregation opened a competition, and hundreds submitted varied proposals and suggestions. Although the deck was undoubtedly stacked in Fontana's favor, the lively competition and numerous submissions are significant indications of the discursive engineering culture of late sixteenth-century Rome.

Preparations and Gathering Materials

Once Fontana got control of the project, he vigorously proceeded to make preparations. He dispatched fifty men to excavate the area where the foundation for the pedestal of the obelisk was to be constructed in the center of St. Peter's Square. Fontana provided the exact measurements of the area to be dug out. The ground at the bottom was found to be muddy with clay and water, making it necessary to insert pilings with beams of oak and bark-stripped chestnut. Symbolic support was provided as well. The excavation was carried out on September 25, 1585, the same propitious date in previous years that Felice Peretti had been made a bishop, and then a cardinal, and then the pope. Bronze medals were fabricated and twelve each placed in two small travertine boxes on which images of the pope were carved along with elaborate symbolic images.[28]

Sixtus V gave Fontana sweeping powers to gather materials and equipment as he carried out the highly public, high-risk project, as can be seen from a document dated October 5, 1585, that Fontana reproduced in his book. He could hire all workers, skilled and unskilled; acquire all necessary tools, instruments, work animals, and materials; appropriate all the timber he needed, including huge logs not available in the city itself. Further, he could purchase whatever he needed without paying any taxes or customs of any kind, including animals in Rome or elsewhere without license or receipt. He could seize and carry away capstans, ropes, and pulley blocks for which, however, he would pay rent and return. He could use any and all equipment belonging to the construction site (*fabrica*) of St. Peter's, the dome of which was in the process of being constructed, and he could command the officials of the basilica to clear the piazza when necessary. He could destroy houses and other structures near the obelisk (and compensate the owners). He and his men could bear arms (except those prohibited) anywhere and anytime of the day or night. All magistrates and officials of the Ecclesiastical State were to "help, obey, favor, and assist" him "notwithstanding any other commands whatever." All others were ordered not to obstruct him in any way, under the penalty of a fine of 500 ducats.[29]

Fontana immediately began gathering equipment and materials. In the Umbrian city of Foligno, a center of hemp production, he collected the fibers to have made (in Rome) the massive amounts of hemp rope needed. This included the three 200-*canne*-long ropes (about 1,100 ft.) made for the double pulley blocks, each block equipped with twelve heads, each head corresponding to two capstans. Fontana itemized the other materials and equipment—the pulleys and pulley blocks and their various parts, bolts, the capstans that would be turned by men and horses, even nails, hammers, and other tools. Huge beams of chestnut, oak, and elm were found in storage houses (presumably in Rome) to make the framework of the *castello*. Fontana ordered ironwork to be manufactured both at Ronciglione (a town near Viterbo with a metalworking industry) and Subiaco thirty miles from Rome. He sent workers to Campomorto to collect a large number of long, thick oak beams, which were transported to Rome on huge two-wheel carts, each pulled by seven pairs of buffalo (fig. 8.5). Lumber was also collected at Terraccina (the seacoast city at the southern border of the Ecclesiastical State)—namely, elm boards for the "bed" of the obelisk—and from Santa Severa north of Rome in territory belonging to the Camera Apostolica, where oak shafts were cut for the capstans, rollers, and crossbars, and where planks of elm were cut as well.[30]

Figure 8.5 Nicola Zabaglia, *Castelli e ponti* (Rome: Niccolo and Marco Pagliarini, 1743), plate XVI. Building materials—travertine, limestone, and lumber shown being hauled in various kinds of carts pulled by buffalo and horses. Collection of Vincent J. Buonanno.

Filippo Pigafetta and the Obelisk

As Fontana was gathering these materials a few months before the project of moving the obelisk began, in March 1586, Filippo Pigafetta (1533–1604) published a small erudite treatise that in the end supported Fontana's control of the project. Pigafetta was the illegitimate son of a nobleman from Vicenza, Matteo di Camillo. He may have attended the University of Bologna, but in any case his learning is evident in all of his writings as is his technical competence as a military engineer. He followed a diplomatic and military career, undertaking extensive travels through most of his adult lifetime. He traveled to Egypt and to the Sinai Peninsula, became a consultant on fortification for Venice, and then in the early 1580s traveled to France, England, and Portugal. During his travels he wrote extensively, mostly on naval and military matters. Perhaps most importantly, Pigafetta translated the *Mechanicorum liber* (1577) by the nobleman Guidobaldo del Monte (1545–1607) into Italian, thereby also translating "noble" theoretical mechanics into the realm of working machines. A significant aspect of this work included the problem of lifting heavy weights.[31]

Pigafetta arrived in Rome in 1585 accompanying the Venetian Embassy headed by the aristocrat Marcantonio Barbaro (1518–1595) to congratulate Sixtus V on his ascension to the papal throne. The obelisk transport project immediately drew his interest, as his small treatise attests. Having actually visited Egypt, the original home of the obelisks, he was able to discuss Egyptian stones with the authority of one who had personally observed many of their varieties in situ, including obelisks. He assured the reader that "these enormous stones" (obelisks) were not made by putting together a mixture of clay, nor were they formed or stamped in the manner of artillery. Rather, they were "pulled from the quarries naturally, as I have seen with my own eyes"; that is, they were monoliths made from a single stone. He supplemented his own observations with a discussion of ancient references to the monoliths in texts such as those of Ammianus Marcellinus (*Rerum gestarum libri*, 17.4.6–23), and Pliny (*Natural History*, 36.14–16).[32]

Pigafetta described at length how to measure the dimensions of the obelisk using Euclidian geometry and calculating the measurements "with Roman palms and with common feet of geometry," that is, with both ancient Roman and contemporary measures. Once the dimensions and volume were determined, the weight of the monument could be extrapolated by weighing a smaller stone block made of a similar material and multiplying accordingly.[33]

Referring to his translation of Guidobaldo del Monte's *Mechanicorum liber* and in particularly his treatment of pulleys, Pigafetta calculated the proportions between the amount of weight to be lifted and the force (*forza*) or power needed by the pulleys and ropes. Using Guidobaldo's theory of pulleys, he concluded that every rope would sustain 8,000 *libbre*, or pounds, and that thirty-two capstans with four ropes each connected to two-pulley tackles could lift the obelisk.[34] In the end, Pigafetta endorsed Domenico Fontana as the architect to do the job.[35]

Moving the Obelisk

Like Pigafetta, Domenico Fontana in his *Della trasportatione* described the physical characteristics of the obelisk and showed how he measured its weight. First, he had a piece of granite (the stone of the monument) cut into a palm-squared cube and weighed it. He then calculated the number of such cubes in the monolith.[36]

Fontana described exactly how he had constructed the *castello*—a machine that functioned as a combination (with the capstans) of scaffolding and crane. He explained that it could be built, then taken apart, and then built again without damaging the beams. (This ability to assemble and disassemble the *castello* was crucial for the success of the project.) The dramatic images designed by Fontana and incised in copper engravings by Bonifacio made the written descriptions clear. The illustrations on the recto are lettered, and the letter identifications are on the verso. Fontana's book differed from many of the spectacular "Theaters of Machines" that were published in the late sixteenth and early seventeenth centuries in that he was depicting not theoretically possible machines but offering testimony to a completed project, showing the steps from beginning to end that had already been successfully carried out (fig. 8.6).[37]

The operation was accomplished in two distinct phases, the first on the last few days in April 1586, the second in September of the same year. In preparation, the houses near the standing obelisk were demolished as was part of the Old Sacristy of St. Peter's. The obelisk itself was protected by reed mats, which were then covered by two-inch-thick wood planks, which in turn were covered by a sheath of vertical iron rods held by horizontal bands or hoops joined together by key bolts. Ropes were arranged on a complex set of pulley blocks and tackle and then attached to carefully arranged capstans (forty of them), each pulled by three or four horses and controlled by a number of men who were supervised by two masters. Practice sessions were held to perfect the synchronization of the turns of the capstans. Each capstan was given a number that corresponded to a particular pulley. In preparation, to coordinate the tension of the ropes, one by one the capstans were turned, until all the ropes had equal tension. Things were left in this state of preparedness for a day.[38]

On the appointed day, April 30, 1586, huge crowds gathered, along with, Fontana wrote, the pope's immediate family, including his sister Camilla. (The pope himself would not have wanted to be visible in case of failure, which is not to say he might not have been watching from a hidden spot.) Fontana reported that

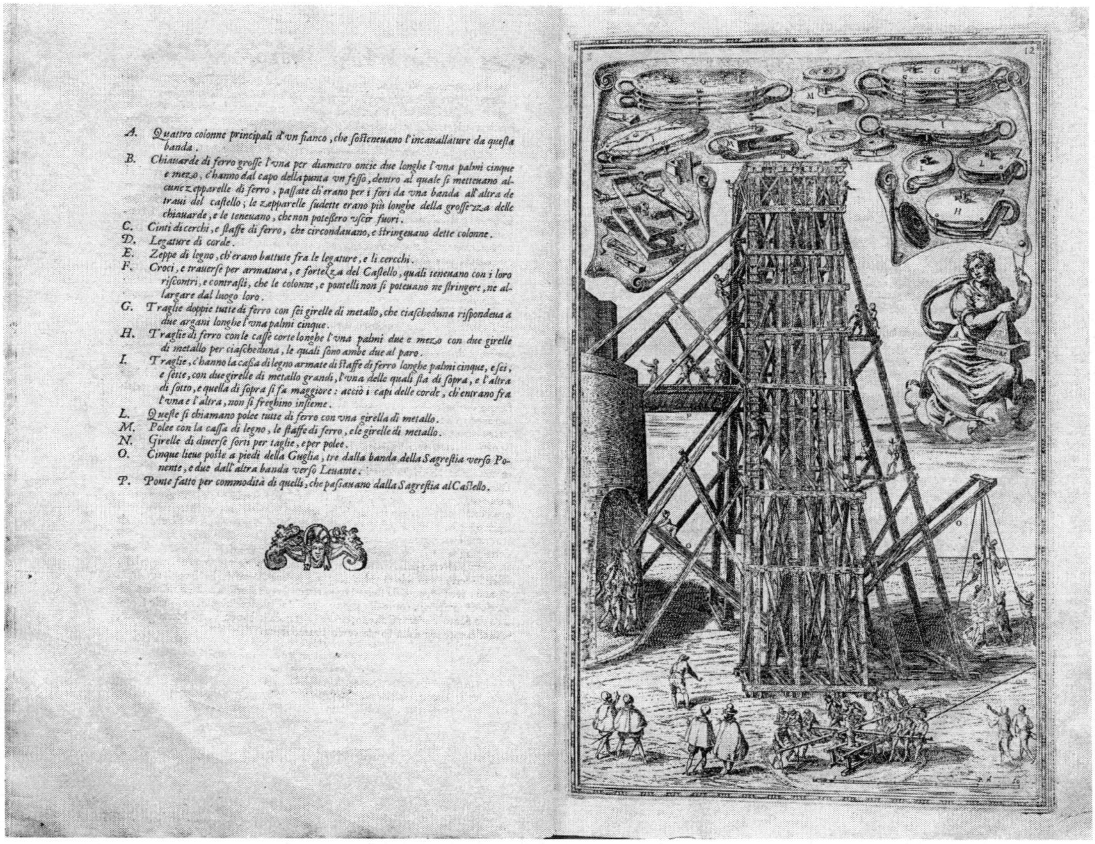

Figure 8.6 The *castello* with pulleys and blocks, parts labeled by letters with descriptions on facing folio. Domenico Fontana, *Della trasportatione dell'Obelisco Vaticano* [. . .] (Rome: Domenico Basa, 1590), fols. 11v and 12r. Courtesy Rosenwald Collection, Library of Congress.

dukes and duchesses attended as did ladies and lords of Rome, ambassadors, "and a very great number of foreigners who had flocked from all parts of Italy to see such a new and wonderful spectacle." The crowds were kept behind barricades and ordered to keep strict silence. Anyone who broke through the barricade was subject to the death penalty. Two masses of the Holy Spirit were said before the start of the work, and all workers were required to take communion. Backup horses and men stood ready in case any of the animals or workers faltered. Another twenty men were assigned to carry any needed supplies such as pulley blocks, cord, and replacement parts to any part of the operation. Twelve carpenters would continuously drive iron and wooden wedges under the obelisk both to help lift it and to provide support. Thirty other men were assigned simply to watch the *castello* at various points so that each part of the operation was observed at all times. Since "the work was directed to the glory of God," Fontana asked all the workers, assistants, and onlookers to kneel in prayer before starting the work. Then the first blast of the trumpet sounded, and the work began.[39]

Immediately the iron hoops began to break. Work was stopped and the hoops were replaced with rope which was found to hold more securely. Finally, the obelisk was lifted off its ancient pedestal. Fontana ordered that the four heavy iron astragals—dadoes or "dumplings" (*dadi o gnoccoli*) is what he calls them, which supported the obelisk on the pedestal—be removed. Two of them were resting on top of the pedestal, the other two attached to the pedestal with lead brackets inserted about 1½ palms (about 13 in.) into the stone. Fontana had one carried to Sixtus V, "who showed very great happiness over it"[40]

Fontana here digressed with a discussion of ancient practices based on physical evidence. He hypothesized "truly" in his opinion that the ancients had used the iron astragals to raise the obelisk. That is, they had fixed them onto the pedestal and then supported the foot of the obelisk against them while they pulled the monolith up by its point. This was obvious, he insisted, because the iron pieces were cracked (*stiacciati,* i.e., *schiacciati*) on the edges where the monument must have pressed against them as it was being raised. His own method, Fontana deduced, was better than that of the ancients because the ancients had put in greater effort and greater expense than he had. Greater effort because by putting the obelisk against the iron pieces and pulling it up, they used one force against the other, whereas he had used capstans and pulleys to raise the bottom as he was raising the top. Greater expense because the ancients must have had to build a scaffold along the whole length of the obelisk, while his own scaffolding extended only about two-thirds of the way up. Fontana further deduced that in antiquity the stone monument had lain on the ground for a long time, since one side was corroded, while the other (the side that had lain on the earth) was smooth.[41]

Fontana suggested that the obelisk had broken at some point during its removal from Egypt, just as the ancient author Pliny had claimed (*Natural History,* 36.15.74). Fontana pointed to three pieces of physical evidence. First, he argued, the point of pyramidion was not proportionate to the shaft in the ratios that existed in the two obelisks found in the Circus Maximus and in a number of smaller obelisks found around Rome. Second, one could see that the pyramidion had been worked by a second master, who had left his side unpolished. Third, the obelisk was shorter in proportion to its width than the other Roman obelisks.[42] In this fascinating archaeological/historical excursion, Fontana addressed humanist and antiquarian interests from a unique vantage point based on physical evidence that was available only temporarily and only to a privileged few.

He then resumed his description of the project at hand. He described the process of lowering the obelisk and placing it on a sledge for its trip about 270 yards to the center of St. Peter's Square (fig. 8.7). The positions of some of the capstans were changed and buttresses prepared in case the work had to pause for any reason. (In this event, Fontana did not want the ropes to bear the full weight of the

Figure 8.7 The Vatican Obelisk on the move. Domenico Fontana, *Della trasportatione dell'Obelisco Vaticano* [. . .] (Rome: Domenico Basa, 1590), fol. 24r. Courtesy Rosenwald Collection, Library of Congress.

obelisk.) The *guglia* was lowered onto the sledge, first by using capstans to pull it down, then by using other capstans to brake it so that it would not crash but descend slowly. After this operation was completed, "all the people felt infinite jubilation, and to show it, they accompanied the architect to his home with drums and trumpets."[43]

The blocks and capstans were dismantled. The entire *castello* was completely taken apart, and each piece—wedges, ropes, pulleys, bolts—was placed separately so that the machine could easily be rebuilt in the center of the piazza.[44]

The center of St. Peter's Square was about thirty feet lower than the original location of the obelisk. Fontana ordered an inclined causeway to be constructed, and the obelisk moved to its new location. A new foundation was built at the center of the piazza, and medals were placed within it, including, on the instruction of Sixtus, two gold medals with effigies of Pius V. The iron astragals were placed on the new pedestal, and then the entire *castello* was rebuilt. By the time the day arrived for raising the obelisk on its new pedestal on September 10, 1586, the capstans had been rearranged. Huge crowds and also the ambassador of France and his retinue arrived to witness the engineering drama. The obelisk raised with fifty-two turns of the capstans "was a beautiful spectacle in many respects." Seven days were taken to move and reinstall the capstans and attach blocks to the faces of the *guglia* to adjust it. The bronze astragals used to support the obelisk were fitted to the base and welded to their brackets. The wedges were knocked out one by one so that gradually the heavy monument came to rest on the astragals. The next day the obelisk was adjusted and set straight. Then, with the monument resting in its new location, the layers of protection were removed, leaving it fully exposed on September 27, 1586.[45]

On that day Sixtus ordered a great procession and ceremony to consecrate the papal insignia (three hills and a star) and gilded cross that had been placed on the pinnacle of the obelisk. Fontana recorded a long list of the high-ranking men in the procession and the order in which they marched. He reported the elaborate ritual of exorcizing the pagan monument to rid it of any lingering pagan spirits. At the conclusion of the ceremony, the Swiss guards fired their artillery. At the same time the mortars at the Castel Sant'Angelo were discharged.[46] Never before had an engineering project also served as such an elaborate urban spectacle and at the same time a deeply solemn occasion for a religious rite.

The spectacular feat and the glory that resulted for both the pope and his architect should not erase the highly discursive context that preceded its execution. As we have seen, men from far and wide proffered diverse proposals. Years later in 1615, the architect Vincenzo Scamozzi (1548–1616), who was in Rome in the 1580s, confirmed that "at various times many elevated minds of the city and from elsewhere toiled to find the most secure and artful way to have it [the obelisk] moved, of which in our youth when we were in Rome we saw part." Scamozzi describes in detail some of the models and plans (which are similar to those shown by Fontana). He further implies that the winner of the competition (i.e., Fontana) was the one "who could achieve the benefice" (i.e., the favor of the pope) and that he "expounded so to speak, the substance of inventions described [by others]"[47] (i.e., he used the ideas of others). The resemblance of the obelisk encasement to Camillo Agrippa's ideas has already been noted, and it could well be that the unknown Sicilian mentioned in Camillo Paleotti's letter provided all or most of the model as well.

Wherever he got the specific ideas for the *castello* and its elaborate mechanisms of pulleys and capstans, the technology in itself was traditional, undoubtedly similar to that used by the ancient Romans. But just as crucial as the *castello* with all of its accoutrements was the fine-tuned coordination of the intricate processes of the complex operation. In addition, the flexibility of the machinery, which could be taken apart and put together again in diverse configurations, was crucial. Fontana's ability to organize and control these complex operations was fundamental to his success.[48]

Following the spectacular engineering feat accomplished by Domenico Fontana, there was a large proliferation of texts and images that described and praised it. The operation was well known by both contemporaries and future generations. No other sixteenth-century engineering project was depicted in such a way—in dramatic and prolific detail, both textual and visual. A relatively fleeting historical moment was made intimately known to all interested contemporaries as well as to future generations. Immediately after the successful completion of the Vatican project, three large-scale copper-engraved and etched prints appeared, filled with details of the operation. They were designed by Sixtus's court painter, Giovanni Guerra, and incised by Natale Bonifacio. Published in 1586, they were made an unusual length with three attached sheets (fig. 8.8). The Vatican Obelisk project also became known by numerous descriptions, dedicatory poems, treatises, and other compositions by contemporaries as well as by Fontana's own dramatically presented book.[49]

We have seen that medals commemorating Pope Pius V were placed in the foundation of the Vatican Obelisk. Yet the transport of the obelisk and its Christianization actually represented a reverse of Pius's policies. He was the pope who had energetically rid Vatican structures of ancient objects such as the antique busts that Pirro Ligorio had assiduously collected and installed in the Casino. Now antiquities—including Rome's largest and most spectacular antiquities, Egyptian obelisks—were decisively brought back into the Christian fold. They were still fascinating ancient objects, but they had also been transformed into proper objects of Christian veneration.

Figure 8.8 Giovanni Guerra and Natale Bonifacio, *Disegno del modo di condurre l'Obelisco detto volgarmente la Guglia*. Rome, 1586. Etching, 30 × 46⅛ in. (76.2 × 117.15 cm). Collection of Vincent J. Buonanno.

Obelisks Moving through Rome

Fontana's spectacular success at the Vatican spurred the transport of three other Roman obelisks that had lain on their sides for centuries and (in the case of two) were buried underground. Fontana oversaw the execution of these three obelisk projects. Although carried out with less fanfare than the Vatican transport, they presented even greater technical problems. The great monolith that became the Lateran Obelisk was the largest obelisk in Rome. It lay buried twenty-four palms (17.32 ft.) underground in the Circus Maximus and had broken into three pieces. Matteo da Castello supervised its excavation, discovering an ancient drain or *cloaca* in the process. His job was hugely difficult because water rose incessantly during the excavation. Five hundred men were needed, three hundred bailing water continuously day and night. Hauling the obelisk to the Lateran was also difficult and required pulling it up the Caelian Hill through narrow streets.[50]

The second obelisk buried in the Circus Maximus, also broken into three pieces, was transported to the Piazza del Popolo. The third obelisk, destined for Santa Maria Maggiore, had long been impeding traffic in a street near the Church of San Rocco and the city river port of Ripetta. This obelisk had originally stood with its twin (now at the Quirinal) near the entrance of the Mausoleum of Augustus. Broken into many pieces, it was moved and reconstructed at Santa Maria Maggiore. There it served a tripartite function: it marked the basilica itself; it defined the starting point of the new street, Via Felice; and it signaled the entrance to Sixtus V's Villa Montalto.[51]

For all three obelisks, broken pieces had to be joined together. Fontana tells us how he finally figured out how to do this, first repairing the Lateran Obelisk. One night, he thought of the ingenious idea of making a socket (*incassatura*) on the broken ends, in which the cut would be made in the form of a swallowtail, large at the bottom and narrower at the top. The ropes could then be put around the ends to move the pieces together. The ropes would be positioned in the wider part, and after being used to pull the obelisk pieces together, they could be pulled out. The spaces would then be filled with lead and stone. Fontana's idea worked. The sections of the obelisk were raised, one by one. After the first was stabilized on the pedestal, the second section was then raised onto and attached to the top of the lower fragment, and the top section subsequently attached to the top of the middle one.[52]

The Power of the Obelisks

Sixtus V's newly positioned obelisks anchored specific streets, piazze, and basilicas within a newly envisioned Christian urbanism. They stand today as spectacular urban markers—the most ancient objects in Rome. Although still standing (Sixtus's four joined by later additions), the obelisks have undoubtedly lost some of the immense attraction that they possessed in the sixteenth century. Michael Cole has described the power of the newly exorcised obelisks with their gilded crosses, the ways in which they not only held off pagan and demonizing foes but mediated divine powers and actively drew celestial benefices to pilgrims honoring the cross. Pilgrims also benefitted from the indulgences (i.e., the remission of temporal punishment in purgatory) of fifteen years that Sixtus V bestowed on those who honored the cross while in a state of grace.[53]

Further power and authority accrued to those obelisks that were covered with hieroglyphs—of Sixtus V's four, the monuments at St. John Lateran and Piazza del Popolo. Egyptian antiquities, including hieroglyphs, had long been a focus of humanist and antiquarian interest. The newly visible obelisks created a flurry of literary, epigraphic, and archaeological activity.[54]

The allure of the obelisks included their promise to provide the deepest knowledge of the world itself, including the natural world that was assumed to have been possessed by the ancient Egyptians, as the papal physician and collector of minerals and fossils, Michele Mercati, explained in his treatise on obelisks. He and his successors associated the hieroglyphs with the corpus of writings by one Hermes Trismegistus, now known to be a mythical personage. The humanist physician Marsilio Ficino had translated the Hermetic corpus from Greek into Latin in the late fifteenth century. Although we now know that the Hermetic corpus is a third-century CE collection of Neoplatonic writings, it was thought in the sixteenth century that they were written by an Egyptian priest, Hermes Trismegistus, who had lived before the time of Moses in Egypt.[55]

It was reasonable to think, Mercati reasoned, that God had given writing to the first man, Adam, along with speech and the ability to give names to His creations. After the biblical flood, people were scattered. All lost the art of writing except for the Egyptians. Egyptian priests, including Hermes Trismegistus, discovered and perfected all the arts and sciences. This included the knowledge of Adam that Moses had divulged to the Egyptian priests. The Egyptian priests wanted to keep this knowledge secret. They recorded the essential core of each of the sciences in hieroglyphs inscribed on obelisks to be read only by other priests. This ancient wisdom would be available to contemporaries, Mercati believed, as soon as hieroglyphic writing could be deciphered.[56]

Although Mercati himself did not pretend to know what the hieroglyphs said, there is an illustration in Fontana's treatise on moving the obelisk in which Mercati undoubtedly had a hand. It is a careful depiction of all four sides of the obelisk that Fontana had moved to the Piazza del Popolo. A legend explains that he had reproduced the hieroglyphs so that learned readers might work on deciphering the lost wisdom of Egypt. Below the insignia of Sixtus V, the pope is named as the dedicatee, and under his insignia is the word *Trismegistos* (fig. 8.9). The obelisk, now proudly standing at the northern gate of the city, thereby not only marked the nearby basilica, Santa Maria del Popolo, but pointed to the ancient wisdom (brought by Sixtus V through Hermes Trismegistus) that would soon (it was implied) be available once again.[57]

The Great Columns

In addition to the four obelisks, in order to serve a triumphant Christianity, Sixtus V appropriated the two great, ancient columns at the center of Rome—Trajan's Column and the Column of Marcus Aurelius (the latter, in the sixteenth century believed to be dedicated to the emperor Antoninus). Trajan's Column had been constructed in 106–113 CE to honor the military victories of the emperor Trajan (53–117 CE) in his Dacian Wars of 101–106 CE.[58] The later Column of Marcus Aurelius was built circa 193 CE in honor of the wars of the emperor Marcus Aurelius (121–180 CE) against the Germanic tribes across the Danube River (170–180 CE). Each column told the story of its emperor's triumphant victories with spectacular continuous helical bands of relief sculpture running in a spiral from bottom to top. The columns originally had been surmounted by the statues of their respective emperors.[59]

Now Sixtus V undertook the repair of the columns and appropriated them for Christian magnificence.[60] Domenico Fontana was put in charge. In his treatise he provided a report on the repair of the Column of Marcus Aurelius. The task was difficult because the 19-palm-high statue (about 14 ft.) had to be lifted to the top of the 185-palm–high column (135 ft.), and once it had been hauled to the top, there was very little space to put it. An "even greater difficulty" had to be overcome. Partly because of its antiquity and partly because it was burned by the Barbarians, the column was in extremely poor condition—"it was reduced to such a point that it seemed impossible, more than difficult to restore it." In many places it was cracked, and large pieces of marble were missing, "to the extent that it frightened he who gazed at it." Fontana described how a *castello* was built around the column and many pieces of marble were added and figures carved on the new marble "with very great diligence" and "with great skill and expense." (The substituted pieces of marble were taken from the destroyed Septizonium.)[61]

Fontana surmounted the two great columns with monumental bronze statues—in 1587 St. Peter and his keys at the summit of Trajan's Column, and in 1589 St. Paul and his sword on the column of Marcus Aurelius (fig. 8.10). Designing and casting the colossal bronze statues (the largest cast statues in late sixteenth-century Rome) was difficult, as was hauling the statues to the top of the columns.

Figure 8.9 Domenico Fontana, *Della trasportatione dell'Obelisco Vaticano* [. . .] (Rome: Domenico Basa, 1590), fol. 75 (two-page spread). Illustration with four sides of the Piazza del Popolo (Flaminio) obelisk and inscriptions associating Sixtus V with Hermes Trismegistus. Courtesy Rosenwald Collection, Library of Congress.

Figure 8.10 Column of Trajan with statue of St. Peter (*left*) and Column of Marcus Aurelius with statue of St. Paul (*right*). Photographs May 2017 © Bob Korn with kind permission.

The huge amount of bronze needed for the statues was scavenged from the doors of the late antique church Sant' Agnese, the doors of the Scala Sancta or Holy Stairs at the Lateran, a piece of ancient pilaster from the Pantheon, and the doors of the old ciborium (the bronze freestanding structure over the altar) of St. Peters.[62]

With the repairs completed on both columns, ceremonies of exorcism were carried out on both to remove the pagan spirits. Now they were ready for their new role as triumphal Christian monuments.[63]

But who was to pay for the great project of repairing and ornamenting the monumental ancient columns? The minutes of the Capitoline Council during May and June 1589 suggest that the familiar struggle over infrastructure expenditures extended to the columns as well. The issue came to a head over the repair of Trajan's Column and a nearby street. In discussions at successive Secret and Public Councils, pressure from the pope is evident. At the Secret Council of April 21, 1589, the wine tax as a source of funding was suggested.[64]

At the Public Council of May 4, it was announced that at the previous Secret Council of May 2, many gentlemen had decided to go to the pope concerning the excessive tax that had been ordered to be collected from four *rioni* nearby in the form of a door-to-door *gettito* for Trajan's Column and a nearby street. The officers of the Commune along with two delegates from each *rione* went to the pope to beg that a different solution be found. The first conservator, Paolo Lancellotti, reported that supplication to the pope mentioned the burdensome taxes resulting from the many papal projects. The delegation advised the pope of "the damage we have seen to the poor, to children, to orphans, and other impoverished people resulting from it [this taxation]." It complained that "in the construction of other streets, the erection of the obelisk, restoration of the Antonine Column, the building of so many and such large buildings dedicated to the divine cult, and [buildings] for public and private use, and in the conducting of the Acqua Felice for public utility of the city and for particular persons," such a great amount of money amounting to many hundreds of thousands of scudi was being demanded. The delegation received "many insults" from the pope and obtained little.[65]

The usual fallback was resorted to on June 5, 1589. It was agreed that the meat tax should be used as well as a small amount collected door-to-door.[66]

Urban Realities

The reality of forced, unwanted, and almost unsupportable tax collection serves as a reminder that the magnificent monuments and straight streets created by Sixtus V and Domenico Fontana do not tell the whole story.

An edict of July 12, 1588, a time well into the years of Sixtus V's urban projects, provides a glimpse of sixteenth-century Roman realities. The edict commanded that "seeing daily through experience how much danger occurs to diverse people, very often with death," from those who lead "unbound and untied buffalo" through the city, this (apparently common) practice was forbidden. The edict forbade herdsmen from leading such animals through the city unless their hoofs were tied and they were yoked. It prescribed severe penalties for violators.[67] Two months later, an edict was issued against goatherds. Goatherds and others who allowed their animals to destroy vineyards, cane thickets, and other places would be fined. If they continued to flagrantly violate the edict, they could be imprisoned, and if they could not pay the fines, they would be put on a public stage and whipped.[68] These edicts are more revealing of urban conditions than of whether and to what extent they were enforced.

While Sixtus V continued to work on the city itself, his efforts extended well beyond it. They included draining the Pontine Marshes—malarial wetlands that extended southeast of Rome to Terracina, the most southern point of the Ecclesiastical State on the Tyrrhenian Sea. Thereby the land could be used for grain crops (grain that was desperately needed by the growing Roman population). Draining the marshes would have the additional effect of removing hideouts used by bandits who plagued the city and its territories. The initial drainage project was a resounding success, celebrated by the pope himself, who led a triumphal procession to Terracina.[69]

That success turned out to be ephemeral. Barely two weeks after the pope's return to the city, the destructive Tiber River flood that occurred on November 4, 1589, destroyed large swathes of the city. Less than a week later, on November 10, another equally devastating flood exacerbated the damage. Because the rains had destroyed many crops, the spring of 1590 was a season of famine in the city. And rainstorms for the most part destroyed the great Pontine drainage project. An *avviso* of April 4, 1590, reported "it is true that the Pontine Marshes that were drained last year [. . .] have returned to their original state because of continuous rain, and it is estimated that it rained for more than 200 days." A few days later, on April 7 an *avviso* again reported that the drained Pontine Marshes, "because of the continuous rain that lasted for four months," had returned to what they were before (marshlands).[70]

Sixtus V died a few months later, on August 27, 1590. In the same year, as reported by the cleric Sanzio Cicatelli, "just in the city of Rome and its region, 60,000 people died, partly through hunger, partly through cold." They were "dying under the benches of the butchers and other shops, reduced to feed on grass like sheep in the fields and to eat even cats and dead dogs and any filthy food they could find." Members of his religious order "recorded more than once a dead body with his mouth full of grass."[71]

Although the number sixty thousand cannot be relied on, it points to a grim reality. It is understandable that the high praise bestowed on Sixtus for his urban renovations by his contemporary advocates and propagandists and by subsequent writers, including historians, was not voiced by many of the Roman people in his own time. Contemporary Romans were hostile to Sixtus because they were forced to pay for his urban innovations even in times of starvation. As early as 1586, an anonymous author wrote that "this year was productive of nothing more than taxes, impositions, new offices, and inventions to raise money." This is "much hated" in every state, the memorialist wrote, but in Rome it is hated "more than [in] any other [state] for not being accustomed to carry that weight," which is usual in the states of other princes. "And hence was born the very bitter hate that his people [the Romans] carried from this time forward." And indeed, Sixtus V had initiated more than thirty-five new taxes.[72]

At the pope's death the anger of the people found immediate expression. The Capitoline Council met and decided to remove the head of the statue of Sixtus V that was on the Capitoline Hill and have it dragged through the city, an action that was ultimately prevented. An *avviso* reported that an armed crowd of almost two thousand men arrived at the Capitoline Hill "demanding the statue of Sixtus in order to drag it around." It was reported that some also wanted to find and kill some men they considered to be associated with Sixtus's oppressions. The poor of Rome were particularly enraged by the tax of one quattrino for every *foglietta* of wine (about a half liter), and the crowd was looking for the heads of "those liars," including the inventor of this new tax, Giovanni Lopez (a Portuguese Jewish banker in Rome)—who hid in the datery (treasury). Nobles were among the crowd and were helping to stir it up, but the cardinals dispatched another nobleman, Filippo Colonna, accompanied by soldiers to protect the statue from the angry Romans. The *avviso* reported that some also wanted to kill the Cavaliere della Guglia, namely Domenico Fontana, among others. Fontana saved himself by hiding in the Palazzo Sforza.[73]

Domenico Fontana was able to avoid bodily harm in the immediate aftermath of his patron's death in 1590 and continued working, but ultimately his career in Rome turned out to be unsustainable. In 1592, the Pope Clement VIII (ruled 1592–1605) removed him from the position of papal architect and demanded that he account for expenses incurred during Sistine building projects. Fontana was accused of mismanaging the construction of a bridge—the Ponte Felice, a bridge that Sixtus V had initiated before his death. The bridge was to be built over the Tiber River in the town of Borghetto near Magliano Sabina, north of Rome (or rather on dry land next to the river, after which the river was diverted).[74]

The bridge was Domenica Fontana's undoing. He was accused of using defective materials to make a poorly constructed bridge. In his lengthy but unsuccessful defense, preserved in the Vatican archives, he cited thefts of supplies and horses by bandits, damage to crops and fields that had to be compensated, and numerous other woes, the costs of which came out of his pocket. He claimed that the Camera Apostolica owed him 17,416 scudi, whereas the Camera Apostolica claimed that he owed them 14,000 scudi. He was forced to leave Rome, while the family workshop he left behind undertook to pay his debt. Fontana subsequently enjoyed a successful career in Naples, and he published an expanded version of his famous treatise in 1604 that included his own account of the Borghetto bridge project as well as some of his Neapolitan projects. Still, the most successful architect/engineer of late sixteenth-century Rome had been forced to leave in disgrace, never to work there again.[75]

Conclusion

A City in Transition

The short time span encompassed by this book, primarily focused on three decades, has enabled an investigation of the processes having to do with civic and hydraulic engineering, urbanization, antiquarian investigations, cartography, and printing. The existence of a still-functioning aqueduct such as the Acqua Vergine—which stands today as a physical structure—is one long-term result of discussions, arguments, bitter dissent, and struggles for contracts and patronage in the sixteenth century. Today, one can still walk down streets that were straightened and widened in that era but without some of the conditions that were very much a part of sixteenth-century reality—the sewage, waste, and mud on the streets, the devastating floods, the bearing down of horse-drawn coaches, and the rumbling of carts laden with stone and lumber.

The ongoing, chronic struggle between the Capitoline Council and the papacy about funding infrastructure projects is no longer visible. Nor are the conflicts over who should receive contracts and make engineering decisions. Such disagreements arose not only from lack of funds, but from a basic lack of agreement about which entity should provide funding for and who should be in charge of what we call public-works projects. Also no longer visible is the simple fact that whenever—frequently as we have seen—the Capitoline Council was forced to resort to the meat tax, it was Roman people, intermittently in a state of borderline starvation, that were forced to pay with a tax on their food. The need to find funding anew for every infrastructure project and even for seemingly routine activities such as street cleaning came from the circumstance that there were no funded bureaucratic structures with permanent salaried employees and designated ongoing authority to carry out tasks. Rome in the late sixteenth century was an early modern city, not a modern one.

Traditional historiography often looks for aspects of the "modern" in early modern cities such as Rome and to view such aspects where they are found as a positive good, suggesting progressive development. From a close-up point of view, what becomes more visible are the conflicts, disagreements, competitions, failures, and lack of funds. What remains today are the great successes—functioning aqueducts and fountains, standing obelisks, and wide, straight streets. The meetings, wide-ranging discussions, arguments, and proposals and counter proposals are no longer apparent but played a crucial role. Also less obvious is the deep interest in antiquarianism, the avid study of physical ruins, and the importance of image making as central to the study of the physical fabric of the city and to its reenvisioning and reconstruction.

What comes into view is the lively culture centered on urbanization, engineering, and construction; the intense interest in certain urban projects and the proliferation of writings providing histories and proposing solutions; the outpouring of urban maps and city images; and the close interrelationships between antiquarian investigation and contemporary problems of urban infrastructure and construction. This vigorous culture was populated by people from a variety of backgrounds and changing identities. Between the rivalries, arguments, learned writings, and the production of spectacular maps and other images, printers and etchers became architects, scholars of ancient Greek or civic law put on their boots and waded into ancient aqueducts, painters designed gardens, and architect/engineers wrote books. This lively, fluid culture of intense argument and interchange, of insult and friendship, had a lasting influence on the development of cities and on a culture of knowledge in which engineering and urban infrastructure projects, topography, antiquarianism, cartography, and image making were understood as intrinsically related to one another.

More broadly, the late sixteenth century was a time when new approaches to the investigation of the natural world proliferated as the values of hands-on practice, observation, measurement, and making came to the fore. These values as applied to both construction and engineering presupposed a complex social matrix. In Rome such a matrix brought together practice and learning, antiquarianism and the study of physical ruins, the planning of infrastructure projects and the discussion of causes, the envisioning of the city as a whole and the creation of cartographical images of that city, the development of plans and proposals, and the composing of writings promoting them. These conditions allowed a flourishing culture of construction and engineering in Rome and also had implications for the culture of knowledge far beyond.

Acknowledgments

The research and writing of this book has had a long trajectory and I thank every one of the grant-giving institutions, libraries, and archives that over the years have assisted me in large and small ways. In addition, there is no way to adequately thank the friends and colleagues who have helped me along the way.

As an independent historian, a requisite task is the annual round of applications for fellowship and grant support. I have had the good fortune to have been supported in my work. Two grants (0240358 and 8812822) from the US National Science Foundation provided resources for crucial research years. In 2003–2004 I spent a wonderful year at the American Academy in Rome as an Andrew W. Mellon postdoctoral Rome Prize Fellow, which allowed me to explore the archives and libraries of Rome and immerse myself in the rich intellectual and artistic life of the academy. Very different but also rewarding fellowships included one at the Shelby Cullom Davis Center for Historical Studies in the Department of History at Princeton University, a fellowship at the Getty Research Institute in Los Angeles, a National Endowment for the Humanities Research Fellowship, and a John Simon Guggenheim Memorial Foundation Fellowship. I spent a fruitful month studying maps at the British Library with a J. B. Harley Research Fellowship in the History of Cartography. A Franklin Research grant from the American Philosophical Society supported two months of research in Turin and Rome. I spent a productive year as a William J. Bouwsma fellow at the National Humanities Center, Research Triangle, North Carolina; a year as a Long-term Fellow at the Folger Shakespeare Library in Washington, DC; three months in the research group led by Sven Dupré, "Art and Knowledge in Premodern Europe," at the Max Planck Institute for the History of Science in Berlin; and six months as a visiting professor at Villa I Tatti, the Harvard Center for Italian Renaissance Studies in Florence. In the final three years of research and writing—what really made possible the completion of this book—was a fellowship from John D. and Catherine T. MacArthur Foundation.

I am indebted to the libraries and archives in which I have worked and to the dedicated scholars, archivists, and librarians who maintain them and assist scholars like me find their way around. Most important for this book have been the Archivio di Stato di Roma, the Archivio di Stato di Torino, the Archivio Segreto Vaticano, the Archivio Storico Capitolino, the Archivium Romanum Societatis Iesu, the Biblioteca Angelica, the Biblioteca dell'Istituto d'Archeologia e Storia dell'Arte, the Arthur and Janet C. Ross Library at the American Academy in Rome, the Biblioteca Apostolica Vaticana, the Biblioteca Casanatensa, the Biblioteca Nazionale Centrale di Roma, the British Library, the Dibner Library at the Smithsonian Institution, the Folger Shakespeare Library, the Library of Congress, Princeton University Libraries, and the Special Collections at the Getty Research Institute.

I thank my colleagues, many of whom over the years have become dear friends. Brian Curran, Tony Grafton, and Ben Weiss were wonderful coauthors and co-explorers in matters concerning obelisks. I am incredibly sad that Brian did not live to see this book completed. I thank my friend Jessica Maier, fellow fellow at the American Academy, who taught me much about maps. I have shared wonderful conversations with Ingrid Rowland and with Paula Findlen, Patricia Osmond, and Carolyn Valone. I thank Libby Cohen, Tom Cohen, Claudia Conforti, Joe Connors, Ronald Delph, David Friedman, Kenneth Gouwens, Ann Huppert, David Karmon, Lynne Lancaster, Fernando Loffredo, Peter Lukehart, Frank Mantello, Craig Martin, Laurie Nussdorfer, Katharine Park, Alina Payne, Michael Roche, Dennis Romano, Pamela Smith, Stefano Villani, and Patricia Waddy. I thank Katherine W. Rinne, who first taught me about Roman aqueducts block by block and who created one of the maps.

Special thanks go to Chiara Bariviera, whose astute knowledge of the archaeology of ancient Rome and expertise in Latin were immensely helpful as she assisted me in the final several years of work on the manuscript. Chiara also created several maps and diagrams for the book.

During my many trips to Rome I stayed in "my attic" near the Ponte Sisto, and

I warmly thank my kindly landlord/friends, Dr. Armando Brunelli and especially Elena Brunelli.

In the last phases of rewriting the manuscript for final submission, my sister Priscilla Long, a brilliant poet and essayist, along with my husband, Bob Korn, videographer and expert editor, helped to improve the coherence of the work and to polish the writing.

Karen Merikangas Darling has been an ideal editor from my first conversations with her some years ago. I warmly thank her, Susannah Marie Engstrom, and the production team at Chicago for making the whole process a seamless and pleasant one. I thank the anonymous reviewers whose comments helped to improve the manuscript. I am especially grateful to C. Steven LaRue, who copyedited the text and a complex set of footnotes with a fine ear and great precision.

Bob Korn has helped with all manner of computer issues, technical difficulties, and travel plans over the years, always taking time from his own work. When I was working in Rome and elsewhere, he came to visit, to take photographs and explore the city. He has been there through the whole thing. This book is dedicated to him and to our daughter Allison Korn, to Marco Yunga, and to Luki and Asha, who make the world go around!

Abbreviations

Arm. = Armadio
ASC = Archivio Storico Capitolino, Rome
ASF = Archivio di Stato di Firenze, Florence
ASR = Archivio di Stato di Roma, Rome
ASRSP = *Archivio della Società Romana di Storia Patria* (journal)
AST = Archivio di Stato di Torino, Turin
ASV = Archivio Segreto Vaticano, Vatican City
b. = busta
BA = Biblioteca Angelica, Rome
BAV = Biblioteca Apostolica Vaticana, Vatican City
BL = British Library, London
BNCF = Biblioteca Nazionale Centrale di Firenze, Florence
BNN = Biblioteca Nazionale di Napoli, Naples
cat. = catena
C.C. = Camera Capitolina
coll. = collezione
cred. = credenzone (storage cabinet)
DA = Turner, Jane, ed. *Dictionary of Art*. 34 vols. New York: Grove Dictionaries, 1996.

DBI = *Dizionario Biografico degli Italiani*. 86– vols. Rome: Istituto della Enciclopedia Italiana, 1960–. http://www.treccani.it/biografie/dizionariobiografico/.

fasc. = fascicolo

fol. = folio

MS Florence, Uffizi = Gabinetto Disegni e Stampe degli Uffizi

maz. = mazzo (bundle or group)

prot. = protocollo (bound group of documents in a busta)

r. = recto

stragr. = stragrande (large format)

tav. = tavola (plate)

tom. = tome (volume)

v. = verso

Notes

INTRODUCTION

1. Essential bibliography includes Mario Caravale and Alberto Caracciolo, eds., *Lo Stato Pontificio da Martino V a Pio IX*, Storia d'Italia, vol. 14 (Turin: Unione Tipografico-Editrice Torinese, 1978); Claudia Conforti and Richard J. Tuttle, eds., *Storia dell'architettura italiana*, vol. 4, *Il secondo Cinquecento* (Milan: Electa, 2001); Jean Delumeau, *Vie économique et sociale de Rome dans la seconde moitié du XVI*[e] *siècle*, 2 vols., Bibliothèque des Écoles française d'Athène et de Rome, 184 (Paris: E. de Boccard, 1957–1959); Luigi Fiorani and Adriano Prosperi, eds., *Roma: La città del papa: Vita civile e religiosa dal giubileo di Bonifacio VIII al giubileo di papa Wojtyla*, Storia d'Italia, 16 (Turin: Giulio Einaudi, 2000); Helge Gamrath, Roma sancta renovata: *Studi sull'urbanistica di Roma nella seconda metà del sec. XVI con particolare riferimento al pontificato di Sisto V (1585–1590)*, Analecta Romana Instituti Danici, supp. 12 (Rome: "L'Erma" di Bretschneider, 1987); Nicoletta Marconi, *Edificando Roma barocca: Macchine, apparati, maestranze e cantieri tra XVI e XVIII secolo* (Città di Castello: Edimond, 2004); Pio Pecchiai, *Roma nel Cinquecento* (Bologna: Licinio Cappelli Editore, 1948); Antonio Pinelli, ed. *Roma del Rinascimento* (Rome: Laterza, 2001); Giorgio Simoncini, *Roma: Le trasformazioni urbane nel Cinquecento*, vol. 1, *Topografia e urbanistica da Giulio II a Clemente VIII* (Florence: Leo S. Olschki, 2008); Simoncini, ed. *Roma: Le trasformazioni urbane nel Cinquecento*, vol. 2, *Dalla città al territorio* (Florence: Leo S. Olschki, 2011); and Gianfranco Spagnesi, *Roma: La basilica di San Pietro, il Borgo e la città* (Milan: Palombi /Jaca Book, 2003).

2. See John W. O'Malley, *Trent: What Happened at the Council* (Cambridge, MA: Belknap Press of Harvard University Press, 2013), 191, for the material

enhancements that the council endorsed. More generally, see Simon Ditchfield, *Liturgy, Sanctity and History in Tridentine Italy: Pietro Maria Campi and the Preservation of the Particular* (Cambridge: Cambridge University Press, 1995), and R. Po-Chia Hsia, *The World of Catholic Renewal, 1540–1770* (Cambridge: Cambridge University Press, 1998).

3. Simon Ditchfield, "Reading Rome as a Sacred Landscape, c. 1586–1635," in *Sacred Space in Early Modern Europe*, ed. Will Coster and Andrew Spicer (Cambridge: Cambridge University Press, 2005), 167–192. And see Ronald K. Delph, "*Renovatio, Reformatio*, and Humanist Ambition in Rome," in *Heresy, Culture, and Religion in Early Modern Italy: Contexts and Contestations*, ed. Ronald K. Delph, Michelle M. Fontaine, and John J. Martin (Kirksville, MO: Truman State University Press, 2006), 73–91; Antonio Pinelli, "Introduzione," in Pinelli, *Roma del Rinascimento*, xi–xxxi; Charles L. Stinger, "*Roma Triumphans*: Triumphs in the Thought and Ceremonies of Renaissance Rome," *Medievalia et Humanistica*, n.s., 10 (1981): 189–201, and *The Renaissance in Rome*, 2nd ed. (Bloomington: Indiana University Press, 1998), 235–291. See also Nicola Courtright, *The Papacy and the Art of Reform in Sixteenth-Century Rome: Gregory XIII's Tower of the Winds in the Vatican* (Cambridge: Cambridge University Press, 2003), esp. 84–103, which emphasizes Gregory XIII's identification with the first Christian emperor, Constantine, and early Christianity.

4. Giovanni Botero, *On the Causes of the Greatness and Magnificence of Cities*, trans. and with an introduction by Geoffrey Symcox (Toronto: University of Toronto Press, 2012). For Botero, see also Luigi Firpo, "Botero, Giovanni," *DBI*, 13:352–362.

5. For the problem of what to call the period, see John W. O'Malley, *Trent and All That: Renaming Catholicism in the Early Modern Era* (Cambridge, MA: Harvard University Press, 2000).

6. See especially, David Edgerton, *The Shock of the Old: Technology and Global History since 1900* (Oxford: Oxford University Press, 2007); and Pamela O. Long, "The Craft of Premodern European History of Technology: Past and Future Practice," *Technology and Culture* 51 (July 2010): 698–714, esp. 706–712. For maintenance, see Andrew L. Russell and Lee Vinsel, "After Innovation, Turn to Maintenance," *Technology and Culture* 29 (January 2018): 1–25.

7. See esp. Nicholas Adams and Laurie Nussdorfer, "The Italian City, 1400–1600," in *The Renaissance from Brunelleschi to Michelangelo: The Representation of Architecture*, ed. Henry A. Millon and Vittorio Magnago Lampugnani (New York: Rizzoli, 1997), 205–230; Mario Bevilacqua, *Il monte dei Cenci: Una famiglia romana e il suo insediamento urbano tra Medioevo ed età barocca* (Rome: Gangemi, 1988), 60–247; and Claudia Conforti, *La città del tardo Rinascimento* (Rome: Laterza, 2005), 19–63.

8. For the engraved, single-sheet prints that proliferated, see esp. Christopher L. C. E. Witcombe, *Print Publishing in Sixteenth-Century Rome: Growth and Expansion, Rivalry and Murder* (London: Harvey Miller, 2008), and for maps of Rome, see Amato Pietro Frutaz, *Le piante di Roma*, 3 vols. (Rome: Istituto di Studi Romani, 1962). For "public information" before newspapers and for printing in Rome, see esp. Brendan Dooley, *The Social History of Skepticism: Experience and Doubt in Early Modern Culture* (Baltimore: Johns Hopkins University Press, 1999), 9–44; Mario Infelise, *Prima dei giornali: Alle origini della pubblica informazione (secoli XVI e XVII)* (Rome: Laterza, 2002); and Gian Ludovico Masetti Zannini, *Stampatori e librai a Roma nella seconda metà del Cinquecento* (Rome: Palombi, 1980).

9. Recent discussions include Elizabeth Merrill, "The *Professione di Architetto* in Renaissance Italy," *Journal of the Society of Architectural Historians* 76 (March 2017): 13–35; and Simon Pepper, "Artisans, Architects and Aristocrats: Professionalism and Renaissance Mili-

tary Engineering," in *The Chivalric Ethos and the Development of Military Professionalism*, ed. D. J. B. Trim (Leiden: Brill, 2003), 117–147.

10. See esp. Vaughan Hart and Peter Hicks, eds., *Paper Palaces: The Rise of the Renaissance Architectural Treatise* (New Haven, CT: Yale University Press, 1998); Pamela O. Long, "The Contribution of Architectural Writers to a 'Scientific Outlook' in the Fifteenth and Sixteenth Centuries," *Journal of Medieval and Renaissance Studies* 15 (Fall 1985): 265–298; *Openness, Secrecy, Authorship: Technical Arts and the Culture of Knowledge from Antiquity to the Renaissance* (Baltimore: Johns Hopkins University Press, 2001), esp. 201–234; *Artisan/Practitioners and the Rise of the New Sciences, 1400–1600* (Corvallis: Oregon State University Press, 2011), 62–93; Pier Nicola Pagliara, "Vitruvio da testo a canone," in *Memoria dell'antico nell'arte italiana*, vol. 3, *Dalla tradizione all'archeologia*, ed. Salvatore Settis (Turin: Giulio Einaudi, 1986), 3–85; and Alina A. Payne, *The Architectural Treatise in the Italian Renaissance: Architectural Invention, Ornament, and Literary Culture* (Cambridge: Cambridge University Press, 1999).

11. Anthony Gerbino, *François Blondel: Architecture, Erudition, and the Scientific Revolution* (London: Routledge, 2010), esp. 1–7. See also Pamela O. Long, "Multi-tasking 'Pre-professional' Architect/Engineers and Other Bricolagic Practitioners as Key Figures in the Elision of Boundaries between Practice and Learning in Sixteenth-Century Europe: Some Roman Examples," in *The Structures of Practical Knowledge*, ed. Matteo Valleriani (Dordrecht: Springer, 2017), 223–246.

12. For a more extended discussion, see Long, *Artisan/Practitioners*, esp. 94–126; "Trading Zones in Early Modern Europe," *Isis* 106 (December 2015): 840–847; and "Multi-tasking 'Pre-professional' Architect/Engineers," 223–246. See also Renata Ago, *Economia barocca: Mercato e istituzioni nella Roma del Seicento* (Rome: Donzelli, 1998), esp. 20–32, 141–143; and Paul A. Anderson, "Master Carpenters in Renaissance and Baroque Rome: The Collaboration of Artists, Architects and Artisans on Monumental Commissions in the Cinquecento and Seicento" (PhD diss., University of California, Santa Barbara, 2008, esp. 1:12–64. And see Pamela H. Smith, *The Body of the Artisan: Art and Experience in the Scientific Revolution* (Chicago: University of Chicago Press, 2004); Lissa Roberts, Simon Schaffer, and Peter Dear, eds., *The Mindful Hand: Inquiry and Invention from the Late Renaissance to Early Industrialization* (Amsterdam: Koninklijke Nederlandse Akademie van Wetenschappen, 2007). For a study of expertise in these centuries, see Eric H. Ash, *Power, Knowledge, and Expertise in Elizabethan England* (Baltimore: Johns Hopkins University Press, 2004), and Ash, ed., *Expertise: Practical Knowledge and the Early Modern State*, Osiris, 2nd ser., vol. 25 (Chicago: University of Chicago Press, 2010).

13. See esp. Antonella Romano, ed., *Rome et la science moderne entre Renaissance et Lumières* (Rome: École française de Rome, 2008); and Maria Pia Donato and Jill Kraye, eds. *Conflicting Duties: Science, Medicine and Religion in Rome, 1550–1750* (London: Warburg Institute, 2009).

14. See esp. Sandro Carocci, "The Papal State," in *The Italian Renaissance State*, ed. Andrea Gamberini and Isabella Lazzarini (Cambridge: Cambridge University Press, 2012), 69–89; Amedeo De Vincentiis, "Papato, stato e curia nel XV secolo: Il problema della discontinuità," *Storica* 8 (2002): 91–115; Rose Marie San Juan, *Rome: A City Out of Print* (Minneapolis: University of Minnesota Press, 2001), 161–186; Maria Antonietta Visceglia, *La città rituale: Roma e le sue cerimonie in età moderna* (Rome: Viella, 2002), esp. 17–117; "Burocrazia, mobilità sociale e *patronage* alla corte di Roma tra Cinque e Seicento: Alcuni aspetti del recente dibattito storiografico e prospettive di ricerca," *Roma Moderna e Contemporanea* 3 (January–April

1995): 11–55; *Morte e elezione del papa: Norme, riti e conflitti, l'età moderna* (Rome: Viella, 2013); and Maria Antonietta Visceglia and Catherine Brice, eds., *Cérémonial et ritual à Rome (XVIe–XIXe siècle)* (Rome: École française de Rome, 1997).

15. Particularly important for urban reconstruction were Pope Nicholas V (ruled 1447–1455) with the help of the humanist and architect Leon Battista Alberti (1404–1472); Pope Paul II (ruled 1464–1471); and then Sixtus IV (ruled 1471–1484) in the fifteenth century, and subsequently in the early sixteenth, Julius II (ruled 1503–1513) and Leo X (ruled 1513–1521). See esp. Charles Burroughs, *From Signs to Design: Environmental Process and Reform in Early Renaissance Rome* (Cambridge, MA: MIT Press, 1990); Elizabeth McCahill, *Reviving the Eternal City: Rome and the Papal Court, 1420–1447* (Cambridge, MA: Harvard University Press, 2013); Anna Modigliani, *Disegni sulla città nel primo Rinascimento romano: Paolo II* (Rome: Roma nel Rinascimento, 2009); Peter Partner, *Renaissance Rome, 1500–1559: A Portrait of a Society* (Berkeley: University of California Press, 1976); Stinger, *Renaissance in Rome*; Manfredo Tafuri, "'Cives esse non licere': The Rome of Nicholas V and Leon Battista Alberti; Elements toward a Historical Revision," *Harvard Architecture Review* 6 (1987): 60–75; Nicholas Temple, Renovatio urbis: *Architecture, Urbanism and Ceremony in the Rome of Julius II* (Abingdon, Oxon: Routledge, 2011); Giorgio Simoncini, *Roma: Le trasformazioni urbane nel Quattrocento*, vol. 1, *Topografia e urbanistica da Bonifacio IX ad Alessandro VI* (Florence: Leo S. Olschki, 2004); Simoncini, ed., *Roma: Le trasformazioni urbane nel Quattrocento*, vol. 2, *Funzioni urbane e tipologie edilizie* (Florence: Leo S. Olschki, 2004); and Carroll William Westfall, *In this Most Perfect Paradise: Alberti, Nicholas V, and the Invention of Conscious Urban Planning in Rome, 1447–55* (University Park: Pennsylvania State University Press, 1974).

16. For an introduction to the large literature on the Sack, see esp. Luigi Guicciardini, *The Sack of Rome*, trans. and ed. James H. McGregor (New York: Italica Press, 1993); and see André Chastel, *The Sack of Rome, 1527*, trans. Beth Archer (Princeton, NJ: Princeton University Press, 1983); Kenneth Gouwens, *Remembering the Renaissance: Humanist Narratives of the Sack of Rome* (Leiden: Brill, 1998); and Judith Hook, *The Sack of Rome, 1527* (London: Macmillan, 1972). For the 1530 flood, see Pietro Frosini, *Il Tevere: Le inondazioni di Roma e i provvedimenti presi dal governo italiano per evitarle* (Rome: Accademia Nazionale dei Lincei, 1977), 160–165; and for Tiber River flooding more generally, see esp. Maria Margarita Segarra Lagunes, *Il Tevere e Roma: Storia di una simbiosi* (Rome: Gangemi Editore, 2004), esp. 69–133.

17. See Chastel, *Sack of Rome*, 207–215; and Sandro Benedetti and Giuseppe Zander, *L'arte in Roma nel secolo XVI*, vol. 1, *L'architettura* (Bologna: Cappelli, 1990), 585–591, for the emperor's entrance into Rome on April 5, 1536 (and see further bibliography in chap. 7, n23). For Paul III, see Léon Dorez, *La cour du Pape Paul III d'après le registres de la trésorerie secrète* (Paris: Librairie Ernest Leroux, 1932), 1:249–291, for his relationship with Charles V. For architecture and building construction from the time of Paul III, see esp. Claudia Conforti, "Roma: Architettura e città," and Federico Bellini, "I grandi cantieri: Campidoglio, San Pietro, *Studium Urbis*," in Conforti and Tuttle, *Storia*, 4:26–65 and 66–93, respectively.

18. For an introduction to the large literature on the Villa Giulia, see esp. David R. Coffin, *The Villa in the Life of Renaissance Rome* (Princeton, NJ: Princeton University Press, 1979), 150–174, and Denis Ribouillault, "La Villa Giulia et l'âge d'or augustéen," in *Le miroir et l'espace du prince dans l'art italien de la Renaissance*, ed. Philippe Morel (Paris: Presses Universitaires François Rebelais and Presses Universitaires de Rennes, 2012), 339–388.

19. For Paul IV, see esp. J. N. D. Kelly, *The Oxford Dictionary of Popes* (Oxford: Oxford

University Press, 1986), 265–266, and Ludwig von Pastor, *History of the Popes from the Close of the Middle Ages*, translated by F. I. Antrobus, R. F. Kerr, et al. (London: Kegan Paul, Trench, Trubner; St. Louis, MO: B. Herder, 1891–1953), 14:56–424. For his policy toward the Jews, see esp. Kenneth R. Stow, *Catholic Thought and Papal Jewry Policy, 1555–1593* (New York: Jewish Theological Seminary of America, 1977), 291–298, for the bull. For a succinct discussion of Paul's rule, see Miles Pattenden, *Pius IV and the Fall of the Carafa: Nepotism and Papal Authority in Counter-Reformation Rome* (Oxford: Oxford University Press, 2013), 8–33.

20. For the Carafa affair, see esp. Pattenden, *Pius IV and the Fall*; and Donata Chiomenti Vassalli, *Paolo IV e il processo Carafa: Un caso d'ingiusta giustizia nel Cinquecento* (Milan: Gruppo Ugo Mursia, 1993).

21. For Pius IV, see Benedetti and Zander, *L'arte in Roma*, 1:343–376; Kelly, *Oxford Dictionary of Popes*, 266–268; Pastor, *History of the Popes*, 15:66–82, on his career before his papal election; and Flavio Rurale, "Pio IV, papa," *DBI*, 83:808–814.

22. See O'Malley, *Trent: What Happened*, 168–247.

23. Eugenio Albèri, ed., *Le relazioni degli ambasciatori veneti al senato durante il secolo Decimosesto* (Florence: Società Editrice Fiorentina, 1857), vol. 10 (2nd. ser., vol. 4), 76, "È naturalmente il Papa inclinato alla privata vita e libera [...] e in tutte le sue azioni mostra piuttosto dolcezza che gravità, lasciandosi vedere da tutti a tutte le ore, e andando a cavallo e a piedi per la città con pochissima compagnia."

24. Onofrio Panvinio, *XXVII Pontificum Maximorum elogia et imagines* [...] *delineatae* (Rome: Antonio Lafreri, 1568), no page no. (opposite portrait of Pius IV), "Multa tota urbe, & per ecclesiasticam ditionem aedificia publica magnifice extruxit." See also John Alexander, *From Renaissance to Counter-Reformation: The Architectural Patronage of Carlo Borromeo during the Reign of Pius IV* (Rome: Bulzoni, 2007), which shows Pius's influence on his famous nephew.

25. See esp. Elena Bonora, *Roma 1564: La congiura contro il papa* (Bari: Laterza, 2011); and Peter Blastenbrei, "Glücksritter und Heilige: Motivstruktur und Täterprofile bei der Accoltiverschwörung gegen Pius IV. im Jahre 1564," *Quellen und Forschungen aus italienischen Archiven und Bibliotheken* 70 (1990): 441–490, both of which contain further bibliography and discussions concerning the motivations of the conspirators; and see Pastor, *History of the Popes*, 16:381–391, for the other assassination attempts as well as the conspiracy of 1564.

26. See esp. Benedetti and Zander, *L'arte in Roma*, 1:377–404; Simona Feci, "Pio V, papa, santo," *DBI*, 83:814–825; Maurilio Guasco and Angelo Torre, eds., *Pio V nella società e nella politica del suo tempo* (Bologna: Il Mulino, 2005); Kelly, *Oxford Dictionary of Popes*, 268–269; Nicole Lemaître, *Saint Pie V* ([Paris]: Fayard, 1994), esp. 219–240; and Pastor, *History of the Popes*, vols. 17 and 18.

27. Guasco and Torre, *Pio V nella società*; Kelly, *Oxford Dictionary of Popes*, 268–269; Lemaître, *Saint Pie V*, esp. 219–240; and Pastor, *History of the Popes*, vols. 17 and 18. Of the large literature on the Battle of Lepanto, still one of the best discussions is Kenneth M. Setton, *The Papacy and the Levant (1204–1571)*, vol. 4, *The Sixteenth Century from Julius III to Pius V* (Philadelphia: American Philosophical Society, 1984), 1045–1104.

28. For Gregory XIII, see Benedetti and Zander, *L'arte in Roma*, 1:405–458; Agostino Borromeo, "Gregorio XIII, papa," *DBI*, 59:204–219; Jack Freiberg, "Pope Gregory XIII, Jurist," in "Art and Science in the Rome of Gregory XIII Boncompagni (1572–1585)," special issue, *Memoirs of the American Academy in Rome* 54 (2009): 41–60; Kelly, *Oxford Dictio-*

nary of Popes, 269–271; and Pastor, *History of the Popes,* vols. 19 and 20. For Gregory's son, see Umberto Coldagelli, "Boncompagni, Giacomo," *DBI,* 11:689–692.

29. For the colleges, see Kelly, *Oxford Dictionary of Popes,* 269–271; and Pastor, *History of the Popes,* 19:234–258; and see Federico Bellini, "I collegi e gli insediamenti nazionali nella Roma di Gregorio XIII (con una nota su Sant'Atanasio dei Greci e la Trinità dei Monti)," *Città e Storia* 2 (2007): 111–130, and Gaspar Dos Reis Souza Lima, "Repertorio delle fabbriche dei nuovi ordini religiosi," in Simoncini, *Roma: Le trasformazioni urbane nel Cinquecento,* 2:57–88, which includes many of the relevant construction projects. And see Evangelia Papoulia, "Unveiling Gregorian Rome: The Urban and Ecclesiastical Patronage of Pope Gregory XIII, 1572–1585" (PhD diss., Courtauld Institute of Art, University of London, 2015).

30. For the enmity between Gregory and Sixtus, see Pastor, *History of the Popes,* 21: 40–42; Italo Insolera, *Roma: Immagini e realtà dal X al XX secolo,* 3rd ed. (Rome: Laterza, 1985), 167–172; and see Papoulia, "Unveiling Gregorian Rome," 230–273.

31. For Sixtus V, see esp. Benedetti and Zander, *L'arte in Roma,* 1:459–521; Marcello Fagiolo and Maria Luisa Madonna, eds., *Sisto V,* 2 vols. (Rome: Istituto Poligrafico e Zecca dello Stato, Libreria dello Stato, 1992); Gamrath, *Roma sancta renovata;* Kelly, *Oxford Dictionary of Popes,* 271–273; Maria Louisa Madonna, ed., *Roma di Sisto V: Le arti e la cultura* (Rome: Edizioni De Luca, 1993); Pastor, *History of the Popes,* vols. 21 and 22; René Schiffmann, Roma felix: *Aspekte der städtebaulichen Gestaltung Roms unter Papst Sixtus V* (Bern: Peter Lang, 1985); Giorgio Simoncini, *"Roma restaurata": Rinnovamento urbano al tempo di Sisto V* (Florence: Leo S. Olschki, 1990).

32. Caravale and Caracciolo, *Lo Stato Pontificio,* esp. 237–413; Paolo Prodi, Il *sovrano pontefice: Un corpo e due anime; La monarchia papale nella prima età moderna* (Bologna: Il Mulino, 1982), translated by Susan Haskins as *The Papal Prince: One Body and Two Souls; The Papal Monarchy in Early Modern Europe* (Cambridge: Cambridge University Press, 1987); and Alberto Caracciolo, "Sovrano pontefice e sovrani assoluti," *Quaderni storici* 18 (April 1983): 279–286. And see Laurie Nussdorfer, *Civic Politics in the Rome of Urban VIII* (Princeton, NJ: Princeton University Press, 1992), esp. 11–15; O'Malley, *Trent and All That,* esp. 81–82 and 85–88; Wolfgang Reinhard, *Freunde und Kreaturen: "Verflechtung" als Konzept zur Erforschung historischer Führungsgruppen Römische Oligarchie um 1600* (Munich: Ernst Vögel, 1979); and Gianvittorio Signorotto and Maria Antonietta Visceglia, "Introduction," in Signorotto and Visceglia, *Court and Politics in Papal Rome,* 1–7.

33. For the importance of the Capitoline government and the urban patriciate classes, see esp. Eleonora Canepari, *La construction du pouvoir local: Élites municipales, liens sociaux et transactions économiques dans l'espace urbain, Rome, 1550–1650* (Rome: École française de Rome, 2017); and Nussdorfer, *Civic Politics.* See also Bevilacqua, *Il monte dei Cenci.* For a summary, see A. D. Wright, *The Early Modern Papacy: From the Council of Trent to the French Revolution, 1564–1789* (Edinburgh Gate, Harlow, Essex: Pearson Education Limited, 2000), esp. 2–14, 231–270, and 272–280. See also Peter Blastenbrei, *Kriminalität in Rom, 1560–1585* (Tübingen: Max Niemeyer, 1995), 8–50, and Irene Fosi, *La giustizia del papa: Sudditi e tribunali nello Stato Pontificio in età moderna* (Rome: Laterza, 2007), translated by Thomas V. Cohen as *Papal Justice: Subjects and Courts in the Papal State, 1500–1750* (Washington, DC: Catholic University of America Press, 2011), 23–46.

34. Stefano Andretta, "Le istituzioni e l'esercizio del potere," in Pinelli, *Roma del Rinascimento,* 93–121; Niccolò Del Re, *La Curia Romana: Lineamenti storico-giuridici,* 4th ed. (Vatican City: Libreria Editrice Vaticana, 1998); *Monsignor Governatore di Roma* (Rome: Istituto

di Studi Romani Editore, 1972); Pietro Palazzini, "Le congregazioni romane da Sisto V a Giovanni Paolo II," in Fagiolo and Madonna, *Sisto V*, 1:19–38; Peter Partner, *The Pope's Men: The Papal Civil Service in the Renaissance* (Oxford: Clarendon Press, 1990); Maria Grazia Pastura Ruggiero, *La Reverenda Camera Apostolica e i suoi archivi (secoli XV–XVIII)* (Rome: Archivio di Stato, Scuola di Archivistica Paleografia e Diplomatica, 1984); Pecchiai, *Roma nel Cinquecento*, 169–207; Marco Pellegrini, "A Turning-Point in the History of the Factional System in the Sacred College: The Power of Pope and Cardinals in the Age of Alexander VI," in Signorotto and Visceglia, *Court and Politics in Papal Rome*, 8–30; and Maria Antonietta Visceglia, "Politica internazionale, fazioni e partiti nella Curia Romana del tardo Cinquecento," *Rivista Storica Italiana* 127 (2015): 721–769.

35. For the Congregation on Streets, Bridges, and Fountains, see Carmen Genovese and Daniela Sinisi, Pro ornatu et publica utilitate: *L'attività della Congregazione cardinalizia super viis, pontibus et fontibus nella Roma di fine '500* (Rome: Gangemi Editore, 2010); and ASR, Congregazione super viis, pontibus et fontibus, Registro 1. For Sixtus's bull, *Immensa aeterni Dei*, see Luigi Tomassetti et al., *Bullarum diplomatum et privilegiorum Sanctorum Romanorum Pontificum Taurinensis editio* (Turin: Seb. Franco and Henrico Dalmazzo, 1857–1872), vol. 9, no. CVII, 985–999.

36. See Andretta, "Le istituzioni e l'esercizio del potere"; Canepari, *La construction du pouvoir local*; Niccolò Del Re, *La Curia Capitolina e tre altri antichi organi giudiziari romani* (Rome: Fondazione Marco Besso, 1993); Nussdorfer, *Civic Politics*; Pecchiai, *Roma nel Cinquecento*, 246–248 for the *caporioni*; and Emmanuel Rodocanachi, *Les institutions communales de Rome sous la papauté* (Paris: Picard, 1901).

37. Fosi, *La giustizia del papa/Papal Justice*. See also Blastenbrei, *Kriminalität in Rom*, 8–50; Nussdorfer, *Civic Politics*, 17–114; and Pecchiai, *Roma nel Cinquecento*, 169–207 for the papal curia and 233–236 for the Capitoline government.

38. The complex relationships of civic officials to the "popular classes" of the *rioni* have been thoroughly explicated by Canepari, *La construction du pouvoir local*. See also Nussdorfer, *Civic Politics*, esp. 60–94, and Nussdorfer, "Il 'popolo romano' e i papi: La vita politica della capitale religiosa," in Fiorani and Prosperi, *Roma, la città del papa*, 239–260. For the composition of the secret and public councils respectively, see Pecchiai, *Roma nel Cinquecento*, 234–238. For the Borgo, see esp. Del Re, *La Curia Capitolina*, 131–155.

39. Blastenbrei, *Kriminalität in Rom*, 51–71; and see Francesco Cerasoli, "Censimento della popolazione di Roma dall'anno 1600 al 1739," *Studi e Documenti di Storia e Diritto* 12 (1891): 169–199; Delumeau, *Vie économique et sociale*, 1:135–220; and Eugenio Sonnino, "The Population in Baroque Rome," in *Rome / Amsterdam: Two Growing Cities in Seventeenth-Century Europe*, ed. Peter van Kessel and Elisja Schulte (Amsterdam: Amsterdam University Press, 1997), 50–70; Sonnino, ed., *Popolazione e società a Roma dal Medioevo all'età contemporanea* (Rome: Editrice "Il Calamo," 1998); and Sonnino, "Le anime dei romani: Fonti religiose e demografia storica," in Fiorani and Prosperi, *Roma, la città del papa*, 327–364. For the Jubilee of 1575, see esp. Barbara Wisch, "The Roman Church Triumphant: Pilgrimage, Penance and Processions Celebrating the Holy Year of 1575," in *Art and Pageantry in the Renaissance and Baroque*, ed. Barbara Wisch and Susan Scott Munshower (University Park: Pennsylvania State University, 1990), 82–117.

40. See esp. Martine Boiteux, "La noblesse romaine sur la scène: Contribution à l'étude de l'évergétisme romain à l'époque moderne," in *Le nobilità delle città capitali*, ed. Martine Boiteux, Catherine Brice, and Carlo M. Travaglini (Rome: Università degli Studi Roma Tre,

CROMA, 2009), 131–180; Canepari, *La construction du pouvoir local,* which focuses on the civic aristocracy and the local basis of its power in the *rioni;* Anna Esposito, "La città e i suoi abitanti," in Pinelli, *Roma del Rinascimento,* 3–47, esp. 13–15 for civic and baronial Roman nobles; Richard J. Ferraro, "Nobility of Rome, 1500–1700: A Study of Its Composition, Wealth, and Investment" (PhD diss., University of Wisconsin–Madison, 1994); Nussdorfer, *Civic Politics,* 32–44; and Maria Antonietta Visceglia, ed., *La nobiltà romana in età moderna: Profili istituzionali e pratiche sociali* (Rome: Carocci, 2001). For the violence and difficulties caused by the feudal nobility, see esp. Irene Fosi, *La società violenta: Il banditismo nello Stato Pontificio nella seconda metà del Cinquecento* (Rome: Edizioni dell'Ateneo, 1985); and Fosi, *La giustizia del papa/Papal Justice,* 77–104.

41. For confraternities, see esp. Claudio Crescentini and Antonio Martini, eds. *Le confraternite romane: Arte storia committenza* (Rome: Fondazione Marco Besso, 2000); Luigi Fiorani, "'Charità et pietate': Confraternite e gruppi devoti nella città rinascimentale e barocca," in Fiorani and Prosperi, *Roma, la città del papa,* 429–476; Lance Gabriel Lazar, *Working in the Vineyard of the Lord: Jesuit Confraternities in Early Modern Italy* (Toronto: University of Toronto Press, 2005); and Vera Vita Spagnuolo, "Fonti per lo studio dell'assistenza e della beneficenza a Roma durante il secolo XVI: Gli Archivi di alcune confraternite," in Fagiolo and Madonna, *Sisto V,* 1:243–260. For the Oratorians and their building project, see Joseph Connors, *Borromini and the Roman Oratory: Style and Society* (New York: Architectural History Foundation; Cambridge, MA: MIT Press, 1980). See also Dos Reis Souza Lima, "Repertorio delle fabbriche." For a study of the houses and urban holdings of one confraternity, see Deborah Nelson Wilde, "Housing and Urban Development in Sixteenth Century Rome: The Properties of the Arciconfraternita della SS.ma Annunziata" (PhD diss., New York University, 1989). On the building projects of the new religious orders, see esp. Clare Robertson, *Rome 1600: The City and the Visual Arts under Clement VIII* (New Haven, CT: Yale University Press, 2015), 182–279;. And for the early Jesuits in Rome, see esp. John W. O'Malley, *The First Jesuits* (Cambridge, MA: Harvard University Press, 1993).

42. D. S. Chambers, *Renaissance Cardinals and Their Worldly Problems* (Aldershot, Hampshire: Variorum, Ashgate, 1997); Both John F. D'Amico, *Renaissance Humanism in Papal Rome: Humanists and Churchmen on the Eve of the Reformation* (Baltimore: Johns Hopkins University Press, 1983), esp. 43–56, and Partner, *The Pope's Men,* provide essential background for the papal curia and cardinalate families in a slightly earlier period. On the cardinals' courts, see also Gigliola Fragnito, "Le corti cardinalizie nella Roma del Cinquecento," *Rivista Storica Italiana* 106 (1994): 5–41; Elena Fumagalli, "La committenza cardinalizia a Roma," in Conforti and Tuttle, *Storia,* 4:94–107; Elena Fasano Guarini, "'Rome, Workshop of All the Practices of the World': From the Letters of Cardinal Ferdinando de' Medici to Cosimo I and Francesco I," in Signorotto and Visceglia, *Court and Politics in Papal Rome,* 53–77; Barbara McClung Hallman, *Italian Cardinals, Reform, and the Church as Property, 1492–1563* (Berkeley: University of California Press, 1985); Wolfgang Reinhard, "Le carriere papali e cardinalizie: Contributo alla storia sociale del papato," in Fiorani and Prosperi, *Roma, la città del papa,* 261–290; Carol M. Richardson, *Reclaiming Rome: Cardinals in the Fifteenth Century* (Leiden: Brill, 2009), esp. 143–313; Clare Robertson, *"Il Gran Cardinale": Alessandro Farnese, Patron of the Arts* (New Haven, CT: Yale University Press, 1992); Patricia Waddy, *Seventeenth-Century Roman Palaces: Use and the Art of the Plan* (New York: Architectural History Foundation; Cambridge, MA: MIT Press, 1990), and "Barberini Cardinals Need Places to Live," in *I Barberini e la cultura europea del Seicento,* ed. Lorenza Mochi Onori, Sebastian

Schütze, and Francesco Solinas, Atti del convegno internazionale Palazzo Barberini alle Quattro Fontane, December 7–11, 2004 (Rome: De Luca, 2007), 487–500.

43. On Roman society see Delumeau, *Vie économique et sociale,* 1:135–220. See also P. Renée Baernstein, "Regional Intermarriage among the Italian Nobility in the Sixteenth Century" in *Marriage in Premodern Europe: Italy and Beyond,* ed. Jacqueline Murray (Toronto: Centre for Reformation and Renaissance Studies, 2012), 201–219; and Partner, *The Pope's Men,* esp. 169–182, both of which address Roman connections with Italian regions outside of Rome. For the significant influence of the Spanish, see Thomas J. Dandelet, *Spanish Rome, 1500–1700* (New Haven, CT: Yale University Press, 2001). For the palace-building activities of elites, see Anna Bedon, "L'orgoglio e la borsa: I palazzi romani della seconda metà del XVI secolo nelle strategie finanziarie delle famiglie municipali," in Curcio, Navone, and Villari, *Studi su Domenico Fontana,* 143–159; Canepari, *La construction du pouvoir local,* esp. 107–152; Joseph Connors, "Alliance and Enmity in Roman Baroque Urbanism," *Römisches Jahrbuch der Bibliotheca Hertziana* 25 (1989): 207–294, translated by Marco Cupellaro as *Alleanze e inimicizie: L'urbanistica di Roma barocca* (Rome: Laterza, 2005); Robertson, *Rome 1600,* esp. 122–181. And see Kimberly L. Dennis, "Camilla Peretti, Sixtus V, and the Construction of Peretti Family Identity in Counter-Reformation Rome," *Sixteenth Century Journal* 43 (2012): 71–101; "Rediscovering the Villa Montalto and the Patronage of Camilla Peretti," in *Wives, Widows, Mistresses, and Nuns in Early Modern Italy: Making the Invisible Visible through Art and Patronage,* ed. Katherine A. McIver (Farnham, Surrey: Ashgate, 2012), 55–73; Carolyn Valone, "Women on the Quirinal Hill: Patronage in Rome, 1560–1630," *Art Bulletin* 76 (March 1994): 129–146; and "Matrons and Motives: Why Women Built in Early Modern Rome," in *Beyond Isabella: Secular Women Patrons of Art in Renaissance Italy,* ed. Sheryl E. Reiss and David G. Wilkins (Kirksville, MO: Truman State University Press, 2001), 317–335.

44. For earlier scholarship, see Antonino Bertolotti, *Artisti bolognesi, ferraresi ed alcuni altri del già Stato Pontificio in Roma nei secoli XV, XVI e XVII* (1885; repr., Bologna: Arnaldo Forni Editore, 1962); *Artisti veneti in Roma nei secoli XV, XVI e XVII* (1884; repr., Bologna: Arnaldo Forni, 1965); *Artisti subalpini in Roma nei secoli XV, XVI e XVII* (1884; Reprint: Bologna: Arnaldo Forni, 1965); *Artisti lombardi a Roma nei secoli XV, XVI, XVII,* 2 vols. (1881; repr., Bologna: Arnaldo Forni, 1969–1970); *Artisti francesi in Roma nei secoli XV, XVI e XVII* (1886; repr., Bologna: Arnaldo Forni, 1975); and Emmanuel Rodocanachi, *Les corporations ouvrières à Rome depuis la chute de l'empire romain,* 2 vols. (Paris: Picard, 1894). See also Ago, *Economia barocca,* esp. 3–79; Eleonora Canepari, "Mestiere e spazio urbano nella costruzione dei legami sociali degli immigrati a Roma in età moderna," in *L'Italia delle migrazioni interne: Donne, uomini, mobilità in età moderna e contemporanea,* ed. Angiolina Arru and Franco Ramella (Rome: Donzelli, 2003), 33–76; Anna Maria Corbo, *Artisti e artigiani in Roma al tempo di Martino V e Eugenio IV* (Rome: De Luca Editore, 1969); Antonio Martini, *Arti, mestieri e fede nella Roma dei papi* (Bologna: Casa Editrice Licinio Cappelli, 1965); Anna Modigliani, "Artigiani e botteghe nella città," in Chiabò et al., *Alle origini della nuova Roma,* 455–477; and for useful summaries, see Esposito, "La città e i suoi abitanti," 23–24, and Luciano Palermo, "L'economia," in Pinelli, *Roma del Rinascimento* 49–91, esp. 56–60. For carpenters, see Anderson, "Master Carpenters in Renaissance and Baroque Rome."

45. The large literature on antiquarianism includes Leonard Barkan, *Unearthing the Past: Archaeology and Aesthetics in the Making of Renaissance Culture* (New Haven, CT: Yale University Press, 1999); Brian Curran, *The Egyptian Renaissance: The Afterlife of Ancient Egypt in Early Modern Italy* (Chicago: University of Chicago Press, 2007); Claudio Franzoni,

"'Urbe Roma in pristinam formam renascente': Le antichità di Roma durante il Rinascimento," in Pinelli, *Roma del Rinascimento,* 291–336; Philip J. Jacks, *The Antiquarian and the Myth of Antiquity: The Origins of Rome in Renaissance Thought* (Cambridge: Cambridge University Press, 1993); Peter N. Miller, ed., *Momigliano and Antiquarianism: Foundations of the Modern Cultural Sciences* (Toronto: University of Toronto Press, 2014); Arnaldo Momigliano, *The Classical Foundations of Modern Historiography* (Berkeley: University of California Press, 1990); Ingrid D. Rowland, *The Culture of the High Renaissance: Ancients and Moderns in Sixteenth-Century Rome* (Cambridge: Cambridge University Press, 1998); Alain Schnapp, *The Discovery of the Past,* trans. Ian Kinnes and Gillian Varndell (New York: Harry N. Abrams, 1997); William Stenhouse, *Reading Inscriptions and Writing Ancient History: Historical Scholarship in the Late Renaissance,* Bulletin of the Institute of Classical Studies, supp. 86 (London: Institute of Classical Studies, School of Advanced Study, University of London, 2005); and Roberto Weiss, *The Renaissance Discovery of Classical Antiquity,* 2nd ed. (Oxford: Basil Blackwell, 1988).

46. See esp. Dominique Julia, "L'accoglienza dei pellegrini a Roma," in Fiorani and Prosperi, *Roma, la città del papa,"* 823–861; Stefania Nanni and Maria Antonietta Visceglia, eds., "La città del perdono: Pellegrinaggi e anni santi a Roma in età moderna, 1550–1750," special issue, *Roma Moderna e Contemporanea* 5 (May–December 1997); and Mario Romani, *Pellegrini e viaggiatori nell'economia di Roma dal XIV al XVII secolo* (Milan: Società Editrice 'Vita e Pensiero,' 1948).

47. See esp. Elizabeth S. Cohen, "'Courtesans' and 'Whores': Words and Behavior in Roman Streets," in "Women in the Renaissance: An Interdisciplinary Forum (MLA 1989)," ed. Ann Rosalind Jones and Betty S. Travitsky, special issue, *Women's Studies* 19 (1991): 201–208; "Seen and Known: Prostitutes in the Cityscape of Late-Sixteenth-Century Rome," *Renaissance Studies* 12 (September 1998): 392–409; Esposito, "La città e i suoi abitanti," 25–26; Laurie Nussdorfer, "Men at Home in Baroque Rome," *I Tatti Studies in the Italian Renaissance* 17 (2014): 103–129; and Tessa Storey, *Carnal Commerce in Counter-Reformation Rome* (Cambridge: Cambridge University Press, 2008). For the ratio of men to women, see esp. Sonnino, "Population in Baroque Rome," esp. 61–63, based on the records of Roman parish registers.

48. See esp. Anna Foa and Kenneth Stow, "Gli ebrei di Roma: Potere rituale e società in età moderna," in Fiorani and Prosperi, *Roma, la città del papa,"* 555–581; Stow, *Catholic Thought and Papal Jewry Policy*; and Serena Di Nepi, *Sopravvivere al ghetto: Per una storia sociale della comunità ebraica nella Roma del Cinquecento* (Rome: Viella, 2013).

49. Canepari, *La construction du pouvoir local,* esp. 67–106.

50. See esp. John Marciari, "Artistic Practice in Late Cinquecento Rome and Girolamo Muziano's Accademia di San Luca"; Laurie Nussdorfer, "Notaries and the Accademia di San Luca, 1590–1630"; and Elizabeth S. Cohen, "The Early Accademia di San Luca and Artists in Rome: A Historian's Observations," all in Lukehart, *Accademia Seminars,* 196–223, esp. 200–203; 54–67, esp. 55–58; and 324–345, esp. 325–328, respectively; and Center for the Advanced Study in the Visual Arts, National Gallery of Art, "History of the Accademia di San Luca, c. 1590–1635: Documents from the Archivio di Stato di Roma," https://www.nga.gov/accademia/en/intro.html. See also Ago, *Economia barocca,* 20–32 and 141–143; Patrizia Cavazzini, *Painting as a Business in Early Seventeenth-Century Rome* (University Park: Pennsylvania State University Press, 2008), esp. 13–48 ("Artists and Craftsmen") and 49–80 ("Training"); and see Long, "Multi-tasking 'Pre-professional' Architect/Engineers."

CHAPTER 1

1. BAV, Urb. lat. 1038, fol. 265r–v (September 18, 1557), "volendo far allegrezza la notte insuberbito si sparse per tutta la citta, crescendo sempre tutto il giorno, et sin alla meza notte sequente, scorrendo per tutte le piazze et vie"; "raffreno il furore"; "piena di fango, et tutta lorda." For the problem of filth in Rome after flooding, see esp. Laura Megna, "'Acque et immonditie del fiume': Inondazioni del Tevere e smaltimento dei rifiuti a Roma tra Cinque e Settecento," *Mélanges de l'École française de Rome, Italie et Méditerranée* 118 (2006): 21–34. For the *avvisi* and their reliability as sources, see Delumeau, *Vie économique et sociale*, 1:25–36; Mario Infelise, "Roman *avvisi*: Information and Politics in the Seventeenth Century," in Signorotto and Visceglia, *Court and Politics in Papal Rome*, 212–228; and Infelise, *Prima dei giornali*.

2. The description of the flood is found within a printed *avviso* by Oldradi on the peace between Pope Paul IV and Philip, the king of Spain—[Angelo] Oldradi, *Aviso della Pace* [. . .] *del diluvio* [. . .] (Rome: Antonio Blado, 1557), fols. 270r–272r, "pose tutto in iscompiglio e in paura" (270v); "insieme con quella bella capelletta di Giulio Terzo che vi era nel meo con tanta arte, e spesa fabricate" (271r); "un miserabile spettacolo, e ogni cosa confusa, mal condotta, e piena di compassione" (271v). The copy that I used of this rare work is in the Biblioteca Casanatense in Rome (Vol. Misc. 1927/18); it is bound with other materials, and I have used the modern folio numbers of this copy. A description of the flood with virtually the same wording along with some suggestions for flood prevention can be found in a manuscript in Rome, BA, Anonimo Manoscritto 153, fols. 30r–31v, "Relatione: dell'inundatione del Tevere in Roma [. . .] nel 1557," and seems to me likely to be Oldradi's original description in his own hand. For Oldradi, see Dennis E. Rhodes, "Ortensia and Hortolana: With Notes on Angelo degli Oldradi," *Gutenberg-Jahrbuch* 71 (1996): 97–98.

3. Oldradi, *Aviso della Pace* [. . .] *del diluvio*, fol. 271r; BAV, Urb. lat. 1038, fol. 275v (October 9, 1557), concerning bread. For mills and other productive activities on the Tiber, see esp. Ivana Ait, "Uno spazio produttivo: Il Tevere nel basso Medioevo," *Rivista Storica del Lazio* 10 (2002): 3–15; "Il Tevere e le attività produttive a Roma nel basso Medioevo," and Paolo Buonora and Manuel Vaquero Piñeiro, "Il sistema idraulico di Roma in età moderna: Assetti di potere e dinamiche produttive," both in Travaglini, *La città e il fiume*, 81–94 and 147–168, respectively; Irene Bevilacqua, "Acque e mulini nella Roma del Seicento," *Città e Storia* 5 (2010): 99–140; Cesare D'Onofrio, *Il Tevere: L'isola tiberina, le inondazioni, i molini, i porti, le rive, i muraglioni, i ponti di Roma*, 4th ed. (Rome: Romana Società Editrice, 1982), 33–64; Maria Margarita Segarra Lagunes, "Le attività produttive del Tevere nelle dinamiche di trasformazione urbana: I mulini fluviali," *Mélanges de l'École française de Rome, Italie et Méditerranée* 118 (2006): 45–52; and Segarra Lagunes, *Il Tevere e Roma*, 249–303. A survey of the flood markers of Rome is found in Rome, BA, MS 153, fols. 50r–51r, which is transcribed and discussed by Enrico Celani, "Alcune iscrizioni sulle inondazioni del Tevere," *Bollettino della Commissione Archeologica Communale* 23 (1895): 283–295. And see Di Martino, Di Martino, and Belati, *Huc Tiber ascendit*, 167–252, and 66–71, for the flood of 1557.

4. For ancient flooding of the Tiber, see Gregory S. Aldrete, *Floods of the Tiber in Ancient Rome* (Baltimore: Johns Hopkins University Press, 2007); and Joël Le Gall, *Le Tibre, fleuve de Rome dans l'antiquité* (Paris: Presses Universitaires de France, 1953). For later periods, see esp. Bevilacqua, "Acque e mulini," 99–140; Silvia Enzi, "Le inondazioni del Tevere a Roma tra il XVI e XVIII secolo nelle fonti bibliotecarie del tempo," *Mélanges de l'École française de Rome,*

Italie et Méditerranée 118 (2006): 13–20; Anna Esposito, "I 'diluvi' del Tevere tra '400 e '500," *Rivista Storica del Lazio* 10 (2002): 17–26; "Le inondazioni del Tevere tra tardo Medioevo e prima età moderna: Leggende, racconti, testimonianze," *Mélanges de l'École française de Rome, Italie et Méditerranée* 118 (2006): 7–12; and "Il Tevere e Roma," in *Le calamità ambientali nel tardo Medioevo europeo: Realtà, percezioni, reazioni,* ed. Michael Matheus, Gabriella Piccinni, Giuliano Pinto, and Gian Maria Varanini, Atti del XII convegno del Centro di Studi sulla civiltà del tardo Medioevo, S. Miniato, May 21–June 2, 2008 (Florence: Firenze University Press, 2010), 257–275; Frosini, *Il Tevere,* 133–233; and Renato Funiciello et al. eds., *La geologia di Roma: Il centro storico,* 2 vols. (Rome: Istituto Poligrafico e Zecca dello Stato, 1995), 119–172. See also Vittorio Di Martino, Roswitha Di Martino, and Massimo Belati, Huc Tiber ascendit: *Le memorie delle inondazioni del Tevere a Roma* (Rome: Arbor Sapientiae, 2017); D'Onofrio, *Il Tevere,* 302–336; and Segarra Lagunes, *Il Tevere e Roma,* 69–133.

5. Aldrete, *Floods of the Tiber,* 54–61; Brian Campbell, *Rivers and the Power of Ancient Rome* (Chapel Hill: University of North Carolina, 2012), 309–320; Frosini, *Il Tevere,* 15–129; Segarra Lagunes, *Il Tevere e Roma,* esp. 11–67; and Strother A. Smith, *The Tiber and Its Tributaries: Their Natural History and Classical Associations* (London: Longmans, Green, 1877), esp. 7–23.

6. For the Ponte Santa Maria, D'Onofrio, *Il Tevere,* 141–165.

7. For the *traghetti,* see esp. Giuseppe Bonaccorso, "Il 'ponte dinamico': I traghetti e l'attraversamento del Tevere a Roma nei secoli XVI–XVIII," in Calabi and Conforti, *I ponti delle capitali d'Europa,* 89–103.

8. See esp. Giorgia Maria Annoscia, *Fonti e strutture per la conoscenza del sistema idrico di Roma nel Medioevo* (Rome: ARACNE, 2007); Buonora and Vaquero Piñeiro, "Il sistema idraulico," esp. 150–154; Emiliano Bultrini, "L'Acqua Crabra: Un fiume scomparso vicende del confine naturale tra Roma e la civitas Tusculana," *ASRSP* 135 (2012): 63–83; Harry B. Evans, *Aqueduct Hunting in the Seventeenth Century: Raffaello Fabretti's* De aquis et aquaeductibus veteris Romae (Ann Arbor: University of Michigan Press, 2002) esp. 72 (for the Marrana) and 246–248 (for the Crabra); Rodolfo Lanciani, *I comentarii di Frontino intorno le acque e gli aquedotti* (Rome: Salviucci, 1880), 3–120; Roberto Luciani, ed., *Roma sotterranea: Porta San Sebastiano, 15 ottobre 1984–14 gennaio 1985* (Rome: Palombi, 1984); Rossella Motta, "La decadenza degli acquedotti antichi e la conduzione dell'Acqua Mariana," in *Il trionfo dell'acqua,* 203–205; Simonetta Pascucci, "L'acqua corrente nel Cinquecento a Roma: Fontane, fontanelle e abbeveratoi," in "Le acque e la città (XV–XVI secolo)," ed. Giuseppe Bonaccorso, special issue, *Roma Moderna e Contemporanea* 17 (January–December 2009): 43–72; and Katherine Wentworth Rinne, *The Waters of Rome: Aqueducts, Fountains, and the Birth of the Baroque City* (New Haven, CT: Yale University Press, 2010), esp. 11–37.

9. BAV, Urb. lat. 1040, fol. 412r. (l'ultimo Maggio [1567]), "li acquaroli, che la condurono, la vendono benissimo." For the *acquaroli,* see D'Onofrio, *Il Tevere,* 77–78; D'Onofrio, *Le fontane di Roma,* 3rd ed. (Rome: Romana Società Editrice, 1986), 144 and (for illustrations) 117–119; Martini, *Arti mestieri,* 212–213; and Rinne, *Waters of Rome,* 34–37.

10. See esp. Renato Sansa, "L'odore del contagio: Ambiente urbano e prevenzione delle epidemie nella prima età moderna," *Medicina e Storia* 2 (2002): 83–108.

11. See esp. Giuseppe Bonaccorso, "Roma e le sue acque potabili nel Cinquecento: La competizione con il Tevere," in Bonaccorso, "Le acque e la città," 73–90; and D'Onofrio, *Le fontane di Roma,* 38–39. For Petroni's writings defending Tiber water, see Alessandro Traiano Petroni, *De Aqua Tiberina* (Rome: Valerius and Aloisius Doricus, 1552), esp. 75–76 (concern-

ing where the water is polluted and where pure), and *De victu romanorum et de sanitate tuenda libri quinque* (Rome: In Aedibus Populi Romani, 1581), esp. 35–39. For Modio, see Gennaro Cassiani, "Modio, Giovanni Battista," *DBI*, 75:238–241; "Patrigno Tevere: Le obiezioni sperimentali di Giovanni Battista Modio al 'dogma' della potabilità dell'acqua del Tevere a metà Cinquecento," *Roma nel Rinascimento* (2014): 357–372; and for Modio's treatise, Giovanni Battista Modio, *Il Tevere* [. . .]. (Rome: Vincenzo Luchino, 1556).

12. Ligorio is the focus of a growing body of scholarship, including a national edition of his works (*Edizione Nazionale delle Opere di Pirro Ligorio*). For a list of his archaeological manuscripts, see Erna Mandowsky and Charles Mitchell, eds., *Pirro Ligorio's Roman Antiquities: The Drawings in MS XIII, B.7 in the National Library in Naples* (London: Warburg Institute, University of London, 1963), 130–140. Most of Pirro's manuscript writings are located in Naples, BNN, Mss. XIII.B.1–10 (10 vols.), and Turin, AST Ms. J.a.III.3.1–Ms. J.a.II.5, an alphabetical dictionary of antiquities (18 vols.), and Ms. J.a.II.6. (XIX)–Ms. J.a.II.17 (XXX), twelve volumes on topics such as coins, medals, and the earthquakes. Fourteen of the eighteen volumes of the alphabetical encyclopedia in Turin were copied for Queen Christina of Sweden in the seventeenth century and are now in the Vatican Library, BAV Ottob. lat. 3364–Ottob. lat. 3377. See also David R. Coffin, *Pirro Ligorio: The Renaissance Artist, Architect, and Antiquarian* (University Park: Pennsylvania State University, 2004), 145–146 for house facades; and see Carmelo Occhipinti, *Pirro Ligorio e la storia cristiana di Roma da Costantino all'Umanesimo* (Pisa: Scuola Normale Superiore, 2007).

13. For the d'Este gardens at Tivoli, see David R. Coffin, *Gardens and Gardening in Papal Rome* (Princeton, NJ: Princeton University Press, 1991), 41–46 and 85–91; *Pirro Ligorio*, 83–105; Claudia Lazzaro, *The Italian Renaissance Garden: From the Conventions of Planting, Design, and Ornament to the Grand Gardens of Sixteenth-Century Central Italy* (New Haven, CT: Yale University Press, 1990), 215–242; Maria Luisa Madonna, "Il Genius Loci di Villa d'Este: Miti e misteri nel sistema di Pirro Ligorio" in Fagiolo, *Natura e artificio*, 190–226; and Denis Ribouillault, *Rome en ses jardins: Paysage et pouvoir au XVI*[e] *siècle* (Paris: CTHS / INHA, 2013), esp. 93–111.

14. For the manuscript encyclopedia on waters, Naples, BNN, Ms. XIII.B.9, and for an erudite edition, Pirro Ligorio, *Libro dei fiumi, dei fonti e dei laghi antichi*, ed. Robert W. Gaston (Rome: De Luca Editori d'Arte, 2015), xvi–xvii, for the controversy concerning drinking Tiber water; and see Marcello Fagiolo, "Il significato dell'acqua e la dialettica del giardino. Pirro Ligorio e la 'filosofia' della villa cinquecentesca," in Fagiolo, *Natura e artificio*, 176–189.

15. C. Paola Scavizzi, "Fonti per uno studio sulla regolazione del Tevere dal Cinquecento al Settecento fra teoria e pratica," *ASRSP* 102 (1979): 237–313, a thorough survey of manuscript and written sources. Scavizzi also emphasizes that the men involved "were of diverse provenance and background," and that there was "intense cultural exchange not only on the theoretical level, but as the transmission and direct acquisition of experience" (243).

16. BA, Anonimo Ms. 153, fols. 2r–29v (Gómez's treatise), 30r–31v (description of 1557 flood), 32r–37r and 48r–50r (flood markers and inscriptions), 52r–v (on a particular canal project from Ponte Milvio to San Paolo), and 57r–69r. See also Celani, "Alcune iscrizioni," 286–295. Gómez's work is Ludovico Gómez, *De prodigiosis Tyberis inundationibus* [. . .] *ad annum M. D. XXXI* [. . .] (Rome: Francesco Minizio Calvo, 1531). The printed copy of Gómez's tract with annotations in the same hand as Ms. 153 is BA, GG. 11.22.

17. Pope Paul IV's creation of the commission is mentioned in the minutes of the meetings of the Capitoline Council, ASC, C.C., cred. I, tom. 37, cat. 37, fols. 14v ("alcuni gentil-

uomini") and 15v. The council meeting occurred on the VIII idus martij 1559. Among the *cives* (citizens) present was Luca Peto, a councillor who played an important role in the Acqua Vergine reconstruction.

18. For the complex ritual destruction of papal statues during the Vacant See, the statue of Paul IV being the first to suffer this fate, see esp. Monika Butzek, *Die kommunalen Repräsentationsstatuen der Päpste des 16. Jahrhunderts in Bologna, Perugia und Rom* (Bad Honnef: Bock und Herchen, 1978), 253–280 (278 for the modern rediscovery of the statue); John M. Hunt, *The Vacant See in Early Modern Rome: A Social History of the Papal Interregnum* (Leiden: Brill, 2016), 178–193; and Pastor, *History of the Popes*, 14:414–416.

19. See esp. Mario Crespi, "Bacci, Andrea," *DBI*, 5:29–30; D'Onofrio, *Il Tevere*, esp. 311–312 and 318–320; and Pamela O. Long, "Hydraulic Engineering and the Study of Antiquity: Rome, 1557–70," *Renaissance Quarterly* 61 (Winter 2008): 1103–1109.

20. Andrea Bacci, *Del Tevere: Della natura et bonta dell'acque & delle inondationi Libri II* (Rome: Vincenzo Luchino, 1558), sig. Aiijv, "L'acqua del Tevere per bontà ha poche pari al mondo."

21. Bacci, *Del Tevere* (1558), esp. fols. 1v–2r, 53r–58r, and 63r–82r, "ò per pioggia ò per natura" (53r).

22. Bacci, *Del Tevere* (1558), fols. 53r–58r. For Aristotle's view of the production of water from air inside the earth, from which Bacci dissents, see Aristotle, *Meteorology*, 1.13.20–27 [349b20–27]. For the broader context of Aritstotle's *Meteorology* in the late Renaissance, see Craig Martin, *Renaissance Meteorology: Pomponazzi to Descartes* (Baltimore: Johns Hopkins University Press, 2011).

23. Bacci, *Del Tevere* (1558), fols. 86r–89r, "non tanto mortifera, quanto horrenda" (87r); "sommersa in un pantano" (89r).

24. Ibid., fols. 96v–100v. For the ancient curators of the river, see Le Gall, *Le Tibre*, 135–183.

25. ASC, C.C., cred. I, tom. 37, cat. 37, fol. 179r (ad quartum idus Junij MD lxiiij). Also in ASC. C.C., cred. I, tom. 22, cat. 22, fol. 72r.

26. See Paolo Agostino Vetrugno, *Antonio Trevisi: Architetto pugliese del Rinascimento* (Fasano: Schena, 1985), which includes an essay concerning Trevisi's Roman activities by Vincenzo Cazzato, "Momenti romani di Antonio Trevisi," 131–140. Cazzato was unaware however of relevant documentation in the ASR.

27. For Carlo Borromeo and his brother Federico in Rome, see Michel De Certau, "Carlo Borromeo, santo," *DBI*, 20:260–269; and Roberto Zapperi, "Borromeo, Federico," *DBI*, 13: 31–33.

28. Antonio Trevisi, *Fondamento del edifitio* [...] *sopra la innondatione del fiume* [...] (Rome: Antonio Blado, 1560), "Allo illustriss. Et Eccell. Signore Conte Federico Bonromeo [*sic*]," 1–2, "Io indotto non posso una opera dotta dedicarvi"; "quella che dal mero fonte del mio rude e basso ingegno viene." The work is known in only two copies, one in the Biblioteca Nazionale Vittorio Emanuele, and a second in the BA, both in Rome. A facsimile edition (also very hard to come by) is in Vetrugno, *Antonio Trevisi*, 11–22 for a discussion of the work and 83–130 for the facsimile. And see Long, "Contribution of Architectural Writers," 292–295, for the Vitruvian dictum.

29. Trevisi, *Fondamento del edifitio*, 3–11r (the tract changes from pagination to foliation after page 6), "Questa machina che dimandano sphera del mondo"; "quello altissimo Architettore che la fabrico"; "al governo della natura universale."

30. Ibid., fols. 11r–13v.

31. Trevisi, *Fondamento del edifitio,* fols. 16r–26v: "molte & infinite persone affocate" (20r). For the election of Giordano Boccabella and Camillo Pignanello as masters of the streets, see ASC, C.C., cred. I, tom. 20, cat. 20, announced at a public council meeting held on quintum Kal. Januarii MD lvij (modern dating, December 28, 1556), fols. 154v–157v, on 157v.

32. Trevisi, *Fondamento del edifitio,* fols. 21r–25v.

33. BAV, Stampati St. Geogr. I. 620 Riserva, Leonardo Bufalini, *Roma* (1551; repr., Rome: Antonio Blado, 1560). For a facsimile reprint of the map with an extensive introduction, see Francesco Ehrle, *Roma al tempo di Giulio III: La pianta di Roma di Leonardo Bufalini del 1551 riprodotta dall'esemplare esistente nella Biblioteca Vaticana* (1911; repr., Vatican City: Biblioteca Apostolica Vaticana, 2011). For Bufalini, see Giovanni Beltrani, "Leonardo Bufalini e la sua pianta topografica di Roma," *Rivista Europea* 22 (1880): 5–28, 361–387; Long, "Multi-Tasking 'Pre-professional' Architect/Engineers," 227–230; Jessica Maier, *Rome Measured and Imagined: Early Modern Maps of the Eternal City* (Chicago: University of Chicago Press, 2015), 77–118; Cosimo Palagiano, "Bufalini, Leonardo," *DBI,* 14:798–799; and Vetrugno, *Antonio Trevisi,* which includes a reproduction of the map and transcriptions of the letters found on the lower border. For Blado, see Francesco Barberi, "Blado, Antonio," *DBI,* 10:753–757.

34. The three extant copies of the map are BAV, Stampati St. Geogr. I. 620 Riserva; BAV, Barb. lat. 4432 (incomplete copy); and BL, Maps S. T. R. 175. For transcriptions of the letters, see Ehrle, *Roma al tempo di Giuilio III* on the map itself and 56–59 ("A li lettori" and "Alli virtuosi architetti"); and see Vetrugno, *Antonio Trevisi,* 45–56, a transcription of the four letters.

35. BAV, Stampati St. Geogr. I. 620 Riserva; and Vetrugno, *Antonio Trevisi,* 23, 45–46. Orazio Naro was a Roman patrician. For the Naro family and their residences, see esp. Sonia Amadio, "Famiglie in carriera ed artisti nella Roma barocca. 1. Naro," *Studi Romani* 45 (July–December 1997): 314–330. The Naro family papers are in ASV, "Rubiccellone della Famiglia Naro," Indice 726.

36. BAV, Stampati St. Geogr. I. 620 Riserva; Vetrugno, *Antonio Trevisi,* 47–48, "può far perfetto tutto l'imperfetto mio." For Borromeo as Pius IV's frequent representative in matters of architectural patronage, see Alexander, *From Renaissance to Counter-Reformation.*

37. BL, Maps S. T. R. 175; Vincenzo Cazzato, "Momenti romani di Antonio Trevisi," in Vetrugno, *Antonio Trevisi,* 131–150, esp. 136–137, and for a transcription of the letter, 49–50.

38. BL, Maps S. T. R. 175; Cazzato, "Momenti romani di Antonio Trevisi," in Vetrugno, *Antonio Trevisi,* 136–137, and for a transcription of the letter, 52–56.

39. The military engineer was Bartolomeo de Rocchi di Brianza, who was working on Roman fortifications in the 1550s. See Gustina Scaglia, "The 'Sepolcro Dorico' and Bartolommeo de Rocchi da Brianza's Drawing of it in the Aurelian Wall between Porta Flaminia and the Tiber River," *Arte Lombarda.* n.s., 96 (1991): 107–116. For the consistory, ASV, Arch. Concist., Acta Vicecanc. 9, fol. 90v. For the *avviso,* BAV, Urb. lat. 1039, fol. 283v (June 28, 1561), "per metter Borgo in Peninsula et per obviare alle inundationi." For the drawings, Florence, Galleria degli Uffizi, Gabinetto di Stampe e Disegni, 288A and 289A; and see esp. Federico Bellini, "La *Civitas Pia* e le fortificazioni vaticane di Pio IV," *Studi Romani* 61 (January–December 2013): 54–57; and Piero Spagnesi, *Castel Sant'Angelo: La Fortezza di Roma: Momenti della vicenda architettonica da Alessandro VI a Vittorio Emanuele III (1494–1911)* (Rome: Palombi, 1995), 38–47, who identifies the drawings as copies of those from Francesco Laparelli of Cortona working on the fortifications of the Vatican and Borgo in the early

1560s. For such canalization projects more generally, see Segarra Lagunes, *Il Tevere e Roma,* 117–129.

40. Austin, Texas: Harry Ransom Center, University of Texas at Austin, "Scrittura atenente a rimedi per l'inondazione del Tevere che sieguono in Roma fatta nel pontificato di Pio IV," Ranuzzi Family Collection Ph 2827.11, fol. 194r–195r, "era come un sacco à questa città"; "con tutti i valenti huomini esperti di Romà"; "con tutti quelli che capitono, che furno molti."

41. For Pius's fortification and hydraulic projects around Castel' Sant'Angelo, see esp. Bellini, "La *Civitas Pia*"; Henry A. Millon, Craig Hugh Smyth, and Francesca Consagra, "The Project for the Castel Sant'Angelo in the Dyson Perrins Codex," in Striker *Architectural Studies,* 111–117; Cesare D'Onofrio, *Castel Sant'Angelo e Borgo tra Roma e papato* (Rome: Romana Società Editrice, 1978), 289–297; Simoncini, *Roma: Le trasformazioni urbane nel Cinquecento,* 1:185–190, 204–206; and esp. Spagnesi, *Castel Sant'Angelo,* 29–37 for Antonio da Sangallo's plans, 38–55 for work under Pius IV, and 56–87 for work under Urban VIII in the seventeenth century.

42. Bellini, "La *Civitas Pia,*" esp. 65–73; Spagnesi, *Castel Sant'Angelo,* 38–55; and for Laparelli and his friendship with Serbelloni, Gerardo Doti, "Laparelli, Francesco," *DBI,* 63: 699–701; and Edoardo Mirri, ed., *Francesco Laparelli: Cortonese Architect at Malta* (Cortona: Tiphys Edizioni, 2009). For the account books for this large-scale fortification project, see ASR, *Archivio delle soldatesche e galere,* b. 17.

43. For a detailed account of the work of Urban VIII, see esp. Spagnesi, *Castel Sant'Angelo,* 56–87.

44. Harry Ransom Center, University of Texas at Austin, "Avvertimenti dell'Ill. Sig.r Gabrio Serbeloni sopra la fortificazione de i Borghi di Roma da osservarsi per conservazione di quella," Ranuzzi Family Collection Ph 12827.12, fol. 196r–v.

45. BAV, Urb. lat. 1040, fols. 354v–355r, "una piacevole inondatione"; "la gente impaurita cominciò a' levar le robbe de luoghi pericolosi, et retirarsi al sicuro"; "e tutta la notte correre le strade di gent[e] che al lume de torce conducevano robbe cavalli, et persone verso i monti"; "è opinione, che le fosse nove d'intorno al Castello fatte da Pio 4.to habbino à giovare assai contra l'impeto del fiume." (The text mistakenly says Clement VI for Clement VII.) See also ASR, *Congregazione super viis pontibus et fontibus,* Registro 1, fols. 10r–11v; and Genovese and Sinisi, *Pro ornatu et publica utilitate,* 12–13, in a meeting of November 29, 1567, where seven noble Romans (*nobili romani*) were present at a meeting of the Congregation of Streets, Bridges, and Fountains and were deputized to repair the damages from this flood. It is noted that proposals were read out loud and conserved in the acts of the notary.

46. The definitive study of Peto, which includes a transcription of his will, is Niccolò Del Re, "Luca Peto giureconsulto e magistrato capitolino (1512–1581)," in *Scritti in onore di Filippo Caraffa* (Anagni: Istituto di Storia e di Arte del Lazio Meridionale, Centro di Anagni, 1986), 309–337. For the law code, see Del Re, *La Curia Capitolina,* esp. 57–73; and Rodocanachi, *Les institutions communales,* esp. 286–308.

47. Del Re, "Luca Peto," 324–331, for a discussion of Peto's writings. For Peto's short tracts, see "Liber sextus varias lectiones continens," in Luca Peto, *De mensuris et ponderibus romanis et graecis* [...] *Libri quinque* (Venice: [P. Manutius], 1573), 73–76 (on the *Georgics*), 76–77, "De Anni 1569 qualitate" (On the qualities of the year 1569), 77–80, "De Anni 1570. qualitate deque erucis et convoluulis" (Concerning the quality of the year 1570, and concerning the worms and caterpillers).

48. Peto, *De mensuris et ponderibus*. For the reform of weights and measures in the Council, see ASC, C.C., cred. I, tom. 38, cat. 38, fols. 266v–268r (March 10, 1571, the secret council) and fols. 269v–270r (approval in the public council on March 14, 1571). The latter is also recorded in ASC, C.C., cred. I, tom. 25, cat. 25, fols. 39r–41r. and 43r (for approval in the public council).

49. Luca Peto, *Discorso* [. . .] *intorno alla cagione della eccessiva inondatione del Tevere* [. . .] (Rome: Giuseppe degl'Angeli, 1573). The copy I used in the Biblioteca Vaticana Apostolica was bound with other materials and had been given page numbers 185r.—187r. This very rare text is reproduced in D'Onofrio, *Il Tevere*, 338–339.

50. Andrea Bacci, *Del Tevere* [. . .] *libri tre* [. . .] (Venice: [Aldo Manuzio], 1576), "All'illustriss. Senato et inclito Popolo Romano."

51. Bacci, *Del Tevere* (1576), 106–108.

52. Bacci, *Del Tevere* (1576), 268–270, "sua Santità desiderosa della commune salute, e del bene particolarmente di questa alma Città, mandò un editto publico, ad eccitar gli animi ingegnosi a questa salutifera impressa, di rimediare alle inondationi del Tevere"; "io mi ritrovai in alcune consulte, che perciò si fecero innanzi alli Illustrissimi, e Reverendissimi Cardinali Sforza, e Montepolciano, e li Signori Mastri di strada, dove, tra molti nobili architettori, e ingegnieri, e altri huomini di valore in diverse professioni, io intesi una gran diversità di pareri intorno a questa inondatione." As an example of such a meeting held on a later date, November 29, 1567, at the palace of Cardinal Montepulciano with Montepulciano and Cardinal Sforza presiding, see ASR, *Congregazione super viis pontibus et fontibus*, 292, fols. 10r–11v, the agenda of which included the subject of flooding. It is reported that writings with very diverse opinions on the subject were passed around and discussed. The substantive discussions are not reported, nor are the conclusions. See also Genovese and Sinisi, *Pro ornatu et publica utilitate*, 12–13, for Congregation no. 8 on November 29, 1567.

53. Bacci, *Del Tevere* (1576), 269–270, "pochissima risolutione s'udì"; "lasciando le scienze da banda, ne ragionavano senza Methodo, e, a uso de gli empirici, havevan riguardo solamente a rimedij"; "le cause del male."

54. Bacci, *Del Tevere* (1576), 270, "lor disegni per secreti," "tutte le volte, che ei si partiranno dallo stile, che gli antichi tennero ordinariamente in questa cura, sempre s'inganneranno a partito."

55. Bacci, *Del Tevere* (1576), 293, "comincierei da' rimedij communi, e più facili, conformi alle cause, che si son dette concorrere a queste inondationi, si come di allargare, e di nettare l'alveo del Tevere, di fortificar le sponde, & d'addirizzargli il corso, e levargli dinanzi ogni impedimento"; "Vorrei dunque si giudicasse dove fosse più di bisogno."

56. Bacci, *Del Tevere* (1576), 295, "li quali terrebbono le ripe, che non dilamassero alle piene del fiume, si procurerebbe materia di legnami a certi bisogni, che occorressero, e si daria qualche vaghezza al fiume." I thank Joanna Long for discussing this passage about riverbank planting with me.

57. Bacci, *Del Tevere* (1576), 295–296.

58. For his Roman citizenship, see ASC, C.C., cred. I, tom. 1, cat. 1, fol. 175r–v, June 13, 1576 (Idibus Junii 1576); and for the *oncia* of water, see ASC, C.C., cred. I, tom. 27, cat. 27, fols. 6v and 168v, December 20, 1576 (XIII Kal. Januarii 1577). For the location of Bacci's house, see D'Onofrio, *Le fontane di Roma*, 40–42.

59. For Clarante, see Segarra Lagunes, *Il Tevere e Roma*, esp. 119, 120, 130, and 144. For his instructions for the sick and the well during the plague, see Paolo Clarante, "Ricordi di

m. Paolo Clarante da Terni medico in Roma" (BAV, Miscellana R.I. IV. 1551 [int. 99]). For his contribution to Gregory's calendar reform, see Gordon Moyer, "Aloisius Lilius and the 'Compendium Novae Rationis Restituendi Kalendarium,'" in Coyne, Hoskin, and Pedersen, *Gregorian Reform of the Calendar*, 174; and see Paolo Clarante, *Della inondatione del Tevere, e della nuova foce del medesimo* (Perugia: Pietroiacomo Petrucci, 1577) I have used BAV, R.G. Scienze IV.199 [12] of this very rare tract. And see Segarra Lagunes, *Il Tevere e Roma*, esp. 107, 110, 119, 120, 130, 144, 166, and 382.

60. Clarante, *Della inondatione*, "una nuova bocca; e quella un nuovo letto" (7); "impresa certo non da semplice architetto; ma ancho Philosopho per ritrovare le cause: e insieme medico per adattarvi i rimedi: e medico non mediocre per trattare il mondo non piccolo, ch'è l'huomo: ma grande. Il cui capo è Roma" (12); and for his work under Pius V (13). See also Segarra Lagunes, *Il Tevere e Roma*, esp. 144 and 382.

61. Lorenzo Parigioli, *Nuovo discorso sopra il diluvio di Roma* (Rome: Heredi d'Antonio Blado, 1579), "non s'impara ogni cosa in su libri"; "la prattica attuale, e l'osservatione delle cose"; "prima della theorica, e dell'arte"; "piu sicura, piu intelligibile, e migliore." I have used the copy of this very rare tract in the Biblioteca Nazionale Centrale di Firenze, Collocazione 64.2, the original of which is unpaginated. The citation is in the "Prefatione," 30 (in the modern pagination of this copy).

62. For this flood, see especially Frosini, *Il Tevere*, 170–171, and Di Martino, Di Martino, and Belati, *Huc Tiber ascendit*, 72–73, which includes a transcription of the description by I. G. I know of one original copy of I. G.'s printed tract: BAV, R.G. Scienze IV 199 (int. 10), 174r–176v (modern foliation), "l'inondatione grande fatta dal terribil fiume del Tevere."

63. I. G. [BAV, R.G. Scienze IV 199 (int. 10)], 174v–175r, "le gentil donne di questa Città"; "quasi le hà haute à portar tutte alla marina"; "era un piacere à vederle fuggir via chi senza pianelle, chi scalze, chi scapegliate e chi in un modo e chi in un'altro gridando misericordia ad alta voce e correndo per quelle strade e per quell'acqua."

64. Bacci, *Del Tevere* (1576), handwritten, unpaginated on flyleaves of BAV, Aldine II 98.

65. Bacci, *Del Tevere* (1576), BAV, Aldine II 98, unpaginated, first recto-verso folio, "à nostri tempi"; "si navigò tre giorni per Roma in piu luoghi non senza qualche notabile ruina, e danni"; "in quanto all'eseguire mi parvè lasciare il carico alli ingegnieri, e piu essercitati di mè nell'operare."

66. Bacci, *Del Tevere* (1576), BAV, Aldine II 98, unpaginated.

67. Bacci, *Del Tevere* (1576), BAV, Aldine II 98, unpaginated. For the flood of 1598, see esp. Di Martino, Di Martino, and Belati, *Huc Tiber ascendit*, 73–81; D'Onofrio, *Il Tevere*, esp. 155–160; Frosini, *Il Tevere*, 171–175, 212–214; San Juan, *Rome: A City Out of Print*, 129–137; and Segarra Lagunes, *Il Tevere e Roma*, esp. 76–79.

68. Gómez, *De prodigiosis Tyberis inundationibus*; and see Andrea Bacci, *Del Tevere dell'eccell. Dottore medico e filosofo Andrea Baccio libro quarto* (Rome: Stampatori Camerali, 1599).

69. For the modern Tiber River walls, see Aldrete, *Floods of the Tiber*, 247–252; and Denis Bocquet, "Storia urbana e storia della decisioni: L'arginamento del Tevere a Roma (1870–1880)," in Travaglini, *La città e il fiume*, 323–342. For some of the later proposals, see BAV, Codex Chigiano H, II, 43, fols. 163r–175v.

70. Cesare S. Maffioli, *La via delle acque (1500–1700): Appropriazione delle arti e trasformazione delle matematiche* (Florence: Leo S. Olschki, 2010), esp. 151–215.

CHAPTER 2

1. For the earlier centuries, see esp. L. Schiaparelli, "Alcuni documenti dei *Magistri Aedificiorum Urbis* (secoli XIII e XIV)," *ASRSP* 25 (1902): 5–60.

2. Cristina Carbonetti Vendittelli, "Documentazione inedita riguardante i *magistri edificiorum urbis* e l'attività della loro curia nei secoli XIII e XIV," *ASRSP* 113 (1990): 169–188; Emilio Re, "Maestri di strada," *ASRSP* 43 (1920): 5–102; Renato Sansa, "Istituzioni e politica dell'ambiente a Roma: Dalle magistrature capitoline alla presidenza pontificia," in *La legislazione medicea sull'ambiente*, vol. 4, *Scritti per un commento*, ed. Giovanni Cascio Pratilli and Luigi Zangheri (Florence: Leo S. Olschki, 1998), 209–224; Daniela Sinisi, "I *notarii magistrorum stratarum* nel '500: Nascita di un ufficio notarile privativo per le magistrature di acque e strade," *Roma Moderna e Contemporanea* 4 (May–August 1996), 363–378; Orietta Verdi, "Da ufficiali capitolini a commissari apostolici: I maestri delle strade e degli edifici di Roma tra XIII e XVI secolo," in Spezzaferro and Tittoni, *Il Campidoglio e Sisto V*, 54–62; and *Maestri di edifici e di strade a Roma nel secolo XV: Fonti e problemi* (Rome: Roma nel Rinascimento, 1997). The submasters or *sottomaestri* came to be long-standing officers in the city. For their activities in a later period, see Tommaso Manfredi, "L'architetto sottomaestro delle strade," in In urbe architectus: *Modelli, disegni, misure; La professione dell'architetto Roma, 1680–1750*, ed. Bruno Contardi and Giovanna Curcio (Rome: Àrgos, 1991), 281–289.

3. The statute is published in Camillo Scaccia Scarafoni, "L'antico statuto dei 'magistri stratarum' e altri documenti relativi a quella magistratura," *ASRSP* 50 (1927): 267–280. For a discussion and detailed summary, see esp. Simoncini, *Roma: Le trasformazioni urbane nel Quattrocento* 1:70–72, 2:242–248.

4. Oddo Colonna was elected Pope Martin V on November 11, 1417, and was able to return to Rome on September 28, 1420. See esp. Maria Chiabò, Giusi D'Alessandro, Paola Piacentini, and Concetta Ranieri, eds., *Alle origini della nuova Roma: Martino V (1417–1431)*, Atti del Convegno, Roma, March 2–5, 1992 (Rome: Istituto Storico Italiano per il Medio Evo, 1992); Philine Helas and Gerhard Wolf, "'E fece uno granni bene alla città di Roma': Considerazioni sulle opere di Martino V per la città di Roma," in Piatti and Ronzani, *Martino V: Genazzano*, esp. 237–239; Peter Partner, *The Papal State under Martin V: The Administration and Government of the Temporal Power in the Early Fifteenth Century* (London: British School at Rome, 1958); Simoncini, *Roma: Le trasformazioni urbane nel Quattrocento*, 1: 72–80; and Verdi, *Maestri di edifici e di strade*, 14–16. For an example of urban expansion in the fifteenth and sixteenth centuries through the building activities of one family, see Bevilacqua, *Il monte dei Cenci*.

5. See Annoscia, *Fonti e strutture*, esp. 157–173, and for a broader view of waste and waste disposal in numerous European cities, Ercole Sori, *La città e i rifiuti: Ecologia urbana dal Medioevo al primo Novecento* (Bologna: Il Mulino, 2001), esp. 151–190.

6. Tomassetti et al., *Bullarum diplomatum et privilegiorum*, vol. 4, no. XV, 716–718, "grandem deformitatem, seu ruinam potius abhominabilem patiantur et iacturam"; "macellarii videlicet, piscarii, sutores, palamantellarii [i.e., conciatori], diversique artifices"; "viscera, intestina, capita, pedes, ossa, cruores, necnon pelles, carnes et pisces corruptos, resque alias foetidas atque corruptas"; "ausu temerario atque sacrilego" (clause 1). See also Simoncini, *Roma: Le trasformazioni urbane nel Quattrocento*, 2:248–251, which contains an Italian translation of the bull; McCahill, *Reviving the Eternal City*, 35–39; and Verdi, *Maestri di edifici e di strade*, esp. 22–34.

7. Re, "Maestri di strada," esp. 17–30; for the text of the statute, 86–102; for street cleaning, 97–100 (clauses 29–36), "le mundezara del Tevere," 100 (clause 38); and see Verdi, *Maestri di edifici e di strade*, 44–59. For Nicholas's urbanism, see esp. Simoncini, *Roma: Le trasformazioni urbane nel Quattrocento*, 1:99–141; Tafuri, "'Cives esse non licere'"; and Westfall, *In This Most Perfect Paradise*.

8. Verdi, *Maestri di edifici e di strade*, 63–68; and for the register, P. Cherubini, A. Modigliani, D. Sinisi, and O. Verdi, "Un libro di multe per la pulizia delle strade sotto Paolo II (21 luglio–12 ottobre 1467)," *ASRSP* 107 (1984): 51–274 and (for Giganti) 244–274, a publication with commentary and analysis of ASR, Presidenza delle Strade, b. 2. For Paul II's wider urban planning, see also Modigliani, *Disegni sulla città*. For Giganti, in addition to the above, see Anna Modigliani, "Giganti, Girolamo," *DBI*, 54:663–666.

9. For Sixtus IV's urbanizing activities, see esp. Egmont Lee, *Sixtus IV and Men of Letters* (Rome: Edizioni di Storia e Letteratura, 1987), esp. 123–150; Simoncini, *Roma: Le trasformazioni urbane nel Quattrocento*, 1:161–204; and Verdi, *Maestri di edifici e di strade*, 68–81.

10. For Cardinal d'Estouteville, see Anna Esposito, "Estouteville (Tuttavilla), Guillaume (Guglielmo), *DBI*, 43:456–460. For the 1480 edict, see Scaccia Scarafoni, "L'antico statuto dei 'magistri stratarum,'" 248–250, and (for the statute itself) 281–285; and Simoncini, ed., *Roma: Le trasformazioni urbane nel Quattrocento*, 2:269–272. See also Richardson, *Reclaiming Rome*, esp. 302–307.

11. Tomassetti et al., *Bullarum diplomatum et privilegiorum*, vol. 5, no. XXIV, 273–278—for the destruction of porticoes and other structures, 273–274 (clause 1). For a summary and partial translation (by G. Mosconi), see Simoncini, *Roma: Le trasformazioni urbane nel Quattrocento*, 272–278, and (for porticoes) 273–274; and see Scaccia Scarafoni, "L'antico statuto dei 'magistri stratarum,'" 250–257, and Verdi, *Maestri di edifici e di strade*, 68–81.

12. Tomassetti et al., *Bullarum diplomatum et privilegiorum*, vol. 5, no. XXIV, 273–278, and see 275, "magis indigere" (clause 3, for the issue of the proprietors of two houses wanting the ruined house between them); and Simoncini, ed., *Roma: Le trasformazioni urbane nel Quattrocento*, 273–278, 275 for clause 3. See also Bevilacqua, *Il monte dei Cenci*, 23–35, esp. 33.

13. Daniela Sinisi, "Presidenza Strade" in Pastura Ruggiero, *La Reverenda Camera Apostolica*, 102–103, for the presidents of the streets, and "I *notarii magistrorum stratarum* nel '500," 363–378.

14. For a detailed discussion of the urban activities of Julius III, see Simoncini, *Roma: Le trasformazioni urbane nel Cinquecento*, 1:149–179. For Jullius III's tax, see Scaccia Scarafoni, "L'antico statuto dei 'magistri stratarum,'" 257–258, and for the edict itself, ASR, Bandi, Collezione 1, b. 2, no. 5, "vapores, et foetorem Aeris"; "infectio, et pestilentes Corporum infirmitates."

15. Simoncini, *Roma: Le trasformazioni urbane nel Cinquecento*, 1:149, for failure to nominate the masters of the streets in 1552.

16. Bonora, *Roma 1564*, 66–67, "ancora una città profondamente medievale."

17. ASC, C. C., cred. I, tom. 37, cat 37, fol. 181r (XI Kal. Augusti MD lxiiij; July 22, 1564), "Urbs expurgetur ab omnibus immundijs et rebus putridis"; "Necque immundicie abijuntur in vias ex fenestris"; "Ne bestie infectes ad urbem ducentur"; and C. C., cred. I, tom. 22, cat. 22, fols. 76r–77v.

18. Tomassetti et al., *Bullarum diplomatum et privilegiorum*, vol. 7, no. CXX, 386–399.

A classic and still essential study of Pius's urban vision is Marcello Fagiolo and Maria Luisa Madonna, "La Roma di Pio IV. 2. Il sistema dei centri direzionali e la rifondazione della città," *Arte Illustrata* 6 (August 1973): 186–212.

19. We know that Bardo was in Rome in 1557 because of a comment that he makes in a treatise on time. Explaining the difference between contiguous and continuous, he uses the flow of water as an example of a continuum, such as rivers drying up but also rising and falling—just as "I saw on the fifteenth of September 1557 in which the Tiber grew abundantly into the city of Rome." I have used the 1581 edition—Marc Antonio Bardo, *Tractatus de tempore* [. . .] (1565; repr., Cologne: Johannus Gymnicus, 1581), 6, "prout vidi die Decimaquinta Septembris, Anno M.D.LVII. In qua abundanter crevit Tiberis in Ur. Rom."

20. Marc Antonio Bardo, *Tractatus iuris protomiseos sive congrui* [. . .] *Ad interpretationem bullarum* [. . .] *super aedificiis construendis, ad decorum Almae Urbis* (Rome: Antonio Blado, 1565), 1–7 (*Etse de cuntarum*), 8–13 (proem with history) and 13–117 (commentary).

21. Bardo, *Tractatus iuris protomiseos sive congrui*, 118–171, for the *quaestiones*.

22. Marc Antonio Bardo, *Facultates magistratus curatorum viarum* [. . .] (Rome: Antonio Blado, 1565), fols. A–X (for the chapters on the officers). In all copies that I have seen this is bound with the *Tractatus iuris protomiseos*.

23. Tomassetti et al., *Bullarum diplomatum et privilegiorum*, vol. 7, no. CXX, 386–399, 387 (clause 4), "non sine animi nostri displicentia."

24. Ibid., 387–388 (clauses 6–9).

25. Ibid., 389 (clause 10). For the growing demand for palaces and the difficulty of obtaining urban space to build them, see esp. Robertson, *Rome 1600*, 123–125.

26. Tomassetti et al., *Bullarum diplomatum et privilegiorum*, vol. 7, no. CXX, 389–391 (clauses 11–13, 15), "sordes, et immunditiae" (clause 12).

27. Ibid., 391–392 (clause 16), "in Alma Urbe quamplurimi adsunt viculi latidudinis duorum, vel trium palmorum, aut circa, in quibus variae sordes, et immunditiae continuo iniiciuntur"; "foetorem, cum pestiferi, et mali aeris generatione inducunt." See Simoncini, *Roma: Le trasformazioni urbane nel Cinquecento*, 1:260, who describes as very widespread in the oldest part of the *abitato* the very narrow alleys between houses that became filled with the waste of those contiguous houses.

28. Tomassetti et al., *Bullarum diplomatum et privilegiorum*, vol. 7, no. CXX, 392 (clause 17), "quia quamplurimae domus ejusdem Urbis quasdam parvas habent Cloacas, Clavichettas vulgariter nuncupatas, in viis publicis respondentes, et ab ipsis domibus, aquam, et immunditias, in vias, et loca publica evomentes"; "quascumque latrinas, seu cantaras, et siccatoria, vulgo sciacquatoria nuncupata, et piscinas."

29. Ibid., 392–394 (clause 19), reeds, "quamplurima canneta"; (clause 20), Via Angelica; (clause 21), gardens in Borgo; (clause 26), wall enclosure of gardens in the city.

30. Ibid., 393–395 (clauses 22, 24, 25, 27, 28).

31. Ibid., 395 (clauses 29, 30). And see Ago, *Economia barocca*, esp. 9–12, and Esposito, "La città e i suoi abitanti," esp. 38–41.

32. ASR, *Congregazione super viis pontibus et fontibus*, Registro 1, and Genovese and Sinisi, *Pro ornatu et publica utilitate*. For Cardinal Montepulciano, see esp. Gigliola Fragnito, "Ricci, Giovanni," *DBI*, 87:246–249, and for Cardinal Sforza, "Sforza, Alessandro," in Salvador Miranda, *The Cardinals of the Holy Roman Church* (Miami: Florida International University Library, 1998–2015), http://www2.fiu.edu/~mirandas/cardinals.htm.

33. Tomassetti et al., *Bullarum Diplomatum et privilegiorum*, vol. 7, no. CXC, 910–912,

912–913, "diversae lites, quaestiones et controversiae ortae, ac diversa gravamina et laesiones, perturbationes et incommoda diversis personis generata et illata" (911, clause 4).

34. Ibid., 912–913 (clause 6).

35. Tomassetti et al., *Bullarum Diplomatum et privilegiorum*, vol. 8, no. XXXIII, 88–95; and see Conforti, *La città del tardo Rinascimento*, 86–87, and Renato Lefevre, "La costituzione edilizia romana del 1574: Contenuto, valore e applicazione," *Economia e Storia* 19 (1972): 20–39.

36. *Statuta almae urbis Romae* [. . .] (Rome: In Aedibus Populi Romani, 1580), 159–162, esp. 159–160.

37. Ibid., 161–162.

38. ASV, Misc. Armadio IV, busta 74, no. 120 ("Bando generale delli signori maistri di strade,") "le vie e strade di essa Città di Roma, e suo distretto si mantenghino nette e acconcie per la sanità de l'habitanti e bene universale." See clauses 4–10 and 54 for disposal of sewage and manure. The document is a manuscript.

39. ASV, Misc. Armadio IV, busta 74, no. 120, "Bando generale," artisan shops (nos. 11, 21, 61), butchers (12), candlemakers (13), laundry (14).

40. ASV, Misc. Armadio IV, busta 74, No. 120, "Bando generale," construction trades (no. 15–20, 22–23), holes (nos. 41–42), bricklaying and license to build (nos. 44–48), carters and gravel (no. 55).

41. ASV, Misc. Armadio IV, busta 74, no. 120 ("Bando generale"), driving cattle and oxen (no. 35), pigs (no. 34), goats (no. 36), wagons pulled by oxen (no. 38).

42. Canepari, *La construction du pouvoir local.*

43. Tomassetti et al., *Bullarum diplomatum et privilegiorum*, vol. 8, no. LXXV, 807–808, and Simoncini, *Roma: Le trasformazioni urbane nel Cinquecento*, 1:304.

44. Tomassetti et al., *Bullarum diplomatum et privilegiorum*, vol. 8, no. CXVII, 985–999; and see Mario Caravale, "Le istituzioni temporali della Chiesa sotto Sisto V," in Fagiolo and Madonna, *Sisto V*, 1: esp. 49–55, and Simoncini, *"Roma restaurata,"* esp. 123–125.

45. The text of the edict, "Smi D. N. Sixti Pape Quinti electio magistratus offici quattuordecim Magistrorum Viarum Urbis," from a manuscript (MS. 47.3.9) in the Biblioteca Giovardiana of Veroli, is published by Scaccia Scarafoni, "L'antico statuto dei 'magistri stratarum,'" 303–306, "ab huiusmodi officiis omnino destituimus et amovemus" (304). A printed (1588) copy of the edict can also be found in ASR, Bandi, Collezione I, b. 6, no. 10: "Erectio magistratus officii quatuordecim magistrorum viarum Urbis" (Rome: Haeredes Antonii Bladij, 1588).

46. For *monti* and bond issues in general, see esp. Francesco Colzi, *Il debito pubblico del Campidoglio: Finanza comunale e circolazione dei titoli a Roma fra Cinque e Seicento* (Naples: Edizioni Scientifiche Italiane, 1999), 59–99 and 75–76; Ferraro, "Nobility of Rome," 361–391 and 823–859; Maria Luisa Lombardo, "Le gabelle della città di Roma nel quadro dell'attività amministrativo-finanziaria della Camera Urbis nel secolo XV," in *Economia e società a Roma tra Medioevo e Rinascimento: Studi dedicati ad Arnold Esch*, ed. Anna Esposito and Luciano Palermo, 205–228 (Rome: Vielli, 2005); Nussdorfer, *Civic Politics*, 86–94; and Pecchiai, *Roma nel Cinquecento*, 278–281.

47. Simoncini, *"Roma restaurata,"* esp. 123–125. For Clement VIII Aldobrandi and his family's patronage, see esp. Robertson, *Rome 1600.*

48. ASR, *Congregazione super viis pontibus et fontibus*, Registro 1, and Genovese and Sinisi, *Pro ornatu et publica utilitate.*

49. Rinne, *Waters of Rome*, esp. 193–194, and Rinne, "Urban Ablutions: Cleansing Counter-Reformation Rome," in Bradley and Stow, *Rome, Pollution and Propriety*, 182–202. See also Italo Riera, "Classificazione e funzionamento delle fognature romane," in Bianchi, *La Cloaca Maxima*, 4–21, who emphasizes the difficulty of identifying and dating the archaeological remains of sewers and drains and suggests a functional classification system.

50. For sewer backflow during the flood of 1557, see Peto, *Discorso* [...] *intorno alla cagione*, 185r–v.

51. Leslie Tomory, "London's Water Supply before 1800 and the Roots of the Networked City," *Technology and Culture* 56 (July 2015): 704–737, and *The History of the London Water Industry, 1580–1820* (Baltimore: Johns Hopkins University Press, 2017), 1–5, 21–28, 239–247, and passim. Historians have for the most part described the development of the networked city as a development of the second half of the nineteenth century when cities in Europe and North America were "increasingly refashioned by the construction of meshes of large-scale integrated and centralized networks" (Tomory, *London's Water Supply*, 705). For recent discussions of the (modern) networked city, with particular attention to waste disposal, see, for example, the following essays, all in Tarr and Dupuy, *Technology and the Rise of the Networked City*: André Guillerme, "The Genesis of Water Supply, Distribution, and Sewerage Systems in France, 1800–1850," 91–115; Jean-Pierre Goubert, "The Development of Water and Sewerage Systems in France, 1850–1950," 116–136; Joel A. Tarr, "Sewerage and the Development of the Networked City in the United States, 1850–1930," 159–185; and Georges Knaebel, "Historical Origins and Development of a Sewerage System in a German City: Bielefeld, 1850–1904," 186–206. And see Erik Van der Vleuten and Arne Kaijser, eds., *Networking Europe: Transnational Infrastructures and the Shaping of Europe, 1850–2000* (Sagamore Beach, MA: Science History Publications, 2006). The foundational study is Thomas P. Hughes, *Networks of Power: Electrification in Western Society, 1880–1930* (Baltimore: Johns Hopkins University Press, 1983).

52. Pietro Narducci, *Sulla fognatura della Città di Roma*, 2 vols. (Rome: Forzani, 1889), based on his survey of the Roman sewers in the 1880s (and see vol. 1, pp. 39–49, and vol. 2, tav. 6a N. 13, for the Cloaca Maxima). For Narducci and his work, see Roberto Narducci, "La riscoperta della rete fognaria romana e la documentazione grafica di fine '800 illustrata dall' ing. cav. Pietro Narducci," in Bianchi, *La Cloaca Maxima*, 42–51. And see Claudio Mocchegiani Carpano, "Le cloache dell'antica Roma," in Luciani, *Roma sotterranea*, 171–173, and Rinne, *Waters of Rome*, 195. For the Cloaca Maxima, in addition to Pietro Narducci cited above, see esp. Heinrich Bauer, "Cloaca, Cloaca Maxima," in Steinby *Lexicon topographicum*, 1:288–290; Elisabetta Bianchi, "Il primo tratto della Cloaca Maxima e gli adattamenti realizzati per la costruzione dei Fori Imperiali," and Elisabetta Bianchi and Luca Antognoli, "La Cloaca Massima dal Foro Romano al Velabro dagli studi di Heinrich Bauer alle nuove indagini," both in Bianchi, *La Cloaca Maxima*, 82–107 and 108–153, respectively; John N. N. Hopkins, "The Cloaca Maxima and the Monumental Manipulation of Water in Archaic Rome," *Waters of Rome*, no. 4 (March 2007), http://www3.iath.virginia.edu/waters/first.html; "The 'Sacred Sewer': Tradition and Religion in the Cloaca Maxima," in Bradley and Stow, *Rome: Pollution and Propriety*, 81–102; and Lawrence Richardson, *A New Topographical Dictionary of Ancient Rome* (Baltimore: Johns Hopkins University Press, 1992), 91–92 (s.v. "Cloaca Maxima").

53. I warmly thank Katherine Rinne for our day-long discussion of this evidence and for her construction of fig. 2.4.

54. See ASR, Presidenza delle Strade, b. 445, fols. 156r–170v (March 28, 1538), and Re,

"Maestri di strada," 69 (no. 48); Rinne, *Waters of Rome*, 198; and "Hydraulic Infrastructure and Urbanism in Early Modern Rome," *Papers of the British School at Rome* 73 (2005): 207.

55. ASC, C.C., cred. I, tom. 1, cat. 1, fol. 130r, "uscendo fuor del condotto dell'acqua vergine di Treio per un fosso la medema acqua"; "passando per gli horti delle monache delle Convertite et di San Silvestro generando malissimo aere"; "conoscendo per l'avvenire esser necessario porvi diligentemente cura accioche in simil disordine non si cada"; "custode di detto cloaca, fosso et acqua"; "alli padroni delle case orti, et terreni ove passa, ò, passerà detta cloaca fosso et acqua, non debbiano gettare o, far gettare alcuna sorta d'immonditia o altro." The document is also cited by Rodolfo Lanciani, *Storia degli scavi di Roma e notizie intorno le collezioni romane di antichità*, ed. Leonello Malvezzi Campeggi (Rome: Quasar, 1990), 3:261.

56. ASR, *Congregazione super viis pontibus et fontibus*, Registro 1, fol. 12v; and Genovese and Sinisi, *Pro ornatu et publica utilitate*, September 18, 1568 (Congregation no. 10), 14.

57. ASR, *Congregazione super viis pontibus et fontibus*, Registro 1, fol. 14r; and Genovese and Sinisi, *Pro ornatu et publica utilitate*, October 12, 1568 (Congregation no. 12), 15. For the hospital, see esp. John Henderson, "'Mal Francese' in Sixteenth-Century Rome: The Ospedale di San Giacomo in Augusta and the 'incurabili,'" in Sonnino, *Popolazione e società a Roma*, 483–523.

58. ASR, *Congregazione super viis pontibus et fontibus*, Registro 1, fol. 20v, "de maximo fetere et causa mortalitatis et in futurum danni insupportabilis"; and Genovese and Sinisi, *Pro ornatu et publica utilitate*, January 11, 1569 (Congregation no. 16), 19–20.

59. ASR, *Congregazione super viis pontibus et fontibus*, Registro 1, fol. 26v, "che se non si remediava a non netter stabio ne se fosse sotto la trinita et anche finir la chiavica di S. Silvestro che ne verra la solita influentia pestifera"; and Genovese and Sinisi, *Pro ornatu et publica utilitate*, April 19, 1569 (Congregation no. 19), 23–24.

60. ASR, *Congregazione super viis pontibus et fontibus*, Registro 1, fol. 36r; and Genovese and Sinisi, *Pro ornatu et publica utilitate*, July 4, 1569 (Congregation no. 22), 27–28. For the tax, see Re, "Maestri di strade," 78 (no. 128); and ASR, Presidenza delle Strade, b. 445, fols. 489r–490v.

61. See esp. Rinne, *Waters of Rome*, 198–199.

62. Renato Sansa has emphasized (for the seventeenth and eighteenth centuries) the very differential treatment of main streets (often cleaned by carters) and smaller side streets (often neglected). See esp. Renato Sansa, "Le norme decorose e il lavoro sporco. L'igiene urbana in tre capitali europee; Londra, Parigi, Roma tra XVI e XVIII secolo," *Storia Urbana* 112 (2006): 85–112, esp. 103.

63. ASR, *Congregazione super viis pontibus et fontibus*, Registro 1, fol. 76r; and Genovese and Sinisi, *Pro ornatu et publica utilitaten*, September 22, 1574 (Congregation no. 61), 59.

64. Douglas Biow, *The Culture of Cleanliness in Renaissance Italy* (Ithaca, NY: Cornell University Press, 2006).

65. ASR, *Congregazione super viis pontibus et fontibus*, Registro 1, fols. 19v, 26r, 50r, 53r, 57r, 60r, 67v, 71r, 75r, 78v, 122v, 127r, 135r, 136r, 144r, 148r, 150r, 155r, 156r, 158v, 161r, 163r, 167v, and 171r; and Genovese and Sinisi, *Pro ornatu et publica utilitate*, January 11, 1569 (Congregation no. 16), 19–20; April 19, 1569 (Congregation no. 19), 23; February 10, 1571 (Congregation no. 34), 39; June 4, 1571 (Congregation no. 37), 42; July 10, 1572 (Congregation no. 43), 45; April 3, 1573 (Congregation no. 50), 51–52; September 14, 1573 (Congregation no. 54), 54; May 31, 1574 (Congregation no. 59), 58; November 23, 1574 (Congregation no. 63), 61; August 11, 1580 (Congregation no. 89), 96; November 8, 1581 (Congregation

no. 92), 99; April 2, 1582 (Congregation no. 99), 106; April 26, 1582 (Congregation no. 100), 106–107; April 1, 1583 (Congregation no. 107), 113–114; January 7, 1584 (Congregation no. 111), 117–118; July 19, 1584 (Congregation no. 113), 119–121; June 7, 1585 (Congregation no. 117), 124; September 6, 1585 (Congregation no. 118), 125; September 28, 1585 (Congregation no. 120), 127–128; February 6, 1586 (Congregation no. 123), 129; May 18, 1586 (Congregation no. 124), 131–132; September 27, 1586 (Congregation no. 126), 133–134; February 23, 1587 (Congregation no. 127), 135.

66. Genovese and Sinisi, *Pro ornatu et publica utilitate*, 178–193, for a list of all the measures having to do with taxes, many of which involve individuals requesting a reduction of an imposed tax.

67. See Martin V. Melosi, *The Sanitary City: Urban Infrastructure in America from Colonial Times to the Present* (Baltimore: Johns Hopkins University Press, 2000), esp. 17–42 for earlier periods and 103–116 for the development of bacteriology; and see Ann Olga Koloski-Ostrow, *The Archaeology of Sanitation in Roman Italy: Toilets, Sewers, and Water Systems* (Chapel Hill: University of North Carolina Press, 2015), esp. 38–51, for the great variations in attitudes toward sanitation; and see Sansa, "L'odore del contagio," and Sori, *La città e i rifiuti*, 129–147.

68. Richard Krautheimer, *The Rome of Alexander VII, 1655–1667* (Princeton, NJ: Princeton University Press, 1985), 15–36, citation on 15.

69. See esp. Sansa, "Le norme decorose," 102, for the incised marble admonitions; and "La pulizia delle strade a Roma nel XVII secolo: Un problema di storia ambientale," *ASRSP* 114 (1991): 127–160.

CHAPTER 3

1. See Long, "Hydraulic Engineering"; and Rinne, "Hydraulic Infrastructure and Urbanism." For the fountains of the Acqua Vergine, see D'Onofrio, *Le Fontane di Roma*, esp. 13–189; Pio Pecchiai, *Acquedotti e fontane di Roma nel Cinquecento (con documenti inediti)* (Rome: Staderini, 1944), 7–56 and 77–93, and Rinne, *Waters of Rome*, esp. 38–55 and 83–108. For Bartolomeo Gritti or Grippetto, see Bertolotti, *Artisti lombardi a Roma*, 63–66.

2. See Grant Heiken, Renato Funiciello, and Donatella De Rita, *The Seven Hills of Rome: A Geological Tour of the Eternal City* (Princeton, NJ: Princeton University Press, 2005); Funiciello, *La Geologia di Roma*; and Leonardo Lombardi, "Geologia e idrogeologia e idrologia del bacino del lo *Spinon*," in Bianchi, *La Cloaca Maxima*, 68–73.

3. For the ancient aqueducts see Thomas Ashby, *The Aqueducts of Ancient Rome*, ed. I. A. Richmond (Oxford: Clarendon Press, 1935); Carlo Fea, *Storia I: Delle acque antiche sorgenti in Roma perdute, e modo di ristabilirle; II: Dei condotti antico-moderni delle acque, Vergine, Felice, e Paola e loro autori* (Rome: Stamperia della Reverenda Camera Apostolica, 1832); Lanciani, *I commentarii di Frontino*; Pecchiai, *Acquedotti e fontane*; and Esther Boise Van Deman, *The Building of the Roman Aqueducts* (Washington, DC: Carnegie Institution of Washington, 1934). More recently, see A. Trevor Hodge, *Roman Aqueducts and Water Supply*, 2nd ed. (London: Duckworth, 2002), 171–214, for surveying and creating the gradient and 215–245, for the hydraulics of the aqueducts, including the gradient. Other studies include Peter J. Aicher, *Guide to the Aqueducts of Ancient Rome* (Wauconda, IL: Bolchazy-Carducci, 1995), a concise guide to the remains of the ancient aqueducts; Paolo Buonara, "Gli acquedotti del papa: Restitutio o innovazione?" in Antinori, "Le reti dell'acqua," 323–334; Lorena Di Car-

lantonio, "I percorsi degli acquedotti e la trasformazione del paesaggio urbano," in Bonaccorso, "Le acque e la città," 31–41; Harry B. Evans, *Water Distribution in Ancient Rome: The Evidence of Frontinus* (Ann Arbor: University of Michigan Press, 1994), esp. 1–52; *Aqueduct Hunting in the Seventeenth Century;* Pietrantonio Pace, *Gli acquedotti di Roma e il* De Aquaeductu *di Frontino* (Rome: Art Studio S. Eligio, 1983); Lorenzo Quilici, "Gli acquedotti di Roma: Solidità, utilità, bellezza," *Archeo.* 53 (July 1989): 51–97; Rabun Taylor, *Public Needs and Private Pleasures: Water Distribution, the Tiber River, and the Urban Development of Ancient Rome* (Rome: "L'"Erma" di Bretschneider, 2000); and *Il trionfo dell'acqua.*

4. Garrett G. Fagan, *Bathing in Public in the Roman World* (Ann Arbor: University of Michigan Press, 1999), 83, 118–123; Koloski-Ostrow, *Archaeology of Sanitation;* and Alex Scobie, "Slums, Sanitation, and Mortality in the Roman World," *Klio* 68 (1986): 399–433.

5. For aqueduct maintenance see Aicher, *Guide to the Aqueducts,* 12, 23–25, 28–29; and Trevor Hodge, *Roman Aqueducts,* 17, 126–127, 227–232.

6. For maintenance and repairs of the Vergine in the fifteenth and sixteenth centuries, see David Karmon, "Restoring the Ancient Water Supply System in Renaissance Rome: The Popes, the Civic Administration, and the Acqua Vergine," *Waters of Rome,* no. 3 (August 2005), http://www3.iath.virginia.edu/waters/first.html; and see Rinne, *Waters of Rome,* 38–55. For the Trevi Fountain, see John A. Pinto, *The Trevi Fountain* (New Haven, CT: Yale University Press, 1986).

7. For the Acqua Vergine, see Ashby, *Aqueducts of Ancient Rome,* 167–182; D'Onofrio, *Le fontane di Roma,* 13–188; Fea, *Storia,* 9–28, 63–71; Karmon, "Restoring the Ancient Water Supply System"; Lanciani, *I comentarii di Frontino,* 120–130; Vittorio Nicolazzo, *Acqua Vergine a Roma: Acquedotti e fontane* (Rome: Colosseo Grafica, 1999); Pascucci, "L'acqua corrente nel Cinquecento a Roma," 49–59; Pecchiai, *Acquedotti e fontane,* 11–37; Lorenzo Quilici, "Sull'acquedotto Vergine dal Monte Pincio alle sorgenti," *Studi di Topografia Romana* 5 (1968) 125–160; Katherine Wentworth Rinne, "Between Precedent and Experiment: Restoring the Acqua Vergine in Rome (1560–70)," in Roberts, Schaffer, and Dear, *Mindful Hand,* 94–115; Rinne, *Waters of Rome,* esp. 38–55; and Van Deman, *Building of the Roman Aqueducts,* 166–178.

8. See Aicher, *Guide to the Aqueducts,* 39–41, 68–74, and Karmon, "Restoring the Ancient Water Supply System." For sinter, see Trevor Hodge, *Roman Aqueducts,* esp. 227–232.

9. See Delph, "*Renovatio, Reformatio*"; and Karmon, "Restoring the Ancient Water Supply." A record of Paul III urging the Capitoline Council to renovate the Acqua Vergine is ASC, C.C., cred. I, tom. 36, cat. 36, fol. 319r (sub die Vegisima septima mensi Novembris [27 November 1535]). And see Fea, *Storia,* 63–64.

10. An anonymous manuscript, BAV, Vat. lat. 7246, 17r–25v, has the same handwriting as Vat. lat. 3587, "Augustinus Steucus de restituenda navigatione Tiberis," which confirms D'Onofrio, *Le fontane di Roma,* 54–59, in his argument that it is by Steuco; and see Agostino Steuco, *De via Pauli et de fontibus inducendis in eam* (Rome: Balthasar Cartularius [n.d., not after 1543]); *De Aqua Virgine in urbem revocanda* (Lyon: Sebastianus Gryphius, 1547); and *De restitudenda navigatione Tiberis* (Lyon: Sebastianus Gryphius, 1547). For an English translation of Steuco's tract on the Acqua Vergine, see Chiara Bariviera and Pamela O. Long, "An English Translation of Agostino Steuco's *De Aqua Virgine in Urbem revocanda* (Lyon: Gryphius, 1547) [*On Bringing Back the Acqua Vergine to Rome*]" *Waters of Rome* 8 (August 2015), http://www3.iath.virginia.edu/waters/Journal8BarivieraLong.pdf. For Steuco, see Ronald K. Delph, "Polishing the Papal Image in the Counter-Reformation: The Case of Agostino

Steuco," *Sixteenth Century Journal* 23 (Spring 1992): 35–47; "From Venetian Visitor to Curial Humanist: The Development of Agostino Steuco's 'Counter'-Reformation Thought," *Renaissance Quarterly* 47 (Spring 1994): 102–139; "*Renovatio, Reformatio*"; and Theobald Freudenberger, *Augustinus Steuchus aus Gubbio, Augustinerchorherr und päpstlicher Bibliothekar (1497–1548)und sein literarisches Lebenswerk* (Münster in Westfalen: Aschendorffschen, 1935).

11. Steuco, *De Aqua Virgine in urbem revocanda,* 7; Bariviera and Long, *English Translation,* 7.

12. Steuco, *De Aqua Virgine in urbem revocanda,* 18 (for oxen falling in the shafts), 9 and 17 (for hydraulic mortar); Bariviera and Long, *English Translation,* 16 (for oxen), 9 and 16 (hydraulic mortar); and see Delph, "Polishing the Papal Image"; "From Venetian Visitor to Curial Humanist"; and "*Renovatio, Reformatio.*"

13. Steuco, *De Aqua Virgine in urbem revocanda,* 5–6, "non ipsa modo ratio, ac rei contemplatio, sed item experientia, et usus, probataque oculis res"; "minori dispendio"; and Bariviera and Long, *English Translation,* 5.

14. Steuco, *De Aqua Virgine in urbem revocanda,* 13–14, "Quae omnis expensa ipsis architectis autoribus, qui ad haec visenda, solerterque perpendenda saepius convenere vix quindecim millium aureorum summam accredit"; "prudentibus, rerumque peritis" (i.e., by consultants and experts on the matter); and Bariviera and Long, *English Translation,* 13.

15. See Karmon, "Restoring the Ancient Water Supply System"; Pecchiai, *Acquedotti e fontane,* 11–38; Quilici, "Sull'acquedotto Vergine"; Rinne, "Between Precedent and Experiment"; and Maria Grazia Tolomeo, "L'Acquedotto Vergine (sec. XVI–XVIII)" in *Il trionfo dell'acqua,* 205–208. For Cardinal Montepulciano's appointment, see also Sylvie Deswarte-Rosa, "Le cardinal Giovanni Ricci de Montepulciano" in Chastel and Morel, *La Villa Médicis,* 2:145.

16. See "Ligorio, Pirro," *DBI,* 65:109–114; Coffin, *Pirro Ligorio,* 5–81; Robert W. Gaston, ed., *Pirro Ligorio: Artist and Antiquarian* (Milan: Silvania, 1988); Long, "Multi-Tasking 'Pre-professional' Architect/Engineers," 230–233; Beatrice Palma Venetucci, "Pirro Ligorio and the Rediscovery of Antiquity," in *The Rediscovery of Antiquity: The Role of the Artist,* ed. Jane Fejfer, Tobias Fischer-Hansen, and Annette Rathje (Copenhagen: Museum Tusculanum Press, University of Copenhagen, 2003), 63–68; Anna Schreurs, *Antikenbild und Kunstanschauungen des neapolitanischen Malers, Architekten und Antiquars Pirro Ligorio (1513–1583)* (Cologne: Walther König, 2000); Caterina Volpi, ed., *Il libro dei disegni di Pirro Ligorio all'Archivio di Stato di Torino* (Rome: Edizioni dell'Elefante, 1994); "Il teatro del mondo: Pirro Ligorio 'architetto' papale nelle Logge Vaticane," *Bollettino d'Arte,* 6th ser., 107 (January–March 1999): 83–102. There is a large literature on the Casino and its decorative programs. Recent studies include Maria Losito, *Pirro Ligorio e il Casino di Paolo IV in Vaticano: L'"essempio" delle "cose passate"* (Rome: Palombi, 2000); and Daria Borghese, ed., *La Casina di Pio IV in Vaticano* (Turin: Umberto Allemandi, 2010).

17. See Coffin, *Pirro Ligorio,* 72–81, for Ligorio's last days in Rome.

18. AST, Cod. J.a.III.3, fol. 14v, s.v. "Aqua Vergo, ò Aqua Vergine," "istesso Pyrrho Ligorio che ha scritto questa opera ha proposto à Papa pio quarto, accioche si conducesse il proprio fonte."

19. Ibid., "peto gran' dottore fusse contrario à questa opera"; "fù cagione quasi di precipitare tanto degna opera, et con negare che l'acqua non poteva venire a Roma, come non fusse gia stata in Roma"; "affino veduto la cosa riuscita scrisse l'opera in cui egli s'avanta d'avere fatto ogni cosa tanta sua potentia la suasione e la bugia."

20. For his trip to Rome, see Mandowsky and Mitchell, *Pirro Ligorio's Roman Antiquities*, 5–6.

21. MS Florence, Uffizi, Arch. no. 4236A, first attributed to Ligorio by Christian Hülsen, "Piante icnografiche incise in marmo," *Mittheilungen des kaiserlich deutschen archäologischen Instituts: Römische Abtheilung* 5 (1890): 57–58, n2, an attribution that cannot be doubted, given that the hand in both writing and drawing is identical with that of Ligorio's archaeological encyclopedia in Turin. For a reproduction of the whole map, see Coffin, *Pirro Ligorio*, 54–55.

22. See AST, Cod. J.a.II.1, fols. 7v–23v, s.v. "Piscina." For the distribution tank, Ligorio uses the Latin word *castellum*.

23. AST, Cod J.a.II.1, fol. 14v, "L'acqua vergine quantunque habbiamo con curiosita ricercato, et di dentro, et di fuori della citta, et nelli suoi fonti."

24. Ibid., "havemo voluto porre qui, alcune poche parole le quali sebene serrono tardamente poste in luce, spero che cosi tarde piacerenno ai curiosi di odire la verità in conoscere le bugiarde"; "particolarmente si è avantato lui essere stato lo inventore et deduttore dell'Acqua Vergine à Roma"; "Luca peto, acompagnato con un altra animuccia"; "per toccare fresca pecunia." Trevisi's disastrous attempt to repair the aqueduct is discussed in greater detail on 73–78. Onofrio Panvinio's interest in aqueducts is indicated by his edition and commentary on Frontinus's *De aquae ductu urbis Romae*; Onofrio Panvinio, *Antiquae urbis imago* [. . .] (Paris: Egidius et Nicolaus Gillius, 1588).

25. AST, Cod. Ja.II.1, fols. 14v–15r. Some scholars have been skeptical of Ligorio's accuracy in his descriptions of structures that are no longer or no longer fully extant. Further, there are some discrepancies between his drawing and his description of the arches over the Marranella. Quilici, "Sull'acquedotto Vergine," 143–144, finds Ligorio's reports largely accurate and is able to explain the discrepancies.

26. AST, Cod. J.a.II.1, fol. 16r–v, "la calamita dell'opera fatta in rappezzare, et rattoppare"; "per diffetto delli usurpatori"; "l'opera di facile serà caduca et frale." For the incident concerning the Goths' attempt to enter Rome through the aqueduct, see Procopius, *History of the Wars*, 6.9.1–11.

27. Luca Peto, *De restitutione ductus Aquae Virginis* (Rome: B. Tosius, 1570), "inde mihi, nescio quare, demandata cura, ut fontes, & loca diligenter inspicerem, & in Senatu referrem"; "ac etiam inspecto opere arcuato, & substructionibus, quae Virgini a Frontino attribuuntur, audacter retuli, etsi subterranei specus quibusdam in locis non apparerent, Virginem esse, & reduci posse" (Sig. A3r). For Peto's appointment to investigate the aqueduct, see ASC, C.C., cred. I, tom. 21, cat. 21 (3ª Junij; ad tertium nonas Junii MD lxi), fol. 87r–v; and also in ASC, C.C., cred. I, tom. 37, cat. 37, fols. 78v–79r. The references are in Frontinus, *De aquae ductu urbis Romae*, 1.5 (Aqua Appia), and 1.10 (Aqua Virgo); and Pliny, *Natural History*, 31.25–42. For the confusion between the Acqua Appia and the Vergine, see Delph, "*Renovatio, Reformatio*," 86; and see Marcello Turci, "Gli acquedotti," and Chiara Bariviera, "Regione XII: *Piscina Publica*," both in Carandini and Carafa, *Atlante di Roma antica*, vol. 1, pp. 92, 94, 95, 96, and vol. 2, tav. 154, and vol. 1, pp. 378, respectively.

28. Peto, *De restitutione ductus Aquae Virginis* (n.p.), "partim equester, ubi multa aqua non erat, partim pedester, ubi sicco pede ambulare licuit, partim vero navicula, ad id accommodata, ab hominibus tracta, ubi rivus navigabilis fuit." For the boundary stones, see Henri Broise and Vincent Jolivet. "Villa Médicis: L'Antiquité," and Glenn M. Andres, "Le jardin du cardinal Ricci," both in Chastel and Morel, *La Villa Médicis*, 2:15–16, and 342–350, respec-

tively. For Fulvio, see Massimo Ceresa, "Fulvio, Andrea," *DBI*, 50:709–712; and Roberto Weiss, "Andrea Fulvio antiquario romano (c. 1470–1527)," *Annali della Scuola Normale Superiore di Pisa: Lettere, Storia e Filosofia*, 2nd ser., 28 (1959): 1–44.

29. For the contract itself, see ASR, Notai RCA, b. 453, fols. 146r–148v (notary Girolamo [Hieron] da Tarano). See also Lanciani, *Storia degli scavi*, 3:260.

30. ASR, Notai RCA, b. 453, fols. 146r–148v (notary Geronimo [Hieron] da Tarano), "con bone et recipiente materie"; "tutte l'altre acque" (fol. 146v); "come si vede che haveva a' tempo delli antichi" (fol. 147r).

31. ASR, Notai RCA, b. 453, fols. 146r–148v (notary Geronimo [Hieron] da Tarano), "metterla a' giuditio di tre architetti"; "esso Antonio" (fol. 147r); "li prezzi honesti et ragionevoli a giuditio della Camera (fol. 147v)." For pozzolana, see Lynne C. Lancaster, *Concrete Vaulted Construction in Imperial Rome: Innovations in Context* (Cambridge: Cambridge University Press, 2005), esp. 51–53 and 54–58.

32. Beltrani, "Leonardo Bufalini," 371–374, provides some archival references to payment authorizations to Trevisi, to which list I have added others. See ASR, Camerale 1, Mandati 907, fols. 147v, 147v–148r, 160v, 164v, 169r, 176r, 181v, 182v, 187v, 207r, 217r, 219v, 222r, 231r, and 234r; Mandati 912, fols. 17v, 20v, 21r, 23v, 33r, 35v, 61r–v, 77v, 78r, 101v, 118r, 121r, 127v, 144r, 156v–157r, 172v, 183v, 194v, 200v, 209v, 217v, 222v, 229v, 231v, 235v–236r, and 237v; Mandati 914, fols. 2r, 12r, 14r, 29r, 33v, 43r, 45r, 49v, 56v, 65r, 68v, 70r, 77v, 81r, 87r, 90r, 99r, 100r, 104v, 110v, 118v, 129v, 130r, 136v, 141v, 149r, 159r, 164r–v, 174r, 185v, 195r, 213v, and 226r; Mandati 916, fols. 113r, 138r, and 232r; Mandati 918, fol. 230r; Mandati 921, fols. 1r, 12v, 24r, 32v, 37r, and 42r. The final authorization noted in the Mandati occurred on September 13, 1566.

33. Peto, *De restitutione ductus Aquae Virginis*, Sig. A2r–v, "satis temerarium"; "temerarijs fabris murarijs."

34. Ibid., Sig A3r, "pecuniae collatae Trivisio illi, & fabris murariis distributae"; "opus ad finem perduci non posse, nisi ab eo Trivisius arceretur."

35. Ibid., Sig. A3r–v, "praetextu quod aequum esset, Trivisium operis auctorem, illius debere esse perfectorem, tantum per nos animadverti, ut opus cito prosequeretur, & recte."

36. The Psylli were an African tribe said to be skilled at snake charming and immune from venom. AST, Cod. J.a.II.1., fol. 16r–v, "a' cottima ò ad opera assoma"; "Antonio da Leccio inciarmatore di serpi"; "con tanta confusione de Maestri Muratori et con tanta lite"; "l'appaltatore Psyllaro ne he morto incarcere." Peto *De restitutione ductus Aquae Virginis* Sig A3v, mentioned only that Trevisi had died (without mentioning prison). I have not thus far been able to find a court proceeding that may have preceded Trivisi's imprisonment or other evidence for Trevisi's death.

37. ASC, C.C., cred. I, tom. 21, cat. 21 (xii Octobris; ad Quartum idus Octobris. MD lx), 48v–49r, "circa dieci cardinali"; "per benefitio et bellezza di questa terra"; "farne conseglio et intendere il parere del Popolo"; "il Popolo assai alegramente ha accetato questa sua offerta, et che pero S. Sta. faccia dar principio et seguire tanto laudabile impresa." Also in C.C., cred. I, tom. 37, cat. 37, fol. 61v. For Cardinal della Cueva y Toledo, see Salvador Miranda, "Cueva y Toledo, Bartolomé de la," in Miranda, *Cardinals*; and for Crispi, who was probably the natural son of Paul III, see Luisa Bertoni, "Crispi, Tiberio," *DBI*, 30:801–803.

38. ASC, C.C., cred. I, tom. 21, cat. 21 (3 Junij; ad tertium nonas Junii), fols. 86r–87v, "ne risulterà grandezza alla citta, et benefitio à questo Popolo." Also in ASC, C.C., cred. I, tom. 37, cat. 37, fols. 78r–79r.

39. ASC, C.C., cred. I, tom. 21, cat. 21 (3 Junij; ad tertium nonas Junii), fol. 87r–v; and also in ASC, C.C., cred. I, tom. 37, cat. 37, fols. 78v–79r.

40. ASC, C. C., cred. I, tom. 21, cat. 21 (20 Junij; ad duodecimum Kal. Julii MD lxi), fols. 93v–94r, "dett' acqua essere alle volte impedita di modo che al publico per doi giorni non viene"; "far un bando contra quelli che impediscono in qualsivoglia modo detta acqua." Also in ASC, C.C., cred. I, tom. 37, cat. 37, fols. 81v–82r.

41. ASC, C.C., cred. I, tom. 21, cat. 21 (5 Julii; ad tertium nonas Juliii MD lxi), fol. 97v. Also in ASC, C.C., cred. I, tom. 37, cat. 37, fol. 83r.

42. ASC, C.C., cred. I, tom. 21, cat. 21 (ad decimum Kal. Septembris MD lxi), fols. 116r–117v and (16 Settembris 1561; ad decimum sextum Kal. Octobris M.D. LXI), fols. 119v–122r (September 16, 1561). The four men chosen to collect money were Cencio Frangipane, Angelo Albertoni, Geronimo Alteri, and Tommaso Cavalieri (fol. 122r.). Also in ASC, C.C., cred. I, tom. 37, cat. 37, fols. 89r–90r, fols. 90v and 91v. For financial policies including the sale of offices, see esp. Clemens Bauer, "Die Epochen der Papstfinanz," *Historische Zeitschrift* 138 (1928): 473–503; Colzi, *Il debito pubblico*; Delumeau, *Vie économique et sociale*, 2:751–843; Nussdorfer, *Civic Politics*, 27–32, 186–199; Peter Partner, "Papal Financial Policy in the Renaissance and Counter-Reformation," *Past and Present* 88 (August 1980): 17–62, esp. 29; Maria Grazia Pastura Ruggiero, "La fiscalità pontificia nel Cinquecento: Aspetti e problemi," in Fagiolo and Madonna, *Sisto V*, 1:211–231; and Pecchiai, *Roma nel Cinquecento*, 267–294. For the *censo* in general, see Ferraro, "Nobility of Rome," 329–342, 859–894.

43. For the Secret Council meeting, ASC, C.C., cred. I, tom. 21, cat. 21 (pro die 13 Octobris MD lxi), fol. 126r–v, and for the Public Council (ad die XV Ottobris Anni MD lxi), fols. 130v–131r. Also in ASC. C.C., cred. I, tom. 37, cat. 37, fol. 93v and fol. 95r.

44. ASC, C.C., cred. I, tom. 21, cat. 21 (ad decimum Kal. Septembris MD lxi), fols. 116r–117v.

45. ASC, C.C., cred. I, tom. 21, cat. 21 (9 Aprilis; ad quintum Idus Aprilis), fol. 163r–v. Also in ASC, C.C., cred. I, tom. 37, cat. 37, fol. 110r–v. The report refers to Cardinal Santa Fiora (Guido Ascanio Sforza (1518–1564), who was the grandson of Pope Paul III and was made cardinal by him in 1534. See Salvador Miranda, "Sforza di Santa Fiora, Guido Ascanio," in Miranda, *Cardinals*.

46. ASC, C.C., cred. I, tom. 21, cat. 21 (28 Aprii; ad quartum Kal. Maii MD lxii), fol. 169r, 170r. Also in ASC, C.C., cred. I, tom. 37, cat. 37, fols. 113r–114r.

47. ASC, C.C., cred. I, tom. 21, cat. 21 (12 Julii; ad quartum Idus Julii MD lxii), fols. 189v–190r. (Secret Council meeting of July 12, 1562), and (Pridie Idus Julii MD lxii), fols. 193r–194r (Public Council meeting of July 14, 1562). Also in ASC, C.C., cred. I, tom. 37, cat. 37, fols. 123r, 124v–125r.

48. ASC, C.C., cred. I, tom. 21, cat. 21 (xv dicembris; ad decimum octarum Kal. Januarij), fol. 211r–v, "non ci pareva honesto non essendo stati pagati sino ad hora li dui terzi delli altri a' chi tocca, ch'il populo: dovesse pagar al pnte l'ultimo terzo." Also in ASC, C.C., cred. I, tom. 37, cat. 37, fol. 135r.

49. ASC, C.C., cred. I, tom. 21, cat. 21 (xxvi Febrij; Ad quartum Kal. Martii MD lxiii), fol. 226v. Also in ASC, C.C., cred. I, tom. 37, cat. 37, fol. 143r–v.

50. Serbelloni wrote on behalf of the captain of the work, Pietro Antonio Tolomei—see ASC, Carte Boccapaduli, Arm. II, maz. IV, b. 104b, fasc. 66, L. 1088, and for a transcription, Pecchiai, *Acquedotti e fontane*, 14. For Giovanni Serbelloni (1519–1591), see Salvador Miranda, "Serbelloni, Giovanni Antonio," in Miranda, *Cardinals*.

51. ASC, C.C., cred. I, tom. 37, cat. 37 (ad XIII Kal. Martij MD lxiiii), fol. 168v, "con grandissima instantia"; "con ogni prestezza"; "opera tanto honorevole et necessaria." Also in C.C., cred. I, tom. 22, cat. 22, fols. 45v–46r.

52. ASC, C.C., cred. I, tom. 37, cat. 37 (Pridie Kal. Novembris MD lxiiii), fol. 186r–v, "novo ordine p[er] condure a'effetto l'opera del condur l'acqua di salone à Roma"; and fol. 187v (Tertius Nonas Novembris MD lxiiii). Also in ASC, C.C., cred. I, tom. 22, cat. 22, fol. 102r–v.

53. ASC, C.C., cred. I, tom 37, cat. 37 (ad decimum octavum Kal. Februarij MD lxv), fols. 191v–192r. Also in ASC, C.C., cred. I, tom. 22, cat. 22, fols. 88v, 91v.

54. ASC, C.C., cred. I, tom. 21, cat. 21 (14 Janarijs 1562; ad XIX Kal. Februarij MD lxii), fols. 147v–148v, "à riveder il lavoro dell'acqua di salone, et visitarlo continovamente accio l'opera si faccia conveniente" (147 v); "ad beneplacitum Popoli Ro" (148 v). Also in ASC, C.C., cred. I, tom. 37, cat. 37, fol. 103r–v. For the papal view of Trevisi as "a sufficient architect," see ASC, C.C., cred. I, tom. 21, cat 21 (ad decimum Kal. Septembris MD lxi), fol. 120v, "un architetto sufficiente."

55. ASC, C.C., cred. I, tom. 21, cat. 21 (9 Aprilis; ad quintum Idus Aprilis), fol. 163r–v. Also in ASC, C.C., cred. I, tom. 37, cat. 37, fol. 110r–v.

56. ASR, Camerale I, Mandati 921, fol. 42r. The final authorization noted in the Mandati occurs on September 13, 1566.

57. ASR, Camerale I, Mandato 918, fol. 230r.

58. Peto, *De restitutione ductus Aquae Virginis,* Sig. A3v, "cum interim Trivisius ille misere admodum diem suum obijsset." For Peto's appointment to the committee on the aqueduct, see ASC, C.C., cred. I, tom. 21, cat. 21, fols. 86r–87v.

59. Many of these statues are now in the Capitoline Museum. See Pastor, *History of the Popes,* 17:110–116; and Lemaître, *Saint Pie V,* esp. 219–224, for his attitude toward antiquities. Concerning the redistribution of antiquities during his pontificate, see also Benedetti and Zander, *L'arte in Roma,* 377–381; Deswarte-Rosa, "Le cardinal Giovanni Ricci de Montepulciano," 158–161; Coffin, *Pirro Ligorio,* 73–74; and Robert W. Gaston, "Pirro Ligorio, the Casino of Pius IV, and Antiques for the Medici: Some New Documents," *Journal of the Warburg and Courtauld Institutes* 47 (1984): 205–209.

60. ASC, C.C., cred. I, tom. 37, cat. 37 (pro die ij Martij MD lxvi), fol. 224r, "Essendo la condutione dell'Acqua di Salone opera importantissima dove si è spesa grossa somma di denari"; "non solo non ne siegue l'effetto per il quale si è speso il danaro, ma anche la spesa et opera gia fatta, si puote dire persa et buttata via." Also in ASC, C.C., cred. I, tom. 22, cat. 22, fols. 173–174r.

61. ASC, C.C., cred. I, tom. 37, cat. 37 (ad VI Idus Junii MD lxvi), fol. 240 v. Also in ASC, C.C., cred. I, tom. 23, cat. 23, fol. 112r. For Pius V's building activities, particularly the aqueduct, see D'Onofrio, *Le fontane di Roma,* 60–68; Karmon, "Restoring the Ancient Water Supply System"; Lanciani, *Storia degli scavi,* 4:19–20; Pecchiai, *Acquedotti e fontane,* 15–20; Rinne, "Between Precedent and Experiment"; and Simoncini, *Roma: Le trasformazioni urbane nel Cinquecento,* 1:223–248, esp. 234 (for the Acqua Vergine).

62. ASC, C.C., cred. I, tom. 37, cat. 37 (Pridie Nonas Julii MD lxvi), fol. 248r, "Sua Sanctita era risoluta, che la detta opera in ogni muodo si conduca alla sua debita perfettione." Also in ASC, C.C., cred. I, tom. 23, cat. 23, fol. 212r–v.

63. For the outbreak of the illness, see Pecchiai, *Roma nel Cinquecento,* 529; Rinne, *Waters of Rome,* 52; and see ASC, C.C., cred. I, tom. 37, cat. 37 (Ad V Idus Septembris. MD lxvi

[September 9, 1566]), fols. 252v–253r, which itemize measures taken by a Secret Council to provide care for the sick. Also in ASC, C.C., cred. I, tom. 23, cat. 23, fols. 28v–29v.

64. The first congregation, the *Congregazione super viis pontibus et fontibus,* met monthly from its creation in 1567 (the first meeting held April 4) until it was abolished by Sixtus V in 1588. The minutes are extant: see ASR, *Congregazione super viis pontibus et fontibus,* Registro 1; and Genovese and Sinisi, *Pro ornatu et publica utilitate.* The second was the *Congregatio super Aqua Salonis* (ASC, C.C., cred. VI, tom. 50, cat. 451 [stragr. 23]. See also Sinisi, "I *notarii magistrorum stratarum* nel '500"; "Lavori pubblici di acque e strade e congregazioni cardinalizie in epoca sistina e presistina," and Verdi, "Da ufficiali capitolini a commissari apostolici," in Spezzaferro and Tittoni, *Il Campidoglio e Sisto V,* 50–52 and 54–62 respectively.

65. For Ricci's biography and his activities as a builder and collector, see especially Deswarte-Rosa, "Le cardinal Giovanni Ricci de Montepulciano," 110–169; and see Glenn M. Andres, *The Villa Medici in Roma,* 2 vols. (New York: Garland, 1976), 1:68–117; Hubert Jedin, "Kardinal Giovanni Ricci (1497–1574)," *Lateranum,* n.s., 15 (1949): 269–358; and Charles-Martial de Witte, *La Correspondance des premiers nonces permanents au Portugal, 1532–1533* (Lisbon: Academia Portuguesa da História, 1986), 1:133–328.

66. Andres, *Villa Medici,* 1:70–72; Deswarte-Rosa, "Le cardinal Giovanni Ricci de Montepulciano," 117–119; and Jedin, "Kardinal Giovanni Ricci," esp. 313–314.

67. Andres, *Villa Medici,* 1:68–205; Deswarte-Rosa, "Le cardinal Giovanni Ricci de Montepulciano," esp. 128–139. Technically, the villa was sold to Ricci's son Giovanni and nephew Giulio Ricci for 2,000 scudi, but he himself was involved in every aspect of its purchase and renovation. An inventory of Sangallo's collection reveals 149 pieces of sculpture and fragments. See Antonino Bertolotti, "Nuovi documenti intorno all'architetto Antonio da Sangallo (il Giovane) ed alla sua famiglia": III," *Buonarroti,* 3rd ser., 4 (1892): 280–281. For the Crescenzi family and their ownership of the area, see Pier Nicola Pagliara, "Le Pincio du XVe au XVIe siècle et la *vigna* de Marcello Crescenzi," in Chastel and Morel, *La Villa Médicis,* 2:92–109.

68. Andres, *Villa Medici,* 1:118–205, and for garden expenditures, 176; Deswarte-Rosa, "Le cardinal Giovanni Ricci de Montepulciano"; Andres, "Le jardin du cardinal Ricci"; and Fritz-Eugen Keller, "Une villa de la Renaissance sur le site d'une villa antique"; all in Chastel and Morel, *La Villa Médicis,* 2:64–77, 2:141–145, and 2:342–350, respectively.

69. For Orsini's gardens, see esp. Coffin, *Gardens and Gardening,* 128–132; and D'Onofrio, *Le fontane di Roma,* 79, 134–138.

70. The council meeting is reported in ASC, C.C., cred. I, tom. 38, cat. 38, fol. 89r. Also in ASC, C.C., cred. I, tom. 23, cat. 23, fol. 176r–v. For the *avviso,* see BAV, Urb. lat. 1040, fol. 597v (14 August 1568), "S'è rissoluto, che la fabrica gia cominciata da Pio. 4 per condursi qua l'aqua di salone si finischi secondo il dissegno à beneficio publico."

71. ASR, *Congregazione super viis pontibus et fontibus,* Registro 1, fol. 12v, "Die 18 settembris 1568, fuit congregatione pro Aqua Salonis,", "in Loco di mastro Nanni Architetto si surroga per la congregatione mastro Jacopino de la porta—che serv' il campidoglio"; and Genovese and Sinisi, *Pro ornatu et publica utilitate,* 14 (Congregation no. 10). For Nanni di Baccio Bigio, see Maria Grazia Ercolino, "Lippi, Giovanni di Bartolomeo, detto Nanni di Baccio Bigio," *DBI,* 65:209–212.

72. See Alessandra Anselmi, "Porta, Giacomo della," *DA,* 25:258–261; Anna Bedon, "Della Porta, Giacomo," *DBI,* 37:160–170; and Vitaliano Tiberia, *Giacomo della Porta: Un architetto tra Manierismo e Barocco* ([Rome]: Bulzoni, 1974). For Della Porta's work on St.

Peter's, see Federico Bellini, *La basilica di San Pietro da Michelangelo a Della Porta* (Rome: Argos, 2011), 1:189–247.

73. ASR, *Congregazione super viis pontibus et fontibus*, Registro 1, "Die 21 settembris 1568," fol. 13r; and Genovese and Sinisi, *Pro ornatu et publica utilitate*, 14–15 (Congregation no. 11). See also Van Deman, *Building of the Roman Aqueducts*, 167–178, which describes the course of the ancient aqueduct in detail, including some of the areas mentioned. The contractors were Michele di Carpi and company, who had taken on the Arcato (perhaps this refers to the arches at the Marranella), the Casa Letta, and the Bocca Lione; Jo. Zebrera Saonese at the Trevi fountain; and Battista da Bologna at the cross of the Trinity (i.e., Trinità dei Monti).

74. ASR, *Congregazione super viis pontibus et fontibus*, Registro 1, "Die 3 novembris 1568," fols. 15r–16v, "che mastro Jacopino de la porta Architetto deputato, è diligentissimo et e andato et fatto molto neli negotii del aggua di salone"; "con degno ale fatighe"; and "Die 15 Decembris 1568," fol. 19r. And see Genovese and Sinisi, *Pro ornatu et publica utilitate*, 16–17 (Congregation no. 13).

75. ASR, *Congregazione super viis pontibus et fontibus*, Registro 1, "Die 14 novembris, 1568," fol. 17r., "per far intender a quello le difficulta circa il negotio de l'aggua della Salone per saper quello comandare sua Beat.ne." See also Genovese and Sinisi, *Pro ornatu et publica utilitate*, 17–18 (Congregation no. 14).

76. ASC, C.C., cred. I, tom 38, cat. 38, fol. 103v. Also in ASC, C.C., cred. I, tom. 23, cat. 23, fol. 199r–v (December 23, 1568). The four men were Pietro Antonio Bandini, Ascanio Caffarelli, Bernardo Olgiati, and Prospero Boccapaduli.

77. Bauer, "Die Epochen der Papstfinanz," 488–491; Colzi, *Il debito pubblico*, esp. 19–57; Delumeau, *Vie économique et sociale*, 2:783–824; Ferraro, "Nobility of Rome," 361–391, 823–859; Partner, "Papal Financial Policy," 23–31; and Pecchiai, *Roma nel Cinquecento*, 282–286.

78. Colzi, *Il debito pubblico*, 32–33.

79. ASC, C.C., cred. I, tom. 38, cat. 38, fol. 128r, "Deputati ad inveniendum pecunias pro Aqua Salonis." Also in ASC, C.C., cred. I, tom. 24, cat. 24, fol. 51v.

80. Peto, *De restitutione ductus Aquae Virginis*, Sig. A3r–A4v.

81. Montepulciano, Ricci Archives, Instrumenti diversi, *Motu proprio* of Pius V, June 28, 1570 (which I have not seen); and ASV, Arm. LII, tome III, fol. 223r–v. See also Andres, *Villa Medici*, 1:146; and Deswarte-Rosa, "Le cardinal Giovanni Ricci de Montepulciano," 146.

82. For Acqua Vergine water given or sold in the 1570s and 1580s, see ASR, Presidenza Acquedotti Urbani, b. 2, 1r–95r; for Peto's award, 1r–1v. Peto's award is dated May 26, 1576. A note on the margin dated August 10, 1610, renews the concession for his successors. Peto was given his water, but in most cases it was sold for eighty scudi for a full *oncia*, forty for a half, and twenty for a fourth. See also ASC, C.C., cred. VI, tom. 53, cat. 454, stragr. 25, fols. 4r–95v, which records grants of water (usually for payment) under Gregory XIII, including the free grant to Peto in appreciation for his work on the aqueduct (fols. 6v–7r). And see BAV, Codex Chigiano H II 43, fol. 2r, May 26, 1576, which also records Luca Peto's *oncia* of water.

83. For Cardinal Orsini's garden and water, see Aloisio Antinori, "L'uso privato dell'Acqua Vergine dalla realizzazione della nuova rete idrica alla revisione generale del 1631," in Antinori, "Le reti dell'acqua," 243 (for a concession to him and the Ospedale degli Incurabili for three *oncie* in 1576); ASC, C.C., cred. I, tom. 38, cat. 38, fols. 369r and 372v on April 26, 1572; C.C., cred. VI, tom. 53, cat. 454, stragr. 25, fols. 5r–6r (for the three *oncie* for him and the Hospital for Incurables); and ASR, Presidenza Acquedotti Urbani, b. 2, fols. 1v–2v. For the hospital, see esp. Henderson, "'*Mal Francese*' in Sixteenth-Century Rome."

84. Grippetto's reward of a quarter of an *oncia* is recorded in ASR, Presidenza Acquedotti Urbani, b. 2, fol. 15r–v. See Rinne, "Hydraulic Infrastructure and Urbanism," 209, for Della Porta's award in 1597 of Acqua Felice water for his gardens on the Quirinal.

85. For a copy, see ASC, C.C., cred. VI, tom. 50, cat. 451 (stragr. 23), fol. 1r–v, "cum maximo totius populi applauso et gaudi ad publicam utilitatem et ipsius urbis ornatum."

86. See Frank Becker, "Intorno all'acquedotto Vergine in Roma," in Conforti and Hopkins, *Architettura e tecnologia,* 158–177; D'Onofrio, *Le fontane di Roma,* esp. 60–123; Pecchiai, *Acquedotti e fontane,* 43–56; and Katherine Wentworth Rinne, "Fluid Precision: Giacomo della Porta and the Acqua Vergine Fountains of Rome," in *Landscapes of Memory and Experience,* ed. Jan Birksted (New York: Spoon Press, 2000), esp. 183–185; *Waters of Rome,* esp. 56–60; and C. Paola Scavizzi, "La rete idrica urbana in età moderna," *Storia della Città,* 29 (1984): 77–96.

87. Genovese and Sinisi, *Pro ornatu et publica utilitate,* 37–39 (Congregation no. 33); and ASR, *Congregazione super viis pontibus et fontibus,* Registro 1 (November 10, 1570), fols. 51r–52v.

88. For the record of this new congregation, see ASC, C.C., cred. VI, tom. 50, cat. 451 (stragr. 23), fols. 1r–3v and 3r–v for the list of proposed new fountains (meeting of November 4, 1570). See also D'Onofrio, *Le fontane di Roma,* 78–86, and Rinne, *Waters of Rome,* 65–66.

89. ASC, C.C., cred. I, tom. 38, cat. 38 (Die XVI Ianarij 1571), fols. 250v–253v, which includes the papal *Motu proprio* that authorized the transaction; fol. 251v for a copy of the contract that was signed by Bonamici; and fol. 270r (xiiii Martij 1571) for the monetary transaction and extraction of the *gabella della carne.* Also in C.C., cred. I, tom. 25, cat. 25, fols. 13r–17r, 44r–v. See also Pecchiai, *Acquedotti e fontane,* 19–21.

90. The relevant contract mentions Giacomo Della Porta and Bartolomeo Grippetto as well as the three masons (*muratori*), Giovanni alias Abramo, Michele da Carpi, and Ludovico da Bellinzona in a contract of June 6, 1571. ASR, *Notai del Tribunale delle Acque e Strade* (Octavius Gracchus), b. 6, prot. II, fols. 215r–216r. See D'Onofrio, *Le fontane di Roma,* 66–67; and Rinne, *Waters of Rome,* 62–64.

91. ASR, *Notai del Tribunale delle Acque e Strade,* b. 6, prot. II, fols. 269r–271r. The masons were Curzio Maccarone and Ludovico Caronica. The members of the congregation named in the contract were the Cardinal Flavio Orsini, Tommaso Cavalieri, a Roman conservator, and Prospero Boccapaduli.

92. For Guglielmo Della Porta, see Carrol Brentano, "Della Porta, Guglielmo," *DBI,* 37: 192–199, and Hanno-Walter Kruft, "Guglielmo della Porta," *DA,* 25:255–257.

93. ASC, C.C., cred. I, tom. 38, cat. 38 (March 5, 1571; III Nonas Martij 1571), fols. 266r–266v, "senza farne parte ne à Noi, ne alli Diputati del Po"; "il voto consultivo"; "di piu lamentandoci"; "si commanda che li Conservatori et Diputati non possano fare cosa alcuna senza intervento delli detti Maestri." The resolution was ratified in the Public Council of xiiii Martij 1571 (fol. 269v). Also in C.C., cred. I, tom. 25, cat. 25, fols. 37v–38v, and for the Public Council ratification, 43v–44r.

94. The contract of May 17, 1571, can be found in ASR, *Notai del Tribunale delle Acque e Strade* (Octavius Gracchus), b. 6, prot. II, fols. 183v–186r, and see Pecchiai, *Acquedotti e fontane,* 23–25.

95. ASR, *Notai del Tribunale delle Acque e Strade* (Octavius Gracchus), b. 6, prot. II, fols. 183v–186r, citation on 183v, "in stipulatione et obligatione huiusmodi contractus noluit

iurare neque se obligare volentes ad effectuationem infrascripti negotii super ductu aque virginis."

96. ASC, C.C., cred. I, tom. 38, cat. 38, fol. 283r–283v (xviij Maii 1571; XV Kal. Junij MD lxxi), "da parte di N.S."; "si desse fine a questo negotio"; "che se fra otto giorni prossimi venisse qualche maestro per fare il medesmo lavoro à migliore et minore prezzo per utilità del Popolo"; "et fatto cassare il contratto stipolato." Also in C.C., cred. I, tom. 25, cat. 25, fol. 65r–v.

97. ASC, C.C., cred. I, tom. 38, cat. 38 (xij Junij 1571; pridie Idus Junij MD lxxi), fol. 286r, "ne volesse pigliar parere di nuovo dalli periti del Arte"; "piu presto con parole brusche"; "homini esperti et iuditiosi"; "che loro facevano fossi ben fatto. Also in C.C., cred. I, tom. 25, cat. 25, fol. 71v.

98. Their contract of June 6, 1571, is in ASR, Notai del Tribunale delle Acque e Strade (Octavius Gracchus), b. 6, prot. II, fols. 216r–217r. The stonemasons were Francesco Schella, Andrea Lucchesini, and Girolamo Caroli da Orta.

99. See D'Onofrio, *Le fontane di Roma*, 65–68, and Rinne, *Waters of Rome*, 74–76. The report is found in ASC, C.C., cred. VI, tom. 59, cat. 460, fols. 19r–20v (modern foliation; earlier foliation, fols. 3r–4r). It is in poor and intermittently illegible condition, making the transcription made by Pecchiai, *Acquedotti e fontane*, 32–34, extremely valuable: "dicto disegnio et inventione del frate non sarebbe stata bona et che averebbe fatto sempre danno senza poterlo mai più rimediare"; "faceva danno come un crivello."

100. ASC, C.C. cred. VI, tom. 59, cat. 460, fols. 19r–20v, "eccetto un condutto di terra cotta dove se ne sono trovati alcuni rotti per difetto del maestro che li metteva in opera"; "il romor dell'acqua de le cantine"; "quanto tiene detto condutto di trevertino"; "fu riprobato da infiniti giuditij di persone inteligenti"; "che riuscirebbe cosi pericolosi et poi sarebbe inrimediabbile, et di questo ve ne sono infiniti testimonij."

101. ASC, C.C., cred. VI, tom. 59, cat. 460, fols. 19r–20v, "l'opera de li condutti di terra cotta"; "li dua Architetti che avevano fatto venir l'acqua de Salone a Roma"; "con li giuditio loro sempre ànno aiutato et riparato alli errori et danni"; "tutto con più avantaggio del popolo."

102. ASC, C.C., cred. I, tom. 38, cat. 38, fol. 289r–v (xxv Junii 1571; ad vij Kal. Julii MD lxxi), "molti potessero godere di questo benefitio"; "la bellezza consiste che siano per la città molte fontane." Also in C.C., cred. I, tom. 25, cat. 25, fols. 76r–77v.

103. For the council's final decision, ASC, C.C., cred. I, tom. 38, cat. 38 (xvi Octobris 1571; xvij Kal. Novembris MD lxxi), fols. 336v–337r, Secret Council, and for the Public Council (xxv Octobris 1571; ad viij Kal. Novembris MD lxxi), fol. 339r. Also in C.C., cred. I, tom. 25, cat. 25, fols. 147v–148v, 152v. For *oncie* and pipe sizes, ASC, Carte Boccapaduli, Arm. II, maz. IV, b. 104b, fasc. 66, L. 1096; and see Antinori, "L'uso privato dell'Acqua Vergine"; and Scavizzi, "La rete idrica," esp. 78–85.

104. ASC, Carte Boccapaduli, Arm. II, maz. IV, b. 104b, fasc. 63, L. 1042, L 1043. This file is dated 1563, but it contains documents from long after. The document being cited, "Capituli, Modi, Prezi, et quantita da concedersi à particolari dell 'acqua di Salone," is undated but is undoubtedly from 1571. The document is present in two copies, one with the amount of scudi left blank, both with images of 3 pipes on 1 *oncia*, ½ *oncia* and ¼ *oncia*, one labeled and the other not.

105. BAV, Codex Chigiano H. II. 43, fol. 9r, "Editto Che le Fontane s'accommodino, e si murino secondo il Livello dato" (Rome: Stampatori Camerali, 1598).

CHAPTER 4

1. Andrea Fulvio, *L'antichità di Roma di Andrea Fulvio antiquario* [...] *con le aggiuntioni e annotationi di Girolamo Ferrucci* [...] (Venice: Girolamo Francini Libraro in Roma, 1588), fols. 74r–75r; and see Weiss, "Andrea Fulvio antiquario romano." See also David Karmon, *The Ruin of the Eternal City: Antiquity and Preservation in Renaissance Rome* (New York: Oxford University Press, 2011), 170–198, 193–196 on the 1561 failure.

2. Fulvio, *L'antichità di Roma,* fols. 74r–75r, "acchioche servisse per la commodità de' carri, & de' cavalli, che molte mercantile da la Ripa à Roma conducono"; "con argani, & funi da un capo all'altro di detto ponte, si spezzò uno de piu grossi canapi, che sosteneva la maggior parte del peso; & la machina cadendo in precipitio nel Tevere, si spezzò e andò il tutto in rovina, con grandissimo danno dell'artefice, che lo fece"; "cadde precipitosamente nel Tevere, fracassandosi tutto & andando talmente in rovina, & perditione, che non se ne puote ricuperare cosa alcuna." The exact nature of the *machina* is not described, and although it could be some kind of constructed element of the bridge being pulled across the gap, I think it more likely a heavy lifting devise or crane. For such machines, see Marconi, *Edificando Roma barocca,* 197–230. I thank Claudia Conforti for discussing this passage with me.

3. For masonry construction foundations by the ancient Romans, see Jean-Pierre Adam, *Roman Building: Materials and Techniques,* trans. Anthony Mathews (London: Routledge, 1999), esp. 125–126 and (specifically for bridges) 284–288; and for a later period, see Roberto Marta, *Tecnica costruttiva a Roma nel Medioevo/Construction Techniques of the Middle Ages in Rome* (Rome: Edizione Kappa, 1989), esp. 53–61. See also Camilla Torre, *Ponti in muratura: Dizionario storico-tecnologico* (Florence: Alinea Editrice, 2003), esp. 80–112 (*fondazione*), and 165–194 (*pila*). See also Nicoletta Marconi, "I ponti: Teoria e pratica nei trattati di architettura tra XVI e XVIII secolo," and Claudia Conforti, "Il cantiere di Michelangelo al Ponte Santa Maria a Roma (1548–49)," both in Calabi and Conforti, *I ponti delle capitali d'Europa,* 38–57 and 75–87, respectively; and Conforti, "Modes and Techniques of Building on Water in 16th-Century Rome," in *Practice and Science in Early Modern Italian Building: Towards an Epistemic History of Architecture,* ed. Hermann Schlimme (Milan: Electa, 2006), 31–42. See also, D'Onofrio, *Il Tevere,* 141–165; Vittorio Galliazzo, *I ponti romani* (Treviso: Canova, 1994), 2:18–20; and Karmon, *Ruin of the Eternal City,* 170–198 (175 for tufa).

4. See Conforti, "Il cantiere di Michelangelo," 38–57; "Modes and Techniques of Building," 31–42; and Karmon, *Ruin of the Eternal City,* 170–198. For Michelangelo's skill in engineering, see William E. Wallace, "Michelangelo Engineer," in Conforti and Hopkins *Architettura e tecnologia,* 96–107.

5. ASC, C.C., cred. I, tom. 21, cat. 21 (Quintus Idus Iulii 1561), fol. 99v (Secret Council); August 6, 1561 (Octavum Idus Augusti 1561), fols. 107v–108r (Public Council); (XIX Kal. Settembris 1561), fols. 110v–111r and (Decimum sextum Kal. Octobris, MD lxi), fol. 120v, "in alcune parti in pietra." Also in C.C., cred. I, tom. 37, cat. 37, fols. 83v, 88v, and 91r. For a discussion, see Conforti, "Il cantiere di Michelangelo," 77–78; D'Onofrio, *Il Tevere,* 147–150; and Karmon, *Ruin of the Eternal City,* 193–196.

6. ASC, C.C., cred. I, tom. 21, cat. 21 (Quintus Idus Iulii 1561), fol. 99v (Secret Council); (Octavum Idus Augusti 1561), fols. 107v–108r, "Che l'Ill.i Sig. Cons.i, e i Priori, Sig[nori] Mastri di strade et quattro gentilhuomini da eleggessi vadino da sua Sta', et fattala capace delle miserie di q[uest]o Po[polo]"; (XIX Kal. Settembris 1561), fols. 110v–111r and 112v–113r; and

(Decimum sextum Kal. Octobris, MD lxi), fol. 120v. Also in C.C., cred. I, tom. 37, cat. 37, fols. 83v, 86v, 87v, 88v and 91r.

7. For the Tiber Island bridges, see esp. D'Onofrio, *Il Tevere*, 94–122; and Galliazzo, *I ponti romani*, 2:10–13 (for Ponte Cestio) and 20–23 (for Ponte Fabricio). For a concise account of the ancient bridges of Rome, see Rabun Taylor, "Tiber River Bridges and the Development of the Ancient City of Rome," *Waters of Rome* 2 (June 2002), http://www3.iath.virginia.edu/waters/Journal2TaylorNew.pdf. "Quattro Capi" can sometimes refer to Ponte Fabricio alone and sometimes to both of the island bridges—see Maffioli, *La via delle acque*, 200–201, who shows that the calculations used by Giovanni Fontana in his study of the currants in 1599 only make sense if they refer to both bridges.

8. ASR, Presidenza delle Strade, b. 445, fols. 385r–397v, "da me Antonio trivisu sotto maestro di stratta"; "alcune case Magazeni molle et fenilli per lo racconciamento del Ponte Quaatro [*sic*] Capi quale e necessarissimo et molto frequentato per la ruina del Ponte di S.ta Maria."

9. For the Ponte Sisto, see D'Onofrio, *Il Tevere*, 203–225; Galliazzo, *I ponti romani*, 2:8–10; Minou Schraven, "Founding Rome Anew: Pope Sixtus IV and the Foundation of Ponte Sisto, 1473," in *Foundation, Dedication, and Consecration in Early Modern Europe*, ed. Maarten Delbeke and Minou Schraven (London: Brill, 2012), 129–151; and Taylor, "Tiber River Bridges," 10–11.

10. ASC, C.C., cred. I, tom. 37, cat. 37 (Septimum Kal. Septembris M.D lxiiii), fol. 182v, "piu necessaria perche minaccia ruina sicome e stato riferito da huomini periti." Also in C.C., cred. I, tom. 22, cat. 22, fol. 80r.

11. ASC, C.C., cred. I, tom. 37, cat. 37 (Septimum Kal. Septembris MD lxiiii), fol. 182v; Secret Council, September 23, 1564 (Nonum Kal. Octobris MD lxiiii), fol. 183r "si truova esausta." Also in C.C., cred. I, tom. 22, cat. 22, fols. 80r, 81v.

12. ASC, C.C., cred. I, tom. 37, cat. 37 (October 31, 1564; Pridie Kal. Novembris MD lxiiii), fol. 186r, "non solo per il Populo: ma anco per il clero et camera apostolica." The discussion of how to raise money continues, turning to the fees collected by customs officials. The delegation included the usual officers of the Commune as well as five nobles, elected in a Public Council of xxii Julii 1564. See C.C., cred. I, tom. 37, fol. 181v; and C.C., cred. I, tom. 22, cat. 22, fol. 77r.

13. ASC, C.C, cred. I, tom. 37, cat. 37 (Tertium Nonas Novembris MD lxiiij), fol. 187v, and (Decimum Octavum Kal. Febrij MD lxv, January 15, 1565), 191v–192r, "a non indugiar piu." Also in C.C. cred. I, tom. 22, cat. 22, fols. 91v–92r, fols. 102r–v, 105r–v.

14. ASC, C.C., cred. I, tom. 37, cat. 37 (VIII Idus Octobris MD lxv), 212r–v, "sei o piu Gentelhuomini." Also in C.C., cred. I, tom. 22, cat. 22, fols. 124v, 148v–149r, and 150r–v.

15. ASC, C.C., cred. I, tom. 37, cat. 37 (XI Kal. Decembris MD lxv), fol. 214v. Also in C.C., cred, I, tom. 22, cat. 22, fols. 154r–155r.

16. ASC, C.C., cred. I, tom. 37, cat. 37 (III Nonas Decembris MD lxv), fol. 215v, "si dovessero fare due Archi et un sperone ò vero pilastro, perche l'opera sarrebbe cosi molto piu Honorevole, sicura, et durabile, et nel Ponte sisto si riparasse quell Pilastro che e pericoloso da Ruvinare"; "al modo di Trovar qualche somma di danari per dar principio à dett'opera, et per hora proveder' al meno di comprare il legname necessario." Also in C.C., cred. I, tom. 22, cat. 22, fol. 157r–v.

17. ASC, C.C., cred. I, tom. 37, cat. 37, fol. 233v, "consultare il modo"; "sodisfare al santo desiderio di N.S. et al benefitio publico"; and again, Public Council, May 10 (VI Idus Maij

MD lxvi), fols. 234v–235r. Also in C.C., cred. I, tom. 22, cat. 22, fol. 191r, and C.C., cred. I, tom. 23, cat. 23, fol. 3r–v.

18. ASC, C.C., cred. I, tom. 37, cat. 37 (ad XIII Kal Junij MD lxvi), fol. 237r; and for repeated requests, June 8, Secret Council (ad vi idus Junij MD lxvi), fol. 241r; June 18, Public Council (ad XIIII Kal. Julii MD lxvi), fol. 242r. Also in C.C., cred. I, tom. 23, cat. 23, fols. 6v–7r, 12r, and 13v–14r.

19. ASC, C.C., cred. I, tom. 37, cat. 37 (ad XIII Kal Julij MD lxvi), fol. 243v, "che sino ad hora non vi si vede alcun preparamento, del che ne restava non puoco maravigliato, et che voleva in tutti modi si ristaurasse." Also in C.C., cred. I, tom. 23, cat. 23, fol. 16r.

20. ASC, C.C., cred. I, tom. 37, cat. 37 (XI Kal. Decembris MD lxvi), fol. 258r, "gia piu Mesi et Anni che tanto sua S.tà quanto suo Antecessore havevano ordinate alli Cons.ri"; "perche chiaramente si vede ch'il Ponte ogni giorno và peggiorando"; "si spenderebbe molto piu et con maggior incommodo del Popolo." Also in C.C., cred. I, tom. 23, cat. 23, fols. 35v–6r.

21. ASC, C.C., cred. I, tom. 37, cat. 37, May 7, 1566, fol. 233v; May 20, 1566 (ad XIII Kal. Junij MD lxvi), fol. 237r. Also in C.C., cred. I, tom. 22, cat. 22, fol. 191r; and for May 20, C.C., cred. I, tom. 23, cat. 23, fols. 6v–7r.

22. ASC, C.C., cred. I, tom. 37, cat. 37 (ad VI Kal. Julii MD lxvi), fol. 246v. Also in C.C., cred. I, tom. 23, cat. 23, fols. 18v–19r.

23. ASC, C.C., cred. I, tom. 37, cat. 37 (V Idus Septembris. MD lxvi), fol. 253r. Also in C.C., cred. I, tom. 23, cat. 23, fol. 30r.

24. ASC, C.C., cred. I, tom 37, cat. 37 (III Nonas Decembris MD lxvi), fols. 261v; approved by the Public Council on December 5, 1566 (Nonis Decembris MD lxvi), fols. 262v–263r. Also in C.C., cred. I, tom. 23, cat. 23, fols. 40v–41r, 43r–v.

25. ASC, C.C., cred. I, tom. 38, cat. 38, fol. 4v. Also in C.C., cred. I, tom. 23, cat. 23, fol. 54r. In the latter (the original record) only Matteo da Castello is mentioned with a blank space for a possible second architect. In the copy (Cred. I, tom. 38), the black space is filled with the well-known architect Vignola. However, extant records do not show Vignola working on the project. For the "assistants of the work," see ASC, C.C., cred. I, tom. 37, cat. 37, December 5, 1566 (Nonis Decembris MD lxvi), fols. 262v–263r, "assistenti all'opera." Also in C.C., cred. I, tom. 23, cat. 23, fol. 43r–v. The men originally chosen included a fourth, Girolamo Pico.

26. For Matteo da Castello, see Anderson, "Master Carpenters in Renaissance and Baroque Rome," 1:104–110 and 452–454; Renata Battaglini Di Stasio, "Bartolini (Bortolino, Bartolani), Matteo, detto—Matteo da Castello," *DBI*, 6:616–661; and Renato Lefevre, "Schede su Matteo Bartolini da Castello architetto in Roma nel tardo Cinquecento," *Rassegna degli Archivi di Stato* 27 (1967): 142–160.

27. ASC, C.C., cred. I, tom. 38, cat. 38 (VII Idus Maij MD lxvii), fol. 14v, "il qual dinaro si ha già per speso in legnami et altri preparamenti necessarij." Also in C.C., cred. I, tom. 23, cat. 23, fols. 67r–68v.

28. ASC, C.C., cred. I, tom. 38, cat. 38 (Idibus Maij MD lxvii), fol. 16r. Also in C.C., cred. I, tom. 23, cat. 23, fols. 69v–70r.

29. ASC, C.C., cred. I, tom. 38, cat. 38 (ad iiij Kal. Maij MD lxix), fol. 120v. Also in C.C., cred. I, tom. 24, cat. 24, fols. 36v–37r.

30. See esp. Karmon, *Ruin of the Eternal City*, 194–196.

31. On visiting the seven churches, see esp. Martine Boiteux, "Parcours rituels romains à l'époque moderne," in Visceglia and Brice *Cérémonial et ritual*, 52–65.

32. ASC, C.C., cred. I, tom. 38, cat. 38 (Quarto Idus Martij MD lxxiij), fol. 429v, "affigantur publica edicta per Urbem." Also in C.C., cred. I, tom. 26, cat. 26, fol. 43v (the folio number used in this and subsequent citations is the archival stamped number).

33. ASC, C.C., cred. I, tom. 38, cat. 38 (Pridie Nonas Aprilis MD lxxiij), fol. 437r–v, "in tutti i muodi voleva," "del Augmento delli 1250 luoghi della Gabella della Carne vendutti dal Po." Also in C.C., cred. I, tom. 26, cat. 26, fols. 51r–52r.

34. ASC, C.C., cred. I, tom 38, cat. 38 (VIII Nonas [sic: Should be Idus] Aprilis MD lxxiij), fols. 438r–439v. Also in C.C., cred. I, tom. 26, cat. 26, fol. 54r–v. The four magistrates were Tommaso Cavalieri, Mario Maffei, Marcello Arberini, and Prospero Boccapaduli.

35. ASC, C.C., cred. I, tom. 38, cat. 38 (VI Nonas Maij MD lxxiii), fols. 440v–441r, "secondo l'ordine datoci dal publico Conseglio habbiamo in piu congregationi visto i desegni delli Architettorì et oditi i loro pareri sopra la ristauratione del Ponte santa Maria, et in somma tutti concorrono che il Ponte se rifaccia con dua Archi, per piu fortezza, piu bellezza, et piu perpetuita di esso, et tutti medesmamente concorrono che ve si spenderà da venticinque milia scudi in quì et che li, mille/25 scudi non passara" (440v). Also in C.C., cred. I, tom. 26, cat. 26, fols. 55v–56r.

36. Ibid. On June 15, 1573, men were chosen to sell the shares at a Secret Council: ASC, C.C., cred. I, tom. 38, cat. 38 (XVII Kal. Julij MD lxxiij), fol. 444r–v. Also in C.C., cred. I, tom. 26, cat. 26, fol. 61r. They were approved at the Public Council on June 23: ASC, C.C., cred. I, tom. 38, cat. 38 (VIIII Kal Julij MD lxxiii), fols. 446v–447r; and for the sale on June 27, fols. 449v–450v. Also in C.C., cred. I, tom. 26, cat. 26, fols. 64r–v, 67v–68v.

37. ASC, C.C., cred. I, tom. 38, cat. 38 (III Nonas Maij MD lxxiij), fol. 442r. Also in C.C., cred. I, tom. 26, cat. 26, fol. 57v.

38. ASC, C.C., cred. IV, tom. 95, cat. 339, fols. 9v–10r (lime), 10r–v (transport of lumber). The four men were Ludovico Mattei, Tommaso Cavalieri, Geronimo Alteri, and Antonio Maccarani.

39. ASC, C.C., cred. I, tom. 38, cat. 38 (III Nonas Augusti MD lxxiiij) fol. 537r; and fols. 538v–539r for the Public Council (VIII Idus Augusti 1574, August 6), in which the measure was approved. Also in C.C., cred. I, tom. 26, cat. 26, fols. 176v–177r, 179r.

40. ASC, C.C., cred. I, tom. 38, cat. 38 (Idibus Octobris 1574), fol. 548v. Also in C.C., cred. I, tom. 26, cat. 26, fols. 192r–193r.

41. ASC, C.C., cred. I, tom. 38, cat. 38 (ad XIIII Kal. Novembris MD lxxiiii), fol. 550r. Also in C.C., cred. I, tom. 26, cat. 26, fol. 195r–v; and see Karmon, *Ruin of the Eternal City*, 116–143, for the Colosseum and a history of its use for building materials, and 193–196, for the Ponte Santa Maria.

42. ASC, C.C., cred. I, tom. 38, cat. 38 (Pridie Idus Novembris MD lxxiiii), fol. 551r–v, "Mastro Matthaeo Architetto"; "valorosamente et bene"; "si è fatigato non solo con l'ingegno, ma anco con la persona"; "l'opera che puol restare à qual si voglia impeto del Fiume." Also in C.C., cred. I, tom. 26, cat. 26, fol. 198v.

43. ASC, C.C., cred. I, tom. 38, cat. 38 (IIII Nonas Decembris MD lxxiiii), fol. 555r–v, "Mastro Mattheo Architetto delle fatighe sue"; "pro recognitione et remuneratione laboris sui"; "cum suis privilegijs." Also in C.C., cred. I, tom. 26, cat., 26, fols. 205v–206r, where it is only recorded at the Public Council meeting of December 2, 1574 (IIII Nonas Decembris MD lxxiiii), and not at the secret meeting. See also fol. 217v of this volume, where sixty *luoghi* of the Monte della Carne are sold (making 6,000 scudi) to pay for the repairs and for Matteo's reward.

44. ASR, Camerale II, Tevere, b. 12, fasc. 106, the copy of the concession of the mill to Matteo. And see Segarra Lagunes, *Il Tevere e Roma*, 292–293; and D'Onofrio, *Il Tevere*, 154–155.

45. ASC, C.C., cred. VI, tom. 59, cat. 460, fol. 310r–v. The letter is bound with a cover sheet labeled "Folium protestationis pro Domino Matteo Bartolino de Castello" (311v), "molte raggioni"; "quali io allegai per provare la necessità secondo il mio parere evidente che si dovessi rimpire quel fondo che è tra il pilastro, qual'hora s'è rifondato verso Trastevere, e l'altro pilastro verso Roma." For the Capitoline Council minutes in which the two officers are recorded, ASC, C.C., cred. I, tom. 26, cat. 26, fol. 235r–v.

46. ASC, C.C., cred. VI, tom. 59, cat. 460, fol. 310r, "che non correva tempo, et che si aspetassi per qualche giorno"; "io, conoscendo la necessità non potrei mancare di dire che mi protestava, et che pregava tutti li Signori Loro che si per caso, che nostro Signore Iddio non lo facessi altro fosse intervento perch'io tenevo per certo che il continuo rodere, e discalsare"; "ch' velocita del Acqua in detto fondo"; "in brevita di tempo"; "far ruinare, et cascare tutto il rimanente, che si era fatto, con tanta spesa di questo inclito Populo."

47. ASC, C.C., cred. VI, tom. 59, cat. 460, fol. 310r-v, "sempre ho detto cioè di farlo riempire di quei massicci del tempio in pace con altezza di cinque in sei palmi"; "con grandissimo mio dispiacere"; "pregando." Karmon, *Ruin of the Eternal City*, 116–143, discusses the use of the Colosseum as a stone quarry. For the Basilica of Maxentius and its erroneous identification, see David Watkin, *The Roman Forum* (Cambridge, MA: Harvard University Press, 2009), 44–52.

48. For the inscription and translation, with further discussion, see Karmon, *Ruin of the Eternal City*, 195, 282n105.

49. For the flood of 1598, which is beyond the purview of this study, see esp. D'Onofrio, *Il Tevere*, esp. 155–165; San Juan, *Rome: A City Out of Print*, 129–137; and Segarra Lagunes, *Il Tevere e Roma*, esp. 76–81.

50. For Matteo's still-standing pier, see esp. Karmon, *Ruin of the Eternal City*, 195.

51. For the Quirinal Palace, see Antonio Menniti Ippolito, *I papi al Quirinale: Il sovrano pontefice e la ricerca di una residenza* (Rome: Viella, 2004). For Sixtus V's Montalto, see esp. Marcella Culatti, *Villa Montalto Negroni: Fortuna iconografica di un luogo perduto di Roma* (Venice: Istituto Veneto di Scienze, Lettere ed Arti, 2009); Dennis, "Camilla Peretti, Sixtus V"; "Rediscovering the Villa Montalto"; Tod A. Marder, "Sixtus V and the Quirinal," *Journal of the Society of Architectural Historians* 37 (December 1978): 283–294; Matthias Quast, *Die Villa Montalto in Rom: Entstehung und Gestaltung im Cinquecento* (Munich: Tuduv, 1991); and for the garden, Coffin, *Gardens and Gardening*, esp. 97–99, 142–145, and 148–149.

52. For the Acqua Felice see Simona Benedetti, "L'acquedotto Felice da Porta Furba alla Mostra del Mosè," in Piera Sette and Benedetti, *Architetture per la città*, 90–129; Elvira Cajano, "L'acquedotto Felice dalla campagna alla città di Roma," in Porzio, *Impronte sistine*, 17–25; D'Onofrio, *Le fontane di Roma*, 193–275; Fea, *Storia*, 28–36; Pascucci, "L'acqua corrente nel Cinquecento a Roma," 59–71; and Rinne, *Waters of Rome*, 122–134.

53. Fea, *Storia*, 316–317 (no. LVIII), a transcription of the document (with some elisions): "modesta impensa"; "sua utilitate, suaque cura, et impensa"; "non majori pretio quingentorum scutorum pro singula uncia"; "pro palatio nostro apostolico sex uncicae [*sic*] in dicta platea gratis haberi possint." I have not located the original document. The vast difference from the eighty scudi per *oncia* of Acqua Vergine water is a question. It could be taking into account the greater height of the springs and thus the faster and more abundant flow of

water in the conduits of the proposed new aqueduct. Or, as D'Onofrio (*Le fontane di Roma,* 195n6) suggested, it could have been using an *oncia* measure that is different from the one used in the Acqua Vergine. Another explanation could be that Gregory XIII gave the three entrepreneurs complete freedom to establish the prices at whatever the market would bear.

54. ASC, C.C., cred. I, tom. 28, cat. 28 (vi Kal. Junii) fols. 165v, 166r–v: "desiderava grandemente, che si conducesse l'acqua di Pantano di griffi"; "quanto sia utile al publico et al privato"; "a tutto loro spese"; "pro uso publico et publicis plateis, et non alias applicandam et convertandam." See also D'Onofrio, *Le fontane di Roma,* 195–198.

55. BAV, Urb. lat. 1053, fol. 20r. (January 12, 1585): "con autorità dal Papa di imponer gravezza sopra molte cose."

56. His "Chirografo," or edict, is published by Fea, *Storia,* 98 (no. XX). I have not been able to locate the original document.

57. For Villa Montalto, in addition to note 51 above, see Domenico Fontana, *Della trasportatione dell'Obelisco Vaticano* [. . .] (Roma: Domenico Basa, 1590), fols. 37r–38r, in which Fontana stresses the beauty of the gardens and fountains as well as the villa.

58. Fea, *Storia,* 98 (no. XX): "et parvo admodum usui sit possidentibus"; "Aquam Felicem de nostro nomine appellari volumus"; "cum Matthaeo Bartolani Civitatis Castelli, quatenus pertinent ad aquam ducendam, quam ille curam, ut accipimus, suscipere paratus est, tractes, et paciscaris." There is a large bibliography on Ferdinando de' Medici, who was a cardinal in Rome until 1587, at which time he returned to Florence to become the grand duke of Tuscany. See as a start Elena Fasano Guarini, "Ferdinando I de' Medici, granduca di Toscana," *DBI,* 46:258–278. Sixtus V had made Giovanni Bernardino Piscina *commissariato* in October 1586. He was charged with care of the documents of the Camera Apostolica (a beginning of a separate archive that would become the Archivio Secreto Vaticano). However, Sixtus soon abandoned the idea, which was only taken up again later. See Vittorio Peri, "Progetti e rimostranze: Documenti per la storia dell'Archivio Segreto Vaticano dall'erezione alla metà del XVIII secolo," *Archivum Historiae Pontificiae* 19 (1981): 191–237, esp. 199–200 (for Piscina).

59. Fea, *Storia,* 99–100; and see BAV, Urb. lat. 1053, fol. 265r (June 8, 1585).

60. BAV, Urb. lat. 1053, fol. 410r. For the account book, see ASR, Presidenza degli acquedotti urbani, b. 82, prot. 677, fol. 1, for the start of the work on September 28, 1585. For Ferdinando de' Medici and the Palazzo Ricci (now known as Palazzo Medici), see Andres, *Villa Medici,* 1:206–456, and Suzanne B. Butters, "Ferdiand et le jardin du Pincio," in Chastel and Morel, *La Villa Médicis,* 2:351–410.

61. For the account books relevant to the Acqua Felice during the rule of Sixtus V, see ASR, Presidenza degli acquedotti urbani, b. 82, 83, and 84.

62. BAV, Urb. lat. 1053, fol. 451r: "un capo d'Acqua tanto grosso che potra servire non solo à Monte cavallo dove dissegna condurerela ma in molti altri luoghi della città"; "grandezza d'animo."

63. ASR, Presidenza degli acquedotti urbani, b. 82, prot. 677, 3–7 (masons), 7–10 (trench diggers), 11 (food and horses), 11 (hiring of boat), 12 (carters of wood), 12–13 (carpenters), 13–16 (carters), 16–17 (water carriers), 17–18 (metals and nails). See also esp. Rinne, *Waters of Rome,* esp. 123–125.

64. BAV, Urb. lat. 1053, fol. 467v. For cardinal Cesi, see Agostino Borromeo, "Cesi, Pier Donato." *DBI* 24:261–266; and for Cardinal Montalto, Simone Testa, "Peretti Damasceni, Alessandro," *DBI,* 82:340–342.

65. For ancient methods of establishing the gradient, see A. Trevor Hodge, "Aqueducts,"

in *Handbook of Ancient Water Technology*, ed. Orjan Wikander (Leiden: Brill, 2000), 49–57; and Hodge, *Roman Aqueducts*, 171–214. See D'Onofrio, *Le fontane di Roma*, 203–208; and Rinne, *Waters of Rome*, 125–127, for surveying the Acqua Felice.

66. For a summary of expenses during Matteo's months of oversight, see ASR, Presidenza degli acquedotti urbani, b. 83, prot. 685 (1585), "Misure dell'aquedutto del Aqua Felice nel tempo amministrato da Matteo da Castello."

67. For the discussion about the springs of Zagarolo, see BAV, Urb. lat. 1054, fols. 202 and 207r (May 17, 1586), and 348r–v (July 26, 1586), the latter of which reports the dismissal of Matteo and the hiring of Giovanni Fontana. See also D'Onofrio, *Le fontane di Roma*, esp. 203–208.

68. BNN, Ms. XII.G.59/60. For Nigrone, see Fernando Loffredo, "La villa di Pedro de Toledo a Pozzuoli e una sicura provenienza per il *Fiume* di Pierino da Vinci al Louvre," *Rinascimento Meridionale* 2 (2011): 105–107; and see D'Onofrio, *Le fontane di Roma*, esp. 203–208. I warmly thank Fernando Loffredo for telling me about this manuscript and providing photocopies of the relevant pages.

69. For the Fontana family and workshop, see Giovanna Curcio, Nicola Navone, and Sergio Villari, eds., *Studi su Domenico Fontana* (Medrisio: Fondazione Archivio del Moderno, 2011); Sabina De Cavi, *Architecture and Royal Presence: Domenico and Giulio Cesare Fontana in Spanish Naples (1592–1627)* (Newcastle upon Tyne: Cambridge Scholars, 2009); Marcello Fagiolo and Giuseppe Bonaccorso, eds. *Studi sui Fontana: Una dinastia di architetti ticinesi a Roma tra Manierismo e Barocco* (Rome: Gangemi Editore, 2008); Alessandro Ippoliti, "Fontana, Domenico," *DBI*, 48:638–643; Tommaso Manfredi, *La costruzione dell'architetto: Maderno, Borromini, i Fontana e la formazione degli architetti ticinesi a Roma* (Rome: Argos, 2008); Leros Pittoni and Gabrielle Lautenberg, *Roma Felix: La città di Sisto V e Domenico Fontana* (Rome: Viviani, 2002); and Matthias Quast, "Domenico Fontana," *DA*, 11:271–274.

70. For Giovanni Fontana, see Marcello Fagiolo, "Intorno a S. Pietro in Montorio e a Giovanni Fontana: Opere viarie, panorami e fontane"; Margherita Fratarcangeli, "Giovanni Fontana e la sua stripe: Edifici d'acque e inondazioni del Tevere"; Silvia Mangiasciutto, "Giovanni Fontana," in Fagiolo and Bonaccorso, *Studi sui Fontana*, 121–140, 339–354, and 419–420, respectively; and Ippoliti, "Fontana, Giovanni," *DBI*, 48:676–677.

71. Fontana, *Della trasportatione dell'Obelisco*, 54r–56r: "impresa certamente, che non cede a quelle degli antichi"; "tale, c'ha bisognato usarvi una diligentia quasi maravigliosa." A reprint edition includes valuable introductory essays: Domenico Fontana, *Della trasportatione dell'Obelisco Vaticano, 1590*, ed. Adriano Carugo with an introduction by Paolo Portoghesi (Milan: Il Polifilo, 1978). There is also a digital reproduction and English translation of the 1590 edition in the Rosenwald Collection of the Library of Congress: Domenico Fontana, *Della trasportatione dell'Obelisco Vaticano, Rome 1590*, trans. David Sullivan, commentary by Ingrid D. Rowland (Oakland, CA: Octavio, 2002). Summaries and translations are my own, while I have also consulted Sullivan's English translation.

72. Francesco Fontana, *Relazione dello stato vecchio, e nuovo dell'Acqua Felice* [. . .] (Rome: Giovanni Francesco Buagni, 1696), 3: "far prova dell'introito dell'Acqua verso Roma, si avvidde, ma troppo tardi per suo rossore";"per non aver tanta forza di salire dove fermamente si dava a credere il di lei Architetto." For introductions to Carlo Fontana and his son, Francesco Fontana, who was made prefect of the Acqua Felice in 1695, see Manfredi, *La costruzione dell'architetto*, 26–55; Saverio Sturm, "Carlo Fontana," and Lorenzo Finocchi Ghersi,

"Francesco Fontana," in Fagiolo and Bonaccorso, *Studi sui Fontana*, 432–438 and 438–439 respectively; and Helmut Hager, "Fontana, Carlo," *DBI*, 48:633–634.

73. Fontana, *Relazione dello stato vecchio*, 4: "fù costretto ad usarvi una meravigliosa diligenza, unita ad un'estrema fatica, come attesta egli medesimo in alcune distinte memorie di manoscritti fatti da lui istesso sopra tal impressa, i quali si conservano in mia casa, come derivante dall'istesso Ceppo."

74. Fontana, *Della trasportatione dell'Obelisco*, fols. 54r–v; and Fontana, *Della trasportatione dell'Obelisco*, trans. Sullivan, 54: "quasi una nuova Roma, e i giardini ricevendo, e bevendo il già tanto tempo desiderato humore, hora crescono freschi, e verdeggianti." For Sixtus's gifts of water from the Acqua Felice, see esp. ASR, Presidenza degli acquedotti urbani, b. 5, fols. 1–37.

75. Giancarlo Palmerio, "Il reimpiego nella costruzione dell'acquedotto Felice a Roma," in *Il reimpiego in architettura: Recupero, trasformazione, uso*, ed. Jean-François Bernard, Philippe Bernardi, and Daniela Esposito (Rome: École française de Rome and "Sapienza," Università di Roma, 2008), 378–379.

76. ASR, Presidenza degli acquedotti urbani, b. 82, prot. 677, 682, for the shafts; and see Palmerio, "Il reimpiego nella costruzione," 379.

77. Palmerio, "Il reimpiego nella costruzione," 380–381.

78. Rinne, *Waters of Rome*, esp. 127–130; Simoncini, *"Roma restaurata,"* 49–55. For the inscription, see Tyler Lansford, *The Latin Inscriptions of Rome: A Walking Guide* (Baltimore: Johns Hopkins University Press, 2009), 278–279. For the Moses Fountain and the sculpture of Moses, see esp. D'Onofrio, *Le fontane di Roma*, 210–220; Tod Marder, "Sisto V e la Fontana del Mosè," in Fagiolo and Madonna, *Sisto V*, 1:519–543, which illuminates the iconography of the fountain as a whole; and Steven F. Ostrow, "The Discourse of Failure in Seventeenth-Century Rome: Prospero Bresciano's *Moses*," *Art Bulletin* 88 (June 2006): 267–291.

79. See Rinne, *Waters of Rome*, 128–129, on which this paragraph depends.

80. For contracts relevant to the Capitoline Fountain and its derivatives, see ASC, C.C., cred. IV, tom. 95, cat. 339, esp. fols. 189r–268r, and Pecchiai, *Acquedotti e fontane*, 63–69, 74–75, and for further documents, 77–100. See also Carla Benocci, "Giovanni e Domenico Fontana ed i 'sistemi di acque e fontane' nei giardini romani in età sistina," in Fagiolo and Madonna, *Sisto V*, 1:545–557; D'Onofrio, *Le fontane di Roma*, esp. 242–275; and Rinne, *Waters of Rome*, 130–134.

81. Luigi Tomassetti et al., *Bullarum diplomatum et privilegiorum*, vol. 9, no. CLXIX, 177–184.

CHAPTER 5

1. See Jerry Brotton, *Trading Territories: Mapping the Early Modern World* (Ithaca, NY: Cornell University Press, 1998); Conforti, *La città del tardo Rinascimento*, 19–63; Surekha Davies, *Renaissance Ethnography and the Invention of the Human: New Worlds, Maps and Monsters* (Cambridge: Cambridge University Press, 2016); John M. Headley, "Geography and Empire in the Late Renaissance: Botero's Assignment, Western Universalism, and the Civilizing Process," *Renaissance Quarterly* 53 (Winter 2000): 1119–1155; Naomi Miller, *Mapping the City: The Language and Culture of Cartography in the Renaissance* (London: Continuum, 2003); Lucia Nuti, *Ritratti di città: Visione e memoria tra Medioevo e Settecento* (Venice: Marsilio, 1996); Jürgen Schulz, *La cartografia tra scienza e arte: Carte e cartografi nel Rinascimento*

italiano, trans. Tessie Doria De Zuliani, rev. ed. (Modena: Franco Cosimo Panini, 2006); and David Woodward, ed., *The History of Cartography*, vol. 3, *Cartography in the European Renaissance*, 2 pts. (Chicago: University of Chicago Press, 2007).

2. For maps of Rome, see Frutaz, *Le piante di Roma*; Christian Hülsen, *Saggio di bibliografia ragionata delle piante icnografiche e prospettiche di Roma dal 1551 al 1748* (1915; repr., Rome: Bardi, 1969); Camillo Scaccia Scarafoni, *Le piante di Roma possedute dalla biblioteca dell'Istituto e dalle altre biblioteche governative della città* (Rome: La Libreria dello Stato, 1939); and more recently, Jean-Marc Besse and Pascal Dubourg Glatigny, "Cartographier Rome au XVIe siècle (1544–1599)," in Romano, *Rome et la science moderne*, 369–414; Marcello Fagiolo, "Piante di Roma antica e moderna: L'ideologia e i metodi di rappresentazione," and Mario Bevilacqua, "L'immagine di Roma moderna da Bufalini a Nolli: Un modello europeo," in Bevilacqua and Fagiolo *Piante di Roma*, 22–61 and 62–95, respectively; and Maier, *Rome Measured and Imagined*. There is a large literature on the relationship of cartography to political power and authority. See, for example, J. B. Harley, *The New Nature of Maps: Essays in the History of Cartography*, ed. Paul Laxton (Baltimore: Johns Hopkins University Press, 2001), and John Marino, "Administrative Mapping in the Italian States," in Buisseret, *Monarchs, Ministers, and Maps*, 5–25. For maps in atlases, see esp. R. V. Tooley, "Maps in Italian Atlases of the Sixteenth Century, Being a Comparative List of the Italian Maps Issued by Lafreri, Forlani, Duchetti, Bertelli and Others, Found in Atlases," *Imago Mundi* 3 (1939): 12–47. For the broader cartographical context, see Leonardo Rombai, "Cartography in the Central Italian States from 1480 to 1680," in Woodward, *History of Cartography*, pt. 1, 909–939.

3. For the early history of the Aurelian Wall, see Paolo Carafa, "Le Mura Aureliane," in Carandini and Carafa, *Atlante di Roma antica*, 1:85–89, and Hendrik W. Dey, *The Aurelian Wall and the Refashioning of Imperial Rome, AD 271–855* (Cambridge: Cambridge University Press, 2011). See also Ann C. Huppert, "Mapping Ancient Rome in Bufalini's Plan and in Sixteenth-Century Drawings," *Memoirs of the American Academy in Rome* 53 (2008): 79–98; and *Becoming an Architect in Renaissance Italy: Art, Science, and the Career of Baldassarre Peruzzi* (New Haven, CT: Yale University Press, 2015). See also Filippo Camerota, "Dal compasso alla lente: Il rinnovamento delle tecniche di cartografia urbana tra Rinascimento e Barocco," in Bevilacqua and Fagiolo, *Piante di Roma*, 108–115; Uta Lindgren, "Land Surveys, Instruments, and Practitioners in the Renaissance," and Hillary Ballon and David Friedman, "Portraying the City in Early Modern Europe: Measurement, Representation, and Planning," both in Woodward, *History of Cartography*, pt. 1, 477–508 and 680–704, respectively; Lucia Nuti, "The Perspective Plan in the Sixteenth Century: The Invention of a Representational Language," *Art Bulletin* 76 (March 1994): 105–128; and Daniela Stroffolino, *La città misurata: Tecniche e strumenti di rilevamento nei trattati a stampa del Cinquecento* (Rome: Salerno, 1999). For urban changes in Rome and the evidence for them in maps, see Insolera, *Roma: Immagini*, 4–206.

4. P. D. A. Harvey, *The History of Topographical Maps: Symbols, Pictures, and Surveys* (London: Thames and Hudson, 1980), 153–168. For the development of city maps in particular, see also, Miller, *Mapping the City*; Eugen Oberhummer, "Der Stadt Plan: Seine Entwickelung und geographische Bedeutung," In Oberhummer, *Verhandlungen des XVI Deutschen Geographentages zu Nürenberg* (Berlin: Dietrich Reimer [Ernst Vohsen], 1907): 66–101; and for the bird's-eye view, Ballon and Friedman, "Portraying the City," 687–696. For maps of Rome, see Maier, *Rome Measured and Imagined*.

5. See Maier, *Rome Measured and Imagined*, esp. 77–118. See also Ballon and Friedman,

"Portraying the City," 683–685; David H. Friedman with Paul Schlapobersky, "Leonardo Bufalini's Orthogonal *Roma* (1551)," *Thresholds* 28 (2003): 10–16; Jessica Maier, "Leonardo Bufalini e la prima pianta a stampa di Roma, 'la più bella di tutte le cose,'" and Allan Ceen, "Bufalini 1551: Distortion and Rectification," both in Bevilacqua and Fagiolo, *Piante di Roma*, 116–127 and 128–133, respectively. George Schelbert, "Gli acquedotti urbani nelle piante e vedute quattrocentesche e cinquecentesche di Roma," in Bonaccorso, "Le acque e la città," 18–19, points out that Bufalini's misplacement of the Acqua Vergine was replicated in maps for a generation despite the knowledge of its actual location developed during its repair in the same period.

6. See John A. Pinto, "Origins and Development of the Ichnographic City Plan," *Journal of the Society of Architectural Historians* 35 (March 1976): 35–50. More generally for the growing status of mixed mathematics in the sixteenth century, see Francesca Fiorani, *The Marvel of Maps: Art, Cartography and Politics in Renaissance Italy* (New Haven, CT: Yale University Press, 2005), 47–48; Anthony Gerbino and Stephen Johnston, *Compass and Rule: Architecture as Mathematical Practice in England* (New Haven, CT: Yale University Press, 2009); Huppert, *Becoming an Architect*; and Alexander Marr, *Between Raphael and Galileo: Mutio Oddi and the Mathemataical Culture of Late Renaissance Italy* (Chicago: University of Chicago Press, 2011).

7. The term *ichnographia* was first used by Vitruvius (*De architectura* 1.1.2) to mean "ground plan." For the development of the ichnographic plan, see Maier, *Rome Measured and Imagined*, esp. 53–54; Pinto, "Origins and Development"; and Stroffolino, *La città misurata*, 113–140.

8. Patrick Gautier Dalché, *La Géographie de Ptolémée en Occident (IVe–XVIe siècle)* (Turnhout, Belgium: Brepols, 2009); and "The Reception of Ptolemy's *Geography* (End of the Fourteenth to Beginning of the Sixteenth Century)," in Woodward, *History of Cartography*, pt. 1, 285–364.

9. Mario Carpo and Francesco Furlan, eds., *Leon Battista Alberti's Delineation of the City of Rome (Descriptio urbis Romae)*, Medieval and Renaissance Texts and Studies, ed. Jean-Yves Boriaud and Francesco Furlan, trans. Peter Hicks (Tempe: Arizona Center for Medieval and Renaissance Studies, 2007); and for the *Ludi*, Kim Williams, Lionel March, and Stephen Wassell, *The Mathematical Works of Leon Battista Alberti* (Basel: Birkhäuser, 2010), 9–140. See also Pierre Souffrin, "La Geometria practica dans les *Ludi Rerum Mathematicarum*," *Albertiana* 1 (1998): 87–104; Luigi Vagnetti, "La 'Descriptio Urbis Romae': Uno scritto poco noto di Leon Battista Alberti (contributo alla storia del rilevamento architettonico e topografico)," *Quaderni dell'Istituto degli Elementi di Architettura e Rilievo dei Monumenti* 1 (October 1968): 25–80; Joan Gadol, *Leon Battista Alberti: Universal Man of the Early Renaissance* (Chicago: University of Chicago Press, 1969), 70–74, and for the broader context, 167–195; Maier, *Rome Measured and Imagined*, 25–31; Anthony Grafton, *Leon Battista Alberti: Master Builder of the Italian Renaissance* (New York: Hill and Wang, 2000), 242–243; and Pinto, "Origins and Development," 36–38. I thank Mario Carpo and Francesco Furlan for providing the Zubov image used in figure 5.1.

10. Gadol, *Leon Battista Alberti*, esp. 73.

11. Pinto, "Origins and Development," 37; and Williams, March, and Wassell, *Mathematical Works*, esp. problems 4 and 5 (measuring the width of a river under various conditions), 16–21 and 88–89; problem 12 (measuring the area of fields), 32–41 and 101–106; problem 16 (circular instrument for measuring angles to draw the map of the city) 50–59 and 122–126; problem 17 (measuring the distance to a remote but visible place), 58–61 and 126–133; and

problem 18 (constructing a device for measuring lengths of distances along a road—i.e., the hodometer), 62–63 and 133; and Stroffolino, *La città misurata*, esp. 14–19.

12. Pinto, "Origins and Development," 37–38.

13. See esp. Mario Baratta, "La pianta d'Imola di Leonardo da Vinci," *Bollettino della Società Geografica Italiana* 12 (1911): 945–967; David H. Friedman, "La pianta di Imola di Leonardo, 1502," in *Rappresentare la città: Topografie urbane nell'Italia di antico regime*, ed. Marco Folin (Reggio Emilia: Edizioni Diabasis, 2010), 121–144; Maier, *Rome Measured and Imagined*, esp. 29 and 89; Carlo Pedretti, "Leonardo architetto a Imola," *Architectura: Zeitschrift für Geschichte der Architektur* 2 (1972): 92–105; and Pinto, "Origins and Development," 38–42.

14. Important studies include Ingrid D. Rowland, "Raphael, Angelo Colocci, and the Genesis of the Architectural Orders," *Art Bulletin* 76 (March 1994): 81–104; John Shearman, *Raphael in Early Modern Sources (1483–1602)* (New Haven, CT: Yale University Press, 2003), 1:500–545; and Francesco P. Di Teodoro, *Raffaello, Baldassar Castiglione e la lettera a Leone X*, rev. ed. (Bologna: Minerva, 2003). For a transcription of the most complete version of the letter (Munich, Bayerische Staatsbibliothek, Codex Italic, 37b), see Renato Bonelli, ed., "Lettera a Leone X," in *Scritti rinascimentali di architettura*, ed. Arnaldo Bruschi et al. (Milan: Il Polifilo, 1978), 459–484, and for an English translation of the same version, Andrea Palladio, *Palladio's Rome: A Translation of Palladio's Two Guidebooks to Rome*, ed. and trans. Vaughan Hart and Peter Hicks (New Haven, CT: Yale University Press, 2006), 177–192. See also Cammy Brothers, "Architecture, History, Archaeology: Drawing Ancient Rome in the Letter to Leo X and in Sixteenth-Century Practice," in *Coming About: A Festschrift for John Shearman*, ed. Lars R. Jones and Louisa C. Mathew (Cambridge, MA: Harvard Art Museums, 2001), 135–140; Maier, *Rome Measured and Imagined*, esp. 49–60; Arnold Nesselrath, "Raphael's Archaeological Method," and Christof Thoenes, "La 'Lettera' a Leone X," both in *Raffaello a Roma: Il convegno del 1983* (Rome: Edizioni dell'Elefante, 1986), 357–371 and 373–381, respectively.

15. See Nicholas Adams and Simon Pepper, "The Fortification Drawings," in *The Architectural Drawings of Antonio da Sangallo the Younger and His Circle*, vol. 1, *Fortification, Machines, and Festival Architecture*, ed. Christoph L. Frommel and Nicholas Adams (New York: Architectural History Foundation; Cambridge, MA: MIT Press, 1994), 61–74; Micaela Antonucci, "Le porte di Roma nei progetti di Antonio da Sangallo il Giovane," *Roma Moderna e Contemporanea* 22 (January–June 2014): 17–35; and Huppert, *Becoming an Architect*, *passim*.

16. BAV, Barb. Lat. 4391 A and Barb. Lat. 4391 B. The map of the walls of Rome is Barb. Lat. 4391 (B), I. The most complete drawing of plans to fortify the Borgo is Barb. Lat. 4391 (B), II. An important study of the collection is Paolo Marconi, "Contributo alla storia delle fortificazioni di Roma nei Cinquecento e nel Seicento," *Quaderni dell'Istituto di Storia dell'Architettura*, 13th ser., fasc. 73–78 (1966): 109–130, which also reproduces some of the relevant maps. See also Bellini, "La *Civitas Pia*," 42–50, and Maier, *Rome Measured and Imagined*, esp. 86–96. I thank Jessica Maier for pointing me to the map collection and to Marconi's article. For Antonio Sangallo the Younger's survey of the walls, see Francesco Paolo Fiore, "Rilievo topografico e architettura a grande scala nei disegni di Antonio da Sangallo il Giovane per le fortificazioni di Roma al tempo di papa Paolo III," in *Il disegno di architettura* ed. Paolo Carpeggiani and Luciano Patetta, 175–180, Atti del Convegno Milano 15–18 febbraio 1988 (Milan: Guerini and Associates, 1989); and Christoph L. Frommel and Nicholas Adams, eds., *The Architectural Drawings of Antonio da Sangallo the Younger and His Circle*, vol. 1, *For-*

tification, Machines, and Festival Architecture (New York: Architectural History Foundation; Cambridge, MA: MIT Press, 1994), 163–164, 350. For the account books of Paul III's fortification project, see ASR, Archivio delle soldatesche e galere, b. 15 (ca. 1537–1549).

17. Although no copies of the 1551 map are extant, three (two complete and one partial) of the 1560 edition published by Antonio Trevisi are. London, British Library, Maps S. T. R. 175., and Vatican City, BAV, Stampe Geogr. I. 620. Riserva, are complete but differ in the letters by Antonio Trevisi, while BAV, Barb. Lat. 4432 is missing four main sheets and four smaller side sheets, "Ad Lectorem, non ad normam, olum [solum], et circinum sed ad pyxidem etiam nauticam." See Frutaz, *Le piante di Roma,* 1:168–170 and 2:189–209; Hülsen, *Saggio di bibliografia ragionata delle piante,* 38–41; and Scaccia Scarafoni, *Le piante di Roma,* 78–82. For Bufalini's life, see the foundational study that accompanies a full-scale reproduction of BAV Stampe Geogr. I.620. Riserva: Ehrle, *Roma al tempo di Giulio III,* 7–30; and also Palagiano, "Bufalini, Leonardo"; Jessica Maier, "Mapping Past and Present: Leonardo Bufalini's Plan of Rome (1551)," *Imago Mundi* 59 (April 2007): 1–23; and Maier, *Rome Measured and Imagined,* 77–118. See also Huppert," Mapping Ancient Rome," and Insolera, *Roma: Immagini,* 112–122.

18. Francesco De Marchi, *Della architettura militare libri tre* (Brescia: Comino Presegni, 1599), fol. 78r (bk. 3, chap. 34), for the debate in front of Paul III, and fols. 42r–v (bk. 2, chap. 82) "valente Architetto, il quale misurò tutta Roma dentro e fuori, e la pose in istampa con tutti li monti e Theatri, e Tempij, strade e altre cose segnalate, alquale io aiutai forse sei mesi per mio piacere, e per più imparare." For De Marchi's life, see Daniela Lamberini, "Francesco De Marchi," *DBI,* 38:447–454; "Francesco De Marchi: Ritratto di un cortigiano del Cinquecento, virtuoso e dilettante di architettura militare," *Storia Architettura* 10 (1987): 69–88; and Morgan Ng, "New Light on Francesco De Marchi (1504–1576) and His Treatise on Fortification," *Mitteilungen des kunsthistorischen Institutes in Florenz* 58 (2016): 403–410.

19. Ehrle, *Roma al tempo di Giulio III,* esp. 22; Lanciani, *Storia degli scavi,* 2: 110–111; and see ASR, *Archivio delle soldatesche e galere,* b 16, fol. 151, where a "Lionardo da Udine" may well refer to Bufalini.

20. For techniques of woodcut printing, see David Woodward, "Techniques of Map Engraving, Printing, and Coloring in the European Renaissance," in Woodward, *History of Cartography,* pt. 1, 591–610. For Bufalini's work as a collaborative project, see Maier, *Rome Measured and Imagined,* 92–96.

21. BL, Maps S. T. R. [175.], Bufalini, *Roma,* "Ad Lectorem," "Quisquis es, Rogat te Leonardus Bufalinus Foroiulien ne tu vile existimes, quod ille tibi iam dat, omnium rerum pulcherrimam se dare credit Romam scilicet, et hanc geminam: neque enim satis tibi factum duxit, redivivam istam unam, quae hodie colitur ante oculus posuisse: nisi veterem etiam, totius olim orbis dominam, summo labore, maxima impensa, diuturna sigilia, quasi e sepulchro excitatam addidisset"; and see Maier, *Rome Measured and Imagined,* 96–108; "Mapping Past and Present," 2–3; "A 'True Likeness': The Renaissance City Portrait," *Renaissance Quarterly* 65 (Fall 2012): 711–752; and "*Roma Renascens*: Sixteenth-Century Maps of the Eternal City," in *Rome: Continuing Encounters between Past and Present,* ed. Dorigen Caldwell and Lesley Caldwell (Surrey: Ashgate, 2011), esp. 40–46.

22. Jean-Louis Ferrary, *Onofrio Panvinio et le antiquité romaines* (Rome: École française di Rome, 1996), 56, "incredibili labore et pertinaci XX annorum studio." The citation is from a manuscript version (BAV, Vat. lat. 6783, fols. 196r–208v) of Panvinio's *Reipublicae Romanae commentariorum libri tres* (Venice: Vincenzo Valgrisio, 1558). Ferrary provides an edition

of the preface of this work using manuscript and print versions. Note that the quotation does not appear in the printed versions of Panvinio's treatise.

23. BL, Maps S. T. R. [175.], Bufalini, *Roma*, "Debeat immensum vult si pensare laborem / Plus tibi quam domino Romula terra suo. Quae flamma fuerat consumpta et armis et annis. / Per te iam nullo est interitura die." I warmly thank Kenneth Gowens for discussing this passage with me.

24. For Marliani's life and work, see Massimiliano Albanese, "Marliani (Marliano), Bartolomeo (Giovanni Bartolomeo)," *DBI*, 70:597–600; Antoninio Bertolotti, "Bartolomeo Marliano, archeologo nel secolo XVI," *Atti e memorie delle RR Deputazioni di Storia Patria per le Provincie dell'Emilia*, 4th ser., pt. 2 (1880): 107–138; Jacks, *Antiquarian and the Myth of Antiquity*, 206–214; Marc Laureys, "Bartolomeo Marliano (1488–1566): Ein Antiquar des 16. Jahrhunderts" in *Antiquarische Gelehrsamkeit und Bildende Kunst: Die Gegenwart der Antike in der Renaissance*, ed. Gunter Schweikhart (Cologne: Walther König, 1996), 151–167; and "Marliano (Bartolomeo) (1488–1566), in *Centuriae Latinae*, vol. 2, *Cent une figures humanistes de la Renaissance aux Lumières*, ed. Colette Nativel (Geneva: Librairie Droz, 2006), 499–503. For Marliani's studies of the plays of Aristophanes, including translations, commentaries, and notes, see BA, Ms. 247. See also Enrico Narducci, "Di alcuni lavori inediti e sconosciuti di Bartolomeo Marliani," *Atti della R. Accademia dei Lincei, Transunti*, 3rd ser., 8 (1884): 188–190.

25. Georg Fabricius, *Roma: Antiquitatum libri due* [. . .] (Basel: Johannes Oporinus, 1550), 9, "Marliani, qui diligentissime omnium scripsit"; "cuius domum cum adiremus, situm montium nobis indicavit ex Capitolio: et ad viarum, que in urbe sunt, initia nos perduxit"; "et aliquoties apud Salamancà et Tramezinum, nostrum sententiam audire visus est per humaniter." For Fabricius, see esp. William Stenhouse, "Georg Fabricius and Inscriptions as a Source of Law," *Renaissance Studies* 17 (March 2003): 96–107. For the print shops as vibrant gathering places for humanists and antiquarians, see David Woodward, *Maps as Prints in the Italian Renaissance: Makers, Distributors and Consumers* (London: British Library, 1996), 41–43.

26. Bartolomeo Marliani, *Antiquae Romae topographia libri septem* (Rome: Antonio Blado in Aedibus D. Ioan Bap. De Maximis, 1534). For Biondo, an initial bibliography would include Riccardo Fubini, "Biondo Flavio," *DBI*, 10:536–559; and Biondo Flavio, *Roma instaurata / Rome restaurée*, ed. and trans. Anne Raffarin-Dupuis, 2 vols. (Paris: Société d'édition Les Belles Lettres, 2005–2012). For Fulvio, see Ceresa, "Fulvio, Andrea," and Weiss, "Andrea Fulvio antiquario romano," 1–44.

27. The activities of the Accademia della Virtù are known primarily through a letter describing it by the humanist Claudio Tolomei (1492?–1556); see "Lettera al Conte Agostino de' Landi," ed. Sandro Benedetti and Tommaso Scalesse in *Trattati con l'aggiunta degli scritti di architettura di Alvise Cornaro, Francesco di Giorgi, Claudio Tolomei, Giangiorgio Trissino, Giorgio Vasari*, ed. Elena Bassi et al. (Milan: Il Polifilo, 1985), 32–61. See also Long, "Contribution of Architectural Writers," 285–287; and Pagliara, "Vitruvio da testo a canone," 67–74.

28. Lucio Fauno, *Delle antichità della citta di Roma* [. . .] (Venice: Michele Tramezzino, 1548), and the useful edition "Alli lettori Lucio Fauno," in *Delle antichità della citta di Roma* [. . .] (Venezia: Michele Tramezzino, 1548), ed. Margaret Daly Davis and Charles Davis, *Fontes* 13, 15 September 2008, http://archiv.ub.uni-heidelberg.de/artdok/580/1/Daly_Davis_Davis_Fontes13.pdf. See also Laura Asor Rosa, "Fauno, Lucio," *DBI*, 45:377–378; Margaret Daly Davis, "Two Early 'Fundberichte': Lucio Fauno and the Study of Antiquities in Farnese Rome," in *Opere e giorni: Studi su mille anni di arte europea dedicati a Max Seidel*, ed. Klaus

Bergdolt and Giorgio Bonsanti (Venice: Marsilio, 2001), 525–532; "Andrea Palladio's 'L'Antichita di Roma' of 1554," *Pegasus: Berliner Beiträge zum Nachleben der Antike* 9 (2007): 151–192; Gennaro Tallini, "Tra studio e bottega. Coordinate bio-bibliografiche per Giovanni Tarcagnota da Gaeta (1499–1566)," *Bibliologia* 6 (2011): 15–42; and "Giovanni Tarcagnota," *Bibliografia* (January 30, 2012), http://www.nuovorinascimento.org/cinquecento/tarcagnota.pdf.

29. Bartolomeo Marliani, *Urbis Romae topographia* [...] (Rome: Valerij, Dorici, and Aloisij fratris, Academiae Romanae Impressorum, 1544), 3, 7, and 12–13.

30. Marliani, *Urbis Romae topographia,* 12–13. A Roman *stadium* is about 186 meters; a *passus,* about 1.48 meters; an *oncia* 2.43 cm. See Frutaz, *Le piante di Roma,* 1:56–57, and vol. 2, pianta XII, tav. 21; and Maier, "*Roma Renascens,*" 36–37; and *Rome Measured and Imagined,* esp. 66–70. The porphyry column with the measure in the Church of the Twelve Holy Apostles was much discussed. For another contemporary notice, see Francesco De Marchi, *Della architettura militare,* n.p. (last folio, far right column of "Tavola delle Misure de Fortificare"). See also the comment by the Vitruvian commentator Guillaume Philandrier in Frédérique Lemerle, ed. and trans., *Les* Annotations *de Guillaume Philandrier sur le* De architectura *de Vitruve, Livres I à IV* (Paris: Picard, 2000), 165–166; and Lemerle, ed. and trans., *Les* Annotations *de Guillaume Philandrier sur le* De architectura *de Vitruve, Livres V à VII* (Paris: Classiques Garnier, 2011): 117. And see Evans, *Aqueduct Hunting in the Seventeenth Century,* 163–165.

31. Marliani, *Urbis Romae topographia,* 12, for the reference to Palatino on lower left. For Palatino, see Claudio Bonacini, *Bibliografia delle arti scrittorie e della calligrafia* (Florence: Sansoni, 1953), 248–254; James Wardrop, "*Civis Romanus Sum*: Giovanbattista Palatino and His Circle," *Signature,* n.s., 14 (1952): 3–39; and Christopher L. C. E. Witcombe, *Copyright in the Renaissance: Prints and the* Privilegio *in Sixteenth-Century Venice and Rome* (Leiden: Brill, 2004), 287. Palatino has no known history of mapmaking; thus, the assumption that he created the map itself is doubtful.

32. Marliani, *Urbis Romae topographia,* 2, "secundum naturalem prospectum"; "aedificia in suo situ locari non potuissent"; "melius putaremus . . . quam inani pictura quosdam oblectare: figuras ipsas planas posuimus."

33. See Ehrle, *Roma al tempo di Giulio III,* 25–28; Frutaz, *Le piante di Roma,* 1:21–22; Laureys, "Marliano (Bartolomeo)," 502; and Maier, *Rome Measured and Imagined,* 66–67.

34. For the confusions concerning the Roman Forum among earlier fifteenth-century humanists, see Frances Muecke, "Humanists in the Roman Forum," *Papers of the British School at Rome* 71 (2003): 207–233; and for an accessible account of the Forum as it is today, Watkin, *Roman Forum.*

35. For Egio, see Franco Pignatti, "Egio, Benedetto," *DBI,* 42:356–357. For Panvinio, see Ferrary, *Onofrio Panvinio.* For Angelo Colocci, "Colocci, Angelo," *DBI,* 27:105–111; and for Orsini, Federica Matteini, "Orsini, Fulvio," *DBI,* 79:649–653. See also Robert W. Gaston, "'La mirabile natura dell'acque correnti: Pirro Ligorio's View of the Waters of the 'Old' World," in *La civiltà delle acque tra Medioevo e Rinascimento,* ed. Arturo Calzona and Daniela Lamberini, Atti del Convegno internazionale, Mantova 1–4 Ottobre 2008 (Florence: Leo S. Olschki, 2010), 1:75–87; Stenhouse, *Reading Inscriptions,* esp. 75–98 on forgery of inscriptions, including Ligorio's. On Ligorio's forgeries, see also Ginette Vagenheim, "La falsificazione epigrafica nell'Italia della seconda metà del Cinquecento: *Renovatio* ed *inventio* nelle *Antichità Romane* attribuite a Pirro Ligorio," in *El monument epigráfico en contextos secun-*

drios: Procesos de reutilización, interpretación y falsificación, ed. Joan Carbonell Manils, Helena Gimeno Pascual, and José Luis Moralejo Álvarez (Belaterra: Universitat Autònoma de Barcelona, 2011): 217–226; "La falsification chez Pirro Ligorio à la lumière des Fasti Capitolini e des inscriptions de Préneste," *Eutopia* 3 (1994): 67–113; and "Bartolomeo Borghesis, Theodor Mommsen et l'édition des inscriptions de Pirro Ligorio dans le *Corpus Inscriptionum Latinarum* (CIL)," *Journal of the History of Collections* 26 (2014): 363–371; and see also "Appunti per una prosopografia dell'Accademia dello Sdegno a Roma: Pirro Ligorio, Latino Latini, Ottavio Pantagato e altri," *Sassoferrato: Studi umanistici piceni* 26 (2006): 211–226.

36. See M. H. Crawford, "Benedetto Egio and the Development of Greek Epigraphy" in *Antonio Agustin between Renaissance and Counter-Reform*, ed. M. H. Crawford (London: Warburg Institute, 1993), 133–153; and Occhipinti, *Pirro Liqorio*, LXXXI–CVI.

37. BAV, Ms. Ross 1204, 38, 44, and 78. The chronology of the dispute is not a matter of consensus. For somewhat varying accounts, see Ferrary, *Onofrio Panvinio*, 82–86; Jacks, *The Antiquarian and the Myth of Antiquity*, esp. 206–221; Marc Laureys and Anna Schreurs, "Egio, Marliano, Ligorio, and the Forum Romanum in the 16th Century," *Humanistica Lovaniensia* 45 (1996): 385–405; and Schreurs, *Antikenbild und Kunstanschauungen*, 51–120.

38. Laureys and Schreurs, "Egio, Marliano, Ligorio," 391; and BAV, Ms. Ross 1204, 31, 60, 72, 95, and 116 (on their transcription of the inscription).

39. Pirro Ligorio, *Libro di M. Pyrrho Ligori napolitano delle antichità di Roma* [. . .] *con le Paradosse del medesimo autore* [. . .] (Venice: Michele Tramezzino, 1553). Facsimile of the first edition, with an introduction by Margaret Daly Davis, *Fontes* 9, July 16, 2003. http://archiv.ub.uni-heidelberg.de/artdok/volltexte/2008/562, 1r (p. 28 of Daly Davis), "M. Benedetto Egio da Spoleto, huomo cosi nelle lettere Greche, come nelle Latine, eruditissimo." For some of the many other mentions of Egio in Ligorio's manuscript writings and of Ligorio in Egio's writings, see Laureys and Schreurs, "Egio, Marliano, Pirro," esp. 397–398.

40. Fauno, *Delle antichità della citta.*

41. Fauno, "Alli lettori Lucio Fauno," 10 (p. 34 of Daly Davis), "Questa materia delle Antichità di Roma, è per la maggior parte cosi dubbia e varia, per non averne alcuno degli antichi distintamente ragionato, e per essere quasi tutti i luoghi antichi disfatti; che male se ne può ragionando dire appunto, cosi è, cosi non è: perché per lo più bisogna andare dietro à congetture." See also Daly Davis, "Two Early 'Fundberichte,'" 530.

42. Ligorio, *Libro di M. Pyrrho Ligori napolitano delle antichità*; and Daly Davis, "Introduction," http://archiv.ub.uni-heidelberg.de/artdok/volltexte/2008/562.

43. Ligorio, *Libro di M. Pyrrho Ligori napolitano delle antichità*, fols. 25–51, and Ligorio's manuscript in Paris, Bibliotèque Nationale (BnF, ms. ital. 1129, 1–48, Libro I, p. 8), cited in Laureys and Schreurs, "Egio, Marliano, Ligorio," 394, "et anchora che costoro habbiano veduto chiaramente quanto ho detto, non hanno voluto tirarsi a dietro del'haver scritto il falso, et non solo in stampa ma in marmo appariscono il loro mal fondato oppenione, per conseglio del mal consegliato Bartolomeo Marliale, al qual non suaderebbe la verità, si ben venisse Cicerone et Romolo, edificator della città punto li cederebbe."

44. Ligorio, *Libro di M. Pyrrho Ligori napolitano delle antichità*, fols. 26v–31v (pp. 54–60 in Daly Davis ed.). See also Muecke, "Humanists." For the Roman Forum as present-day archaeologists understand it to have developed in antiquity, see Filippo Coarelli, *Il Foro Romano*, 2 vols. (Rome: Edizioni Quasar, 1983–1985); Dunia Filippi, "Regione VIII: *Forum Romanum Magnum*," in Carandini and Carafa, *Atlante di Roma antica*, 1:143–206; and G. Tagliamonte, "Forum Romanum (fino alla prima età repubblicana," N. Purcell, "Forum

Romanum (the Republican Period)," Purcell, "Forum Romanum (the Imperial Period)," and C. F. Giuliani and P. Verduchi, "Forum Romanum (età tarda)," in Steinby, *Lexicon topographicum*, 2:313–325, 325–336, 336–342, and 342–343, respectively; and Richardson, *New Topographical Dictionary*, 170–174.

45. For an edition of the Regionary Catalogs, see Roberto Valentini and Giuseppe Zucchetti, eds., *Codice topografico della città di Roma* (Rome: Tipografia del Senato, 1940–1953), 1:63–258, 193–258 for editions with late interpolations. For further discussion, see Howard Burns, "Pirro Ligorio's Reconstruction of Ancient Rome: The *Anteiquae Urbis Imago* of 1561" in Gaston, *Pirro Ligorio*, 19–91, at 23–25; and see Vincent Jolivet, "La localisation des toponymes de la Rome antique à partir des Régionnaires: Une étude de cas," in Leone, Palombi, and Walker, *Res bene gestae*, 103–125; and see Muecke, "Humanists," esp. 217–220. For Pomponio Leto and his academy, see Chiara Cassiani and Myriam Chiabò, eds. *Pomponio Leto e la prima Accademia Romana* (Rome: Roma nel Rinascimento, 2007); and D'Amico, *Renaissance Humanism*, esp. 91–102.

46. Woodward, *Maps as Prints*, 42–43; and see Daly Davis, "Introduction," in *Libro di M. Pyrrho Ligori napolitano delle antichità*, 5, and for the privilege, 26; and Burns, "Pirro Ligorio's Reconstruction," 49–50n37, where the petition for the *privilegi*, Archivio di Stato di Venezia, Senato Terra, Filze Filza 16 (1552), is cited in full along with the results of the Senate vote. For a detailed study of the Roman circus and Ligorio's understanding of it, see Silvia Tomasi Velli, "Gli antiquari intorno al circo romano: Riscoperta di una tipologia monumentale antica," *Annali della Scuola Normale Superiore di Pisa: Classe di Lettere e Filosofia*, 3rd ser., 20 (1990): 61–168. For the Tramezzino brothers, see Pier Silverio Leicht, "L'editore veneziano Michele Tramezino ed i suoi privilegi," in *Miscellanea di scritti di bibliografia ed erudizione in memoria di Luigi Ferrari* (Florence: Leo S. Olschki, 1952), 357–367; Alberto Tinto, *Annali tipografici dei Tramezzino* (Venice: Istituto per la Collaborazione Culturale, 1966); and Witcombe, *Copyright in the Renaissance*, 112–122. The brothers lived in the Rione Parione before the Sack, and each was listed as a *libraro* (bookseller) in the 1527 census. See Egmont Lee, ed., Descriptio Urbis*: The Roman Census of 1527* (Rome: Bulzoni, 1985), 88.

47. BL, Maps* 155. (5.), Pirro Ligorio, *Urbis Romae situs cum iis quae adhuc conspiciuntur veter. monument. reliquiis Pyrrho Ligorio Neap. invent.* (Venice: Michaelis Tramezini, 1552). See also Frutaz, *Le piante di Roma*, 1:170–171, and vol. 2, pianta CXI (tav. 222), who reproduces the example in the Biblioteca Nazionale di Roma, P.R. 13; Hülson, *Saggio di bibliografia ragionata delle piante*, 41–43; Robert W. Karrow Jr. *Mapmakers of the Sixteenth Century and Their Maps* (Chicago: Speculum Orbis Press for the Newberry Library, 1993), 349–350; Scaccia Scarafoni, *Le piante di Roma*, 82 (no. 143); and for a recent assessment, see Maier, *Rome Measured and Imagined*, 122–124.

48. BL, Maps* 155. (5.), Ligorio, *Urbis Romae situs* (1552). For George Lily, see Karrow, *Mapmakers*, 349–350; and see Roberto Almagià, "Pirro Ligorio cartografo," *Rendiconti delle Sedute dell'Accademia Nazionale dei Lincei: Classe di Scienze, Morali, Storiche e Filologiche*, 8th ser., 11: (1956): 49–61.

49. BL, Maps* 155. (35.), Pirro Ligorio, *Urbis Romae* [...] (Venice: Michele Tramezzino, 1553), "Michael Tramizinus Lectoribus": "Pyrri Ligorii industria pictam et nostris denum aeneis tabellis excusam nuper emisimus"; "Id interim ne diutius cum ineptis et leberide caecioribus antiquariis ceu Socraticulis Strepsiades suos nil nisi nugas docentibus fallamini, libenter accipite, ceterum quod hic pleraque contra perversam veter(um) antiquariorum sententiam nominantur et locantur, id haud temere factum ex Pyrrhi Ligorii paradoxis

propediem in lucem prodituris clarissime perspicietis." The map, published in Rome, was engraved by Julius de Musis, as is indicated on the lower right under Tramezzino's typographical mark—"Giulio de' Musi / in Aes incidit / MDLIII." See also Frutaz, *Le piante di Roma,* 1:60–61, and vol. 2, pianta XVI (tav. 25); Hülson, *Saggio di bibliografia ragionata delle piante,* 43–44; and Scaccia Scarfoni, *Le piante di Roma,* 23–24. And see Maier, *Rome Measured and Imagined,* 124–126.

50. BL, Maps* 155. (35.), Ligorio, *Urbis Romae situs* (1553), "Urb. In Ambitu Habet .pas XIII.mill. DL." For the region *Transtiberim* in the amended Regionary Catalog, see Valenti and Zucchetti, eds., *Codice Topografico,* 1:247–250.

51. Bartolomeo Marliani, *Urbis Romae topographia: Nuper ab ipso auctore nonnullis erroribus sublatis emendate, Addita etiam interpretatione nominum* [. . .] (Rome: Valerij, Dorici, & Aloisij, fratris, Academia Romanae Impressorum, 1544), "De praenomine, nomine, et cognomine," "et praesertim ex ijs, quibus in Foro Romano nuper repertis." . The letter to readers is added to the example of the 1544 edition in the Rubenstein Library of Duke University, call no. A-17 fm348U c. 1, and it is also in the copy in Rome in the Biblioteca Casanatense, Vol. misc. 1001 4. For the *Fasti* and their significance, see William McCuaig, "The Fasti Capitolini and the Study of Roman Chronology in the Sixteenth Century," *Athenaeum* 79 (1991): 141–159; Lily Ross Taylor, "Degrassi's Edition of the Consular and Triumphal Fasti," *Classical Philology* 45 (April 1950): 84–95; Stenhouse, *Reading Inscriptions,* 103–112; and Vagenheim, "La falsification chez Pirro Ligorio," 67–113, which details Pirro's arguments with Marliani over the original location of the *Fasti.* And see Bartolomeo Marliani, *Consulum, dictatorum censorumque series* [. . .] (Rome: R. Ian, 1549), and *Annales consulum, dictatorum, censorumque Romanorum* [. . .] (Rome: Antonio Blado, 1560). For the *Fasti* in the ancient world, see Mary Beard, *The Roman Triumph* (Cambridge, MA: Harvard University Press, 2007), 61–68 and 72–75.

52. Marliani, *Urbis Romae topographia. / Nuper ab ipso auctore nonnullis erroribus,* "Lectori," 121–122, citation on 121, "Id admoneo ne per errorem factum esse existimes, quemadmodu[m] Strepsiadem quendam latrare accepimus. Qui etiam de quodam suo fo. Ro. contra omnium doctissimorum virorum opinionem, somnia quaedam balbutit, et eorum, quae non intelligit, se magistrum arroganter profitetur. [. . .] Et quanquam huiusmodi nebuloni respondere pudet, tamen ne ij, qui Topographiae urbis parum erudite sunt, istius nugis in errorem inducantur, polliceor propediem demo[n]straturum, quae à me de foro scripta sunt, Pythio oraculo esse veriora."

53. For the "Nuper adiecta," Marliani, "B. Marliani topographiae urbis Romae haec nuper adiecta" in Marliani, *Urbis Romae topographia. / Nuper ab ipso auctore nonnullis erroribus.* See Laureys and Schreurs, "Egio, Marliano, Ligorio," esp. 395–396 and note 35; and Ferrary, *Onofrio Panvinio,* 83–86, who refers (84n164) to two examples—Paris, BnF, J 457, and London, BL, C.48.i.2)—and 169–170. Laureys and Schreurs used an example held in Rome at the British School (shelf mark 600.605) (Marc Laureys, personal communication, March 31, 2013). I warmly thank Marc Laureys for discussing the "Nuper adiecta," providing further information about the location of the copy that he used, and lending me his own transcription. Subsequently, I discovered an edition of the *Urbis Romae topographia* that includes the "Nuper adiecta" at the Getty Research Institute, Special Collections, ID/accession Number 84-B16413, which provides a digitalized version online: http://archive.org/details/vrbis romaetopogr00marl. It is this copy that I have used here.

54. Marliani, *B. Marliani topographiae urbis Romae haec nuper adiecta* [1544] (http://

archive.org/details/vrbisromaetopogr00marl), title page, "contra novam, & stultam opinionem cuiusdam Strepsiadis/ Argumentum nebularum Aristophanis admodum ridiculum"; "figuras quasdam sub nomine Urbis eode[m] Strepsiade Magistro impressas, prorsus falsas esse"; sig. A iij, "postquam melle saturatus est, reliqua stercore contaminat."

55. See Bertolotti, "Bartolomeo Marliano," 118–138, who reproduces much testimony and discusses the will in detail.

56. Ibid. Ruscone was called an *orpellaro*, that is, a *ricamatore*, or embroiderer. For Marliani's translations and comments on the plays of Aristophanes, see BA, Ms. 247, translation and commentary on the *Clouds* (*Nebulae*), fols. 38r–83v. The Cavalieri di San Pietro was an honorary knighthood first created by Pope Leo X in 1520 that could be acquired by a payment which then accrued interest—see Hallman, *Italian Cardinals*, 131.

57. BL, Maps C.25.d.9(1.). Pirro Ligorio, *Anteiquae urbis imago accuratissime ex vetusteis monumenteis formata* (Rome: M. and Fr. Tramezini, 1561). See Frutaz, *Le piante di Roma*, 1:61–63, vol. 2, pianta XVII (tav. 26–32), and vol. 3, tav. 671–684, and Maier, *Rome Measured and Imagined*, esp. 108 and 126–129.

58. Burns, "Pirro Ligorio's Reconstruction."

59. BL, Maps C.25.d.9(1.), Ligorio, *Anteiquae urbis imago*; and Frutaz, *Le piante di Roma*, 1:61–63, vol. 2, pianta XVII (tav. 26–32), and vol. 3, tav. 671–684.

60. BL, Maps C.25.d.9(1.), Ligorio, *Anteiquae urbis imago*, "Effigies Antiquae Romae Ex vestigiis Aedificiorum Ruinis Testimonio / Veterum Auctorum Fide Numismatum Monumentis / Aeneis Plumbaeis Saxeis / Tiglinisque Collecta atque in hanc Tabellam Redacta atque descripta a Pyrrho Ligorio Romano per XIIII Regiones in quas urbem divisit Impo. Caesar. Aug."

CHAPTER 6

1. For printmaking and its development, see Paolo Bellini, "Stampatori e mercanti di stampe in Italia nei secoli XVI e XVII," *I Quaderni del Conoscitore di Stampe* 26 (1975): 19–45, translated as "Printmakers and Dealers in Italy during the Sixteenth and Seventeenth Centuries," *Print Collector*, no. 13 (May–June 1975): 17–45; Maria Antonietta Bonaventura, "L'industria e il commercio delle incisioni nella Roma del '500," *Studi Romani* 8 (July–August 1960): 430–436; Michael Bury, *The Print in Italy, 1550–1620* (London: British Museum Press, 2001); "The Taste for Prints in Italy to c. 1600," *Print Quarterly* 2 (March 1985): 12–26; David Landau and Peter Parshall, *The Renaissance Print, 1470–1550* (New Haven, CT: Yale University Press, 1994); and Parshall, "Prints as Objects of Consumption in Early Modern Europe," *Journal of Medieval and Early Modern Studies* 28 (Winter 1998): 19–36. For Roman printmaking in particular, see Eckhard Leuschner, *Antonio Tempesta: Ein Bahnbrecher des römischen Barock und seine europäische Wirkung* (Petersberg: Michael Imhof, 2005), esp. 131–245; Evelyn Lincoln, *The Invention of the Italian Renaissance Printmaker* (New Haven, CT: Yale University Press, 2000); San Juan, *Rome: A City out of Print*; and Witcombe, *Print Publishing*, esp. 7–18.

2. See Bury, *Print in Italy*, 13–32; Coolie Verner, "Copperplate Printing," in *Five Centuries of Map Printing*, ed. David Woodward (Chicago: University of Chicago Press, 1975), 51–75; and Woodward, "Techniques of Map Engraving," in Woodward, *History of Cartography*, pt. 1, 591–610.

3. Besse and Dubourg Glatigny, "Cartographier Rome," esp. 380–382.

4. For the Tramezzino brothers, see Tinto, *Annali tipografici*, esp. xv–xvi and xviii. Michele's apprenticeship in Rome in 1522 and his desire to set up his own shop are known from letters he sent to the Venetian publisher Giovanni Bartolomeo Gabiano—see Angela Nuovo, "Una lettera di Michele Tramezino a Giovanni Bartolomeo Gabiano (1522)," *La Bibliofilia* 115 (2013): 147–156; and *The Book Trade in the Italian Renaissance* (Leiden: Brill, 2013), esp. 77–79, 95–96, 109–110, 135, 213–214, 249–250, and 418.

5. For the document in which the goods of the brother are divided, see Tinto, *Annali tipografici*, 102–103, "viver senza travaglio"; and Nuovo, *Book Trade*, 129–130.

6. Tinto, *Annali tipografici*, xxii–xxiii; and for the book privilege system in general, Nuovo, *Book Trade*, 195–257.

7. For the print shops as vibrant gathering places for humanists and antiquarians, see Woodward, *Maps as Prints*, 41–43; and for the Tramezzino shop, Nuovo, *Book Trade*, 418–419. For Salamanca, see Sylvie Deswarte-Rosa, "Les gravures de monuments antiques d'Antonio Salamanca à l'origine du *Speculum Romanae Magnificentiae*," *Annali di Architettura* 1 (1989): 47–62; and Maria Cristina Misiti, "Antonio Salamanca: Qualche chiarimento biografico alla luce di un'indagine sulla presenza spagnola a Roma nel '500," in *La stampa in Italia nel Cinquecento*, ed. Marco Santoro, Atti del Convegno, Rome, October 17–21, 1989 (Rome: Bulzoni, 1992), 1:545–563.

8. The essential documents for the shop were first collected by Francesco Ehrle, *Roma prima di Sisto V: La pianta di Roma du Pérac-Lafréry del 1577 riprodotta dall'esemplare esistente nel Museo Britannico* (Vatican City: Biblioteca Apostolica Vaticana, 1908); and see F. Roland, "Un Franc-Comptois éditeur et marchand d'estampes a Rome au XVIe siècle: Antoine Lafrery (1512–1577)," *Mémoires de la Société d'Emulation du Doubs*, 8th ser., 5/6 (1910–1911): 320–370. For Salamanca, see Valeria Pagani, "Documents on Antonio Salamanca," *Print Quarterly* 17 (June 2000): 148–155; and "Guido Roberti and Francesco Salamanca," *Print Quarterly* 30 (September 2013): 259–272. See Witcombe, *Print Publishing*, 67–105, for Lafreri, including his partnership with Salamanca, 107–221, and esp. 290–301 for Duchetti. For Lefreri's engagement with humanist investigations of antiquities, see Birte Rubach, "Three Prints of Inscriptions: Antonio Lafreri and His Contact with Jean Matal," in Rebecca Zorach, ed., *The Virtual Tourist in Renaissance Rome: Printing and Collecting the* Speculum Romanae Magnificentiae (Chicago: University of Chicago Press, 2008), 25–35, who shows that Lafreri was concerned with antiquarian studies as well as the market. See also Witcombe, *Copyright in the Renaissance*, 129–134.

9. As with the *Speculum* (see below), each contained a diverse selection of maps, most depicting a variety of locales and regions—but not Rome itself. See Bury, *The Print*, 49–50 and fig. 1; Roland, "Antoine Lafrery," 343–350; and Tooley, *Maps in Italian Atlases*.

10. Bury, *Print in Italy*, 49–50, fig. 2.

11. It is important to note that the *Speculum* does not represent a single printed edition—each collection contains a diverse group of prints that share Dupérac's famous title page. Many collections thus titled contain some prints that were not even published by Lafreri. Further, the collections were often augmented well beyond the sixteenth century. See Bury, *Print in Italy*, 50, 59–60; Christian Hülsen, "Das *Speculum Romanae Magnificentiae* des Antonio Lafreri," in *Collectanea variae doctrinae Leoni S. Olschki* (Munich: Jacque Rosenthal, 1921), 121–170; Bates Lowry, "Notes on the *Speculum Romanae Magnificentiae* and Related Publications," *Art Bulletin* 34 (March 1952): 46–50; Peter Parshall, "Antonio Lafreri's *Speculum Romanae Magnificentiae*," *Print Quarterly* 23 (March 2006): 3–28; Zorach, *Virtual Tourist*;

and University of Chicago Library, The *Speculum Romanae Magnificentiae* Digital Collection, http://speculum.lib.uchicago.edu; Stefano Corsi and Pina Ragionieri, eds., *Speculum Romanae Magnificentiae: Roma nell'incisione del Cinquecento* (Florence: Mandragora, 2004); and Lawrence R. McGinniss and Herbert Mitchell, *Catalogue of the Earl of Crawford's "Speculum Romanae Magnificentiae" now in the Avery Architectural Library* (New York: Avery Architectural Library, Columbia University, 1976). See also, more generally, Fabia Borroni Salvadori, *Carte, piante e stampe storiche delle raccolte lafreriane della Biblioteca Nazionale di Firenze* (Rome: Istituto Poligrafico e Zecca dello Stato, 1980).

12. Witcombe, *Print Publishing*, esp. 4–7.

13. Woodward, *Maps as Prints*, 43–45; Gian Ludovico Masetti Zannini, "Rivalità e lavoro di incisori nelle botteghe Lafréry-Duchet e de la Vacherie," in *Les foundations nationals dans la Rome pontificale* (Rome: École française de Rome, 1981), 547–566, who reproduces a partial transcription of the criminal investigation; Witcombe, *Print Publishing*, esp. 223–301; and Witcombe, *Copyright in the Renaissance*, 138–139 (for Vaccari) and 139–141 (for Claudio Duchetti).

14. Cristina Acidini [Luchinat], "Dosi, Giovanni Antonio (Giovan Antonio, Giovannantonio), detto Dosio," *DBI*, 41:516–523; Franco Borsi, Cristina Acidini, Fiammetta Mannu Pisani, and Gabriele Morolli, eds., *Giovanni Antonio Dosio: Roma antica e i disegni di architettura agli Uffizi* (Rome: Officina Edizioni, 1976); Antonio Ernesto Denunzio, "Brevi considerazioni intorno a due ambiti della committenza dosiana: La corte farnesiana di Roma e quella vicereale di Napoli," and Antonella Marciano, "Disegno e rappresentazione tra antico e moderno," both in Emanuele Barletti, ed., *Giovan Antonio Dosio da San Gimignano: architetto e scultore fiorentino tra Roma, Firenze e Napoli* (Florence: Edizione Firenze, 2011), 75–113 and 139–145, respectively; Christian Hülsen, ed., *Das Skizzenbuch des Giovannantonio Dosio im staatlichen Kupferstichkabinett zu Berlin* (Berlin: Heinrich Keller, 1933); Antonella Marciano, *Giovanni Antonio Dosio: Fra disegno dell'antico e progetto* (Naples: La Scuola di Pitagora, 2008); and Carolyn Valone, "Giovanni Antonio Dosio: The Roman Years," *Art Bulletin* 58 (December 1976): 528–541. For Gamucci, see Laura Asor Rosa, "Gamucci, Bernardo," *DBI*, 52:132–133.

15. See Hülsen, *Saggio di bibliografia ragionata delle piante*, 51–52; Hülsen, *Das Skizzenbuch*, v–vi; Frutaz, *Le piante di Roma*, 1:176–177, and vol. 2, pianta CXVII, tav. 229; and Saccia Scarafoni, *Le piante di Roma*, 85–86.

16. Cristina Acidini [Luchinat], "Roma antica," in Borsi et al., *Giovanni Antonio Dosio*, 49–63 for the drawings for Gammucci; 50–51 for drawings of the Forum (no. 2523/A, the basis for the engraving of Gamucci, and a copy, 2522/A); and Valone, "Dosio: The Roman Years."

17. Giovanni Antonio Dosio, *Urbis Romae aedificiorum illustriumquae supersunt reliquiae* [...] *ut hodie cernuntur descriptae* [...] (Rome, 1569), which was reprinted as Dosio, *Le antichità di Roma aedificiorum illustriumquae supersunt reliquiae* [...], ed. Franco Borsi (1569; reprint, Rome: Edizioni Colombo Ristampe, 1970). For Cavalieri, see Paola Pizzamano, ed. *Giovanni Battista Cavalieri: Un incisore trentino nella Roma dei papi del Cinquecento* (Rovereto, Trent: Nicolodi, 2001); Bruno Passamani, "Cavalieri (Cavalleri, Cavallieri, De Cavalieri, De Cavalleri, De Cavaleriis), Giovanni Battista," *DBI*, 22:673–675; and Witcombe, *Print Publishing*, esp. 392–393 and *passim*. And see Valone, "Dosio: The Roman Years," 528.

18. For Conti, see Irene Fosi, "Conti, Torquato," *DBI*, 28:479–480; and see Valone, "Dosio: The Roman Years," 535–536.

19. For the *Forma Urbis Romae,* see Gianfilippo Carettoni, Antonio M. Colini, Lucos Cozza, and Guglielmo Gatti, *La pianta marmorea di Roma antica: Forma Urbis Romae,* 2 vols. (Rome: Comune di Roma, 1960), the fundamental modern reference work on the fragments; and the Stanford Digital *Forma Urbis Romae* Project, http://formaurbis.stanford.edu/, the website of the ongoing digital research project that includes a detailed history, bibliography, and digital scans of all 1,186 surviving fragments. See also Roberto Meneghini, "La cartografia antica e il catasto di Roma imperiale," in Leone, Palombi, and Walker *Res bene gestae,* 205–218; John Pinto, "*Forma Urbis Romae*: Fragment and Fantasy," in Striker, *Architectural Studies,* 143–146; and William Stenhouse, "The *Forma Urbis Romae* before Nolli: Antiquarian Scholarship in the Sixteenth and Seventeenth Centuries," in *Giambattista Nolli and Rome: Mapping the City before and after the Pianta Grande,* ed. Ian Verstegen and Allan Ceen (Rome: Studium Urbis Rome Center, 2013), 15–26. Sixteenth-century copies of some of the fragments that no longer exist today can be found in BAV Vat. lat. 3434 and may have been drawn by Dosio. Rowland, "Raphael, Angelo Colocci," 88, points out that some fragments from the marble plan had already begun to emerge in the early sixteenth century.

20. Bernardo Gamucci, *Libri quattro dell'antichità della città di Roma* [...] (Venice: G. Varisco, 1565), 33, "M. Giovanni Antonio Dosi da San Gimignano giovane virtuoso, architetto & antiquario di non poca espettatione." Other editions of Gamucci's book followed in 1569, 1580, and 1589. There is a facsimile edition of the 1569 edition: Bernardo Gamucci, *Le antichità della città di Roma* [...] (1569; Rome: Centro Editoriale Internazionale, 1993).

21. See Gamucci, *Libri quattro dell'antichità della città di Roma,* 193–201. The "divine Michelangelo" (il divin Michelagnol [*sic*] Buonarruoti) is first mentioned with reference to the stairs of the Senate building with their niches containing the Nile and Tiber river gods (16), and for Bramante's *Tempietto,* see 176–177.

22. Ibid., 3, "Io non voglio nè negar, nè affermare che sia cosi; perche non è mia intentione in tutta questa opera senon di mostrare quelle cose per vere, che con l'autorità de gli scrittori, o con la certezza de gli edificij si possono ancora nè tempi nostri accertare, rimettendomi in questo sempre al giudicio di coloro, che ne saranno piu di me intendenti."

23. Ibid., 36–37. For the basilica, including its misidentification, see Watkin, *Roman Forum,* 44–52.

24. Gamucci, *Libri quattro dell'antichità della città di Roma,* 55 (Trajan's Column), 78 (Circus Maximus), 76, 79, 152–153, 195 (obelisks), 162 (the Pantheon), "M. Raffael Gamucci Geometra, & Aritmetico de' nostri tempi in Roma non inferiore ad alcuno altro."

25. Ibid., 18–21. An illustration of the Forum with identifying letters shows its exact location in Gamucci's view (20–21), "soggetto veramente degno & da molti antichi & moderni scrittori con autorità, & studio descritto" (18); "tante controversie"; dall'historie, & le regole & ordini, che si traggono dall'archittura [*sic*]"; "s'è ritrovato ogni giorno, nel cavar fra le ruine" (19).

26. The definitive study of Duperac, including a complete catalog of his prints and drawings, is Emmanuel Lurin, "Etienne Dupérac, graveur, peintre et architecte (vers 1535?–1604): Une artiste-antiquaire entre l'Italie et la France" (PhD diss., L'Université Paris IV, Sorbonne, 2006). See also Long, "Multi-Tasking 'Pre-professional' Architect/Engineers," 233–236.

27. See Ehrle, *Roma prima di Sisto V,* 8–11.

28. See Lurin, "Etienne Dupérac, graveur, peintre et architecte," 13–15 for a summary biography and 39–237 for his extensive collaboration with Onofrio Panvinio. See also Cristina Bragaglia Venuti, "Etienne Dupérac e i paesaggi della Loggia di Pio IV," *Rivista dell'Isti-*

tuto Nazionale d'Archeologia e Storia dell'Arte, 3rd ser., 57 (2002): 279–3010; "Etienne Dupérac and Pirro Ligorio," *Print Quarterly* 23 (December 2006): 408–413, citation on 412; and Ferrary, *Onofrio Panvinio*, 3–38 and *passim*.

29. For Panvinio's life, see Stefan Bauer, "Panvinio, Onofrio," *DBI*, 81:36–39; Karl A. Gersbach, "The Books and Personal Effects of Young Onorfrio Panvinio, OSA, in Vat. Lat. 7205," *Analecta Augustiniana* 52 (1989): 51–76; "Onofrio Panvinio's Brother, Paolo, and His Role in the Posthumous Edition of the 'De primatu Petri et Apostolicae sedis potestate' and the Purchase of Onofrio's Manuscripts for the Vatican Library," *Analecta Augustiniana* 56 (1993): 241–264; "Onofrio Panvinio, OSA, and His Florentine Correspondents Vincenzio Borghini, OSB, Pietro Vettori, Francesco de' Medici," *Analecta Augustiniana* 40 (1997): 207–280 (a selection of Gersbach's many studies on aspects of Panvinio's life and work); Lurin, "Etienne Dupérac, graveur, peintre et architecte," 39–237; and P. Davide Aurelio Perini, *Onofrio Panvinio e le sue opere* (Rome: Tipografia Poliglotta, 1899). The two editions of the *Fasti* are Onofrio Panvinio, *Fasti et triumphi romani* [. . .] (Venice: Jacopo Strada, 1557), and his approved version, *Fastorum libri V* [. . .] (Venice: Vicenzo Valgrisio, 1558); and see McCuaig, "Fasti Capitolini," 153–154. For Jacopo Strada, see Dirk J. Jansen, "Strada, (1) Jacopo Strada," in *DA*, 29:737–740.

30. Ferrary, *Onofrio Panvinio*, esp. 26, 32–33, 37, and 141 for BAV, Vat. lat. 3439; and Lurin, "Etienne Dupérac, graveur, peintre et architecte," 39–237. And see Panvinio, *XXVII Ponitificum Maximorum elogia et imagines*.

31. Lurin, "Etienne Dupérac, graveur, peintre et architecte," 39–237, for his extensive collaboration with Onofrio Panvinio, and 77–89, for Panvinio's ideas about images; and for the same, see also Lurin, "Les restitutions de scènes antiques: Onofrio Panvinio iconographe et inventeur d'images," in *Programme et invention dans l'art de la Renaissance*, ed. Michele Hochmann, Julian Kliemann, Jérémie Koering, and Philippe Morel (Rome: Académie de France à Rome, 2008), 153–173. See also William Stenhouse, "Panvinio and *Descriptio*: Renditions of History and Antiquity in the Late Renaissance," *Papers of the British School at Rome* 80 (2012): 233–256. And see Ferrary, *Onofrio Panvinio*, 26–38, for Panvinio and Duperac, and 55, for Panvinio's friendship with Ligorio.

32. Frutaz, *Le piante di Roma*, 1:65–66, and vol. 2, pianta XX, tav. 35; and see Ferrary, *Onofrio Panvinio*, 33–35; and Lurin, "Etienne Dupérac, graveur, peintre et architecte," esp. 185–186, 1042–1045.

33. BSISA, Roma XI.16.3.10, *Specimen, seu perfecta urbis antiquae imago* (Rome: Antonio Lafreri, 1573), a small 23½ × 18 inch (60 × 46 cm) image of the city. See Frutaz, *Le piante di Roma*, 1:66–67, and vol. 2, pianta XXI, tav. 36, "accuratissime delineate"; Hülsen, *Saggio di bibliografia ragionata delle piante*, 58–60; Emmanuel Lurin, "Étienne Dupérac vedutista e cartografo: La costruzione della pianta di Roma del 1577," in *Le città dei cartografi: Studi e ricerche di storia urbana*, ed. Cesare De Seta, Brigitte Marin, and Marco Iuliano (Naples: Electa Napoli, 2008), 51–52; Lurin, "Etienne Dupérac, graveur, peintre et architecte," 1222–1225, where Lurin suggests that the engraver was not actually Dupérac but someone less skilled; and Scaccia Scarafoni, *Le piante di Roma*, 26–27.

34. BL, Maps* 23810. (4.) for the fragment with the privilege from Gregory XIII; and for the complete map, BL, Maps* 155. (7.), Stefano Duperac, *Urbis Romae sciographia ex antiquis monumentis accuratiss. delineata* (Rome: Lorenzo Vaccari, 1574; repr., Francesco Villamena, [1590s]). Hülsen, *Saggio di bibliografia ragionata delle piante*, 60–61, identifies this example in the British Library as the unique example of the first 1574 printing. However, as Frutaz, *Le*

piante di Roma, 1:67–68, and vol. 2, pianta XXII, tav. 37–50 on 1:67, points out, the phrase "Romae excudebat Franciscus Villamena," on the upper-right corner of this example, indicates that it is a reprint, since Francesco Villamena (or Villamoena, ca. 1565–1624) did not arrive in Rome until the late 1580s during the reign of Sixtus V. Frutaz reproduces a second reprint by Giovanni Giacomo De Rossi from the mid to late seventeenth century. See also Giovanna Aragozzini and Marco Nocca, *Le piante di Roma dal Cinquecento all'Ottocento* (Rome: Dino Audino, 1993), 38–39 and tav. 8, who identify the original publisher as Vaccheria; Lurin, "Etienne Dupérac, graveur, peintre et architecte," 1226–1233, 1230, for the later addition of the Roman coat of arms; and Maier, *Rome Measured and Imagined,* 134–135. It can be assumed that no example of the first printing is extant. The plates are conserved at the Calcografia Nazionale di Roma (now the Istituto Nazionale per la Grafica). See Carlo A. Petrucci, *Catalogo generale delle stampe tratte dai rami incisi posseduti dalla Calcografia Nazionale* (Rome: La Libreria dello Stato, 1953), 126n1439.

35. BL, Maps* 155. (7.), Duperac, *Urbis Romae sciographia*: "magno cum labore"; "diligenti pervestigatione, acri animadversione, acurata descriptione"; "perpetuo adnotans / quo quidque loco repertum atque erutum esset"; "adhibito eruditissimorum hominum iuditio"; "magno mihi auxilio."

36. BL, Maps* 23805. (8.), Stefano Duperac, *Nova urbis Romae descriptio* (Rome: Antonio Lafreri, 1577). See Bury, *Print in Italy,* 136–137; Ehrle, *Roma prima di Sisto V,* for a facsimile reprint of the map and 25–33 for an introduction that includes a complete list of labeled items that are on the map; Frutaz, *Le piante di Roma,* 1:186, and vol. 2, tav. 247–255; Hülsen, *Saggio di bibliografia ragionata delle piante,* 66–68; Lurin, "Etienne Dupérac, graveur, peintre et architecte," 1450–1454; "Étienne Dupérac vedutista e cartografo," 50–51; and Maier, *Rome Measured and Imagined,* 139–143. For a summary, see Aragozzini and Nocca, *Le piante di Roma,* 50–51, and tav. 7.

37. For etching, see Landau and Parshall, *Renaissance Print,* 27–28; and see Lurin, "Étienne Dupérac vedutista e cartografo," esp. 51–57. For the *Vestiges,* Stefano Duperac [Étienne Dupérac], *I vestigi dell'antichità di Roma* [. . .] (Rome: Lorenzo della Vaccheria, 1575); and see Lurin, "Etienne Dupérac, graveur, peintre et architecte," 397–465 and 1312–1450 (E98–E138).

38. Lurin, "Etienne Dupérac, graveur, peintre et architecte," 599–695.

39. See R. Almagià, "Intorno a un cartografo italiano del secolo XVI," *Rivista Geografica Italiana* 20 (February–March 1913): 99–112; Fabia Borroni Salvadori, "Cartaro, Mario," *DBI,* 20:796–799; Annalisa Cattaneo, "Mario Cartaro, incisore viterbese del XVI secolo," *Grafica d'Arte* 9, no. 35 (1998): 2–9, 11, no. 41 (2000): 6–14, and 11, no. 42 (2000): 3–11; V. Federici, "Di Mario Cartaro, incisore viterbese del secolo XVI," *ASRSP* 21 (1898): 535–552; Long, "Multi-Tasking 'Pre-professional' Architect/Engineers," 236–239; Maier, *Rome Measured and Imagined,* esp. 143–152; and Witcombe, *Copyright in the Renaissance,* 166–169; *Print Publishing,* 226–227, 279–281, 283–285, and 292–295. The signed celestial globe is in the Istituto e Museo di Storia della Scienza in Florence (inventory no. 123), http://www.mhs.ox.ac.uk/epact/catalogue.php?ENumber=87909, and the terrestrial is at the Observatory of Monte Mario in Rome. For Cartaro's role in assessing Lafreri's estate, see Ehrle, *Roma prima di Sisto V,* 45–47. For Cartaro's activity in Naples and his work with Sigliola, see Marino, "Administrative Mapping," 12–15; Ernesto Mazzetti, ed., *Cartografia generale del Mezzogiorno e della Sicilia,* 2 vols. (Naples: Edizioni Scientifiche Italiane, 1972); Vladimiro Valerio, "The Neapolitan Saxton and His Survey of the Kingdom of Naples," *Map Collector* 18 (March

1982): 14–17; and "Cartography in the Kingdom of Naples during the Early Modern Period," in Woodward, *Cartography*, pt. 1, 962–970.

40. BL, Maps* 23805. (7.), Mario Cartaro, *Urbis Romae descriptio* (Rome: Mario Cartaro, 1575), measuring ca. 36 by 22 inches (41 × 55.2 cm). See Frutaz, *Le piante di Roma*, vol. 1, p. 184, and vol. 2, tav. 237; and Hülsen, *Saggio di bibliografia ragionata delle piante*, 62.

41. Mario Cartaro, *Novissimae urbis Romae accuratissima descriptio* (Rome: Mario Cartaro, 1576), "Urbis Romae situm [. . .] diligenter a se descriptum et aeneis formis accurate repraesentatum." There are two extant examples, both in Rome: BIASA, Roma: X. 648, and BNC, P. A. 1. See Frutaz, *Le piante di Roma*, 1: 185 and 2: tav. 238–246; Hülsen, *Saggio di bibliografia ragionata delle piante*, 65–66; and E. Rocchi, *Le piante icnografiche e prospettiche di Roma del secolo XVI* (Turin: Roux e Viarengo, 1902), vol. 1, pp. 80–100, and vol. 2, tav. XVI, XVI bis. See also Scaccia Scarafoni, *Le piante di Roma*, 88; Maier, "A 'True Likeness,'" esp. 738–747, who points out (n68) that "Roma Renasces" is known from ancient coins and that often classical Latin dropped the "n" of a gerund when it was preceded by an "s"; and Maier, "*Roma Renascens*," 46–50.

42. BL, Maps* 155. (10.), Mario Cartaro, *Celeberrimae urbis antiquae fidelissima topographia post omnes alias aeditiones accuratissime delineata* (Rome: Mario Cartaro, 1579); and see Frutaz, *Le piante di Roma*, 1:68–69, and vol. 2, tav. 51–55; Hülsen, *Saggio di bibliografia ragionata delle piante*, 68–70; Maier, "A 'True Likeness,'" esp. 742–747; *Roma Renascens*, 46–50; and E. Rocchi, *Le piante icnografiche e prospettiche*, 1:121–172, and vol. 2, tav. XXIII.

43. BL, Maps* 155. (10.), Cartaro, *Celeberrimae urbis antiquae*. Upper-right cartouche: "Studioso Lectori"; "eadem [i.e., aedificia] suis propriis locis eo pacto situaverimus, ut verissimis intervallis quantum ex apparentibus hodie vestigiis colligi potuit"; "Habes itaque Romam et Antiquam et Novam duobus tabellis eiusdem magnitudinis a nobis excussam"; "plurimorum doctorum hominum studio suffulti et diligenti inspectione omnium que extant antiquorum fragmentorum roborati, mathematicorum instrumentorum usu."

44. For a comprehensive study, see Leuschner, *Antonio Tempesta*. For a catalog of hundreds of his images, see Sebastian Buffa, ed., *Antonio Tempesta: Italian Masters of the Sixteenth Century*, vols. 35–37 of *The Illustrated Bartsch*, ed. Walter L. Strauss (New York: Abaris Books, 1983–1984), and for a succinct summary, see C. Höper, "Tempesta, Antonio," *DA*, 30:428–429. For van Aelst, see Leuschner, *Antonio Tempesta*, 155–163; Loredana Lorizzo, "Nicolas van Aelst's Will and a List of his Plates," *Print Quarterly* 31 (2014): 3–20; Witcombe, *Copyright in the Renaissance*, esp. 272–276; and Witcombe, *Print Publishing*, esp. 360–370.

45. Newberry Library, Novacco 4F 256 (PrCtO): Antonio Tempesta, *Recens prout hodie iacet almae urbis Romae cum omnibus viis aedificiisque prospectus accuratissime delineatus* (Rome, 1593). The 1606 version owned by the Vatican Library, BAV, Stampe Geogr.1.621. Riserva, is published in a facsimile edition in Francesco Ehrle, *Roma al tempo di Clemente VIII: La pianta di Roma di Antonio Tempesta del 1593 riprodotta da una copia vaticana del 1606* (Vatican City: Biblioteca Apostolica Vaticana, 1932). See also Stefano Borsi, *Roma di Sisto V: La pianta di Antonio Tempesta, 1593* (Rome: Officina Edizioni, 1986); Hülsen, *Saggio di bibliografia ragionata delle piante*, 74–76; Frutaz, *Le piante di Roma*, 1:192–194, and vol. 2, tav. 262–274; Eckhard Leuschner, "Prolegomena to a Study of Antonio Tempesta's "Map of Rome," in Bevilacqua and Fagiolo, *Piante di Roma*, 158–167; Scaccia-Scarafoni, *Le piante di Roma*, 92; and Witcombe, *Copyright in the Renaissance*, 242.

46. Leuschner, "Prolegomena to a Study," 161–162. Hülsen, *Saggio di bibliografia ragionata delle piante*, transcription on 74: "Urbem non illam veterem, sed quam hodie sub sanctis

Pontificibus florentem aspicimus"; "haud parvis meis sumptibus et vigiliis." And see Giacomo Bosio, *Dell'istoria della sacra religione et illma. militia de San Giovanni gierosolimitano*, 3 vols. (Rome: Stamperia Apostolica Vaticana, 1594–1602), and for Bosio's life, Gaspare De Caro, "Bosio, Giacomo," *DBI*, 13:261–264.

47. Borsi, *Roma di Sisto V*, 20.

CHAPTER 7

1. Michel de Montaigne, *Montaigne's Travel Journal*, trans. Donald M. Frame. Repr. ed. (San Francisco: North Point Press, 1983), xiv, for his education, and 73 and 101, for the scrutiny of his books. For the Porta del Popolo and Pius's other new gates, see Federico Bellini, "Le porte romane di Pio IV (1559–1565)," in "Entrare in città: Le porte di Roma," ed. Giuseppe Bonaccorso and Claudia Conforti, special issue, *Roma Moderna e Contemporanea* 22 (January–June 2014): 37–61.

2. Montaigne, *Montaigne's Travel Journal*, 72.

3. Ibid., 72–74.

4. Ibid., 74–77, citations on 76 and 77. See also Umberto Gnoli, *Topografia e toponomastica di Roma medioevale e moderna*, 2nd ed., ed. Livio Jannattoni (Foligno: Edizioni dell'Arquata, 1984), 173; and Simoncini, *Roma: Le trasformazioni urbane nel Cinquecento*, 1:260–261, for Montaigne's residences in Rome.

5. Montaigne, *Montaigne's Travel Journal*, 77–78. The monks accompanying the man on the cart would have been members of the Confraternity of San Giovanni Decollato, whose mission was to provide comfort and assistance to persons condemned to death. For a history of the confraternity, see Andrea Carlino, *Books of the Body: Anatomical Ritual and Renaissance Learning*, trans. John Tedeschi and Anne C. Tedeschi (Chicago: University of Chicago Press, 1999), esp. 92–108; and Vincenzo Paglia, *La morte confortata: Riti della paura e mentalità religiosa a Roma nell'età moderna* (Rome: Edizioni di Storia e Letteratura, 1982), 123–133, on the role of the spectators in public executions.

6. Richard Ingersoll, "The Ritual Use of Public Space in Renaissance Rome" (PhD diss., University of California, Berkeley, 1985). See as well Martine Boiteux, "Fêtes et cérémonies romaines au temps des Carrache," in *Les Carrache et les décors profanes: Actes du Colloque*, Collection de L'École française de Rome (Rome: École française de Rome, 1988), 183–213; "Rivalità festive: Rituali pubblici romani al tempo di Sisto V," in Fagiolo and Madonna, *Sisto V*, 1:357–392; "La noblesse romaine"; Silvia Carandini, "L'effimero spirituale: Feste e manifestazioni religiose nella Roma dei papi in età moderna," in Fiorani and Prosperi, *Roma, la città del papa*," 519–553; Marcello Fagiolo, ed., *La festa a Roma dal Rinascimento al 1870*, 2 vols. (Rome: Umberto Allemandi, 1997); and Gamrath, *Roma sancta renovata*, 123–163.

7. See Setton, *Papacy and the Levant*, 1045–1104. Marcantonio Colonna notably entered Rome by the same gate as the emperor Charles V in his triumphal entry of 1536. Both men arrived from the south on the Via Appia and through the Porta San Sebastiano. Marcantonio then processed in a triumphal march of high pageantry to the Roman Forum and to St. John Lateran. See Francesco Cancellieri, *Storia de' solenni possessi de' sommi pontefici* [...] (Rome: L. Lazzarini, 1802), 112–119.

8. A detailed account of each of the papal *possessi* with the citation of many primary sources is the account of Cancellieri, *Storia de' solenni possessi*; and see Irene Fosi, "Court and City in the Ceremony of the *Possesso* in the Sixteenth Century," in Signorotto and Visceglia,

Court and Politics in Papal Rome, 31–52; and Marcello Fagiolo, "L'effimero di Stato: Dal conclave al possesso," in Fagiolo, *La festa a Roma*, 2:8–25. For Via Papale, see Valeria Cafà, "The Via Papalis in Early Cinquecento Rome: A Contested Space between Roman Families and Curials," *Urban History* 37 (2010): 434–451. For fifteenth-century hostile crowds, see Ingersoll, "Ritual Use of Public Space," 171–223, and for sixteenth-century crowd behaviors, see Hunt, *Vacant See*, 257–263.

9. On the *possesso* and other aspects of papal ceremonial ritual, see Boiteux, "Parcours rituels romains," Irene Fosi, "'Parcere subiectis, debellare superbos': L'immagine della giustizia nelle cerimonie di possesso a Roma e nelle legazioni dello Stato Pontificio nel Cinquecento," and Maria Antonietta Visceglia, "Il cerimoniale come linguaggio politico: Su alcuni conflitti di precedenza alla corte di Roma tra Cinquecento e Seicento," all in Visceglia and Brice, *Cérémonial et ritual à Rome*, 31–45; 89–104; and 117–176, respectively. And see Cancellieri, *Storia de' solenni possessi*; Jennifer Mara DeSilva, "Taking Possession: Rituals, Space, and Authority," *Royal Studies Journal* 3 (2016): 1–17; Fosi, "Court and City in the Ceremony of the *Possesso*"; and Stinger, *Renaissance in Rome*, 53–57. For Pius V's *possesso* procession, see Lemaître, *Saint Pie V*, 202–203. For the arch of Titus, see Richardson, "Arcus Titi (2)," in *New Topographical Dictionary*, 30; and J. Arce, "Arcus Titi (Via Sacra)," in Steinby, *Lexicon Topographicum*, 1:109–111.

10. Fosi, "Court and City in the Ceremony of the *Possesso*."

11. For the master of ceremonies (Giovanni Francesco Firmano, d. 1565), see Donatella Rosselli, "Firmano, Giovanni Francesco," *DBI*, 48:223. The citation is from BAV, Barb. lat. 2800: Firmano, *Caeremoniale Io. Francisci Firmani Maceratensis. Capellae S.mi. D. N. Papae Caerimoniarus Clerici* [. . .] [], fols. 160v–161r: "ceciderunt infiniti homines unus sup alium"; "fuerunt soffocati circa 26."

12. BAV, Urb. lat 1040, fol. 172v, January 19, 1566: "Non gettò denari"; "fece adunar tutti, i poveri in Campo Santo poi uscir per una sola Porta, et à tutti fece dar tre giulij per ciascuno."

13. Cancellieri, *Storia de' solenni possessi*, 119–121, for Gregory XIII, and 121–127, for Sixtus V. For Sixtus V, see also Fosi, "Court and City in the Ceremony of the *Possesso*," 48–52; and Boiteux, "Rivalità festive," 359–363.

14. Cancellieri, *Storia de' solenni possessi*, 110–112, citation of Cornelio Firmano on 111. For Cornelio Firmano (1532/33–1588), who was the nephew of Giovanni Francesco Firmano, mentioned above (n. 11), see Donatello Rosselli, "Firmano, Cornelio," *DBI*, 48:221–223. And see Lemaître, *Saint Pie V*, 97–99 and 202–206; and Pastor, *History of the Popes*, 17:70–71.

15. Fosi, "Court and City in the Ceremony of the *Possesso*," 48–52; and see Boiteux, "Rivalità festive," 359–363, and Cancellieri, *Storia de' solenni possessi*, 121–127.

16. Boiteux, "Parcours rituels romains," 52–59; Ingersoll, "Ritual Use of Public Space," 118–122; Pecchiai, *Roma nel Cinquecento*, 368–370; and Barbara Wisch, "The Matrix: *Le Sette Chiese di Roma* of 1575 and the Image of Pilgrimage," *Memoirs of the American Academy in Rome* 56/57 (2011–2012): 271–303.

17. Gregory Martin, *Roma sancta (1581)*, ed. George Bruner Parks (Rome: Edizioni di Storia e Letteratura, 1969), esp. 17–70; and see esp. Genoveffa Palumbo, "'L'assedio delle reliquie' alla città di Roma: Le reliquie oltre la devozione nello sguardo dei pellegrini," in Nanni and Visceglia, "La città del perdono," 377–403.

18. Boiteux, "Parcours rituels romains," 55–57. For Sixtus's papal bull of February 13, 1586, which made Santa Maria del Popolo one of the seven churches, see Tomassetti et al.,

Bullarium diplomatum et privilegiorum, vol. 8, no. XXXIII, 663–666; and see Antinori, "Le basiliche romane meridionali."

19. For a comprehensive account of urban reconstructions including work on streets in the fifteenth century, organized chronologically and by pope, see Simoncini, *Roma: Le trasformazioni urbane nel Quattrocento,* vol. 1; and see Christoph L. Frommel, "Papal Policy: The Planning of Rome during the Renaissance," in *Art and History: Images and Their Meaning,* ed. Robert I. Rotberg and Theodore K. Rabb (Cambridge: Cambridge University Press, 1988), 39–65; and Maria Letizia Gualandi, "'Roma resurgens': Fervore edilizio, trasformazioni urbanistiche e realizzazioni monumentali da Martino V Colonna a Paolo V Borghese," in Pinelli, *Roma del Rinascimento,* 123–160. For Nicholas V, see also Westfall, *In This Most Perfect Paradise.* For accounts of two medieval streets, see Deborah Robbins, "Via della Lungaretta: The Making of a Medieval Street," and Richard Ingersoll, "Piazza di Ponte and the Military Origins of Panopticism," both in *Streets: Critical Perspectives on Public Space,* ed. Zeynep Çelik, Diane Favro, and Richard Ingersoll (Berkeley: University of California Press, 1994), 165–175 and 177–188, respectively. For the Via Santa and Sixtus IV's intervention, see Maurizio Caperna, *La Lungara,* vol. 1, *Storia e vicende edilizie dell'area tra il Gianicolo e il Tevere* (Rome: Edizioni Quasar, 2013), 21–43. For new streets created by Pope Paul II (1464–1471), see Modigliani, *Disegni sulla città;* and for a broader European context, see Conforti, *La città del tardo Rinascimento,* esp. 65–101.

20. Tomassetti et al., *Bullarum diplomatum et privilegiorum,* vol. 5, no. XII, 377–378. And see Marcello Fagiolo and Maria Luisa Madonna, "Il revival del trionfo classico: Da Alessandro VI alla sfilata dei Rioni," in Fagiolo, *La festa a Roma,* 1:34–49, esp. 34; Simoncini, *Roma: Le trasformazioni urbane nel Quattrocento,* 1:226–229 and 2:281–284 (partial translation and summary by G. Mosconi); and Enrico Guidoni and Giulia Petrucci, *Roma, Via Alessandrina: Una strada "tra due fondali" nell'Italia delle corti (1492–1499)* (Rome: Edizioni Kappa, 1997).

21. See esp. Caperna, *La Lungara,* 59–67; Suzanne B. Butters and Pier Nicola Pagliara, "Il Palazzo dei Tribunali e via Giulia a Roma/ The Palazzo dei Tribunali and Via Giulia in Rome," *Zodiac* 14 (1995–1996): 14–29; Luigi Salerno, Luigi Spezzaferro, and Manfredo Tafuri, *Via Giulia: Una utopia urbanistica del 500* (Rome: Stabilimento Aristide Staderini, 1973), esp. 15–197; Simoncini, *Roma: Le trasformazioni urbane nel Cinquecento,* 1:35; and Temple, *Renovatio Urbis,* esp. 34–93 and 151–159.

22. Simoncini, *Roma: Le trasformazioni urbane nel Cinquecento,* 1:57–58, and 73, for Via Ripetta, and 135, for Via del Babuino.

23. See esp. Gualandi, "'Roma resurgens,'" in Pinelli, *Roma del Rinascimento,* 149–150; Karmon, *Ruin of the Eternal City,* 100–110; Maria Luisa Madonna, "L'ingresso di Carlo V a Roma," in Fagiolo, *La festa a Roma,* 1:50–65; and Visceglia, *La città rituale,* 191–201.

24. Bellini, "Le porte romane," 37–61; "Michelangelo, La Strada e la Porta Pia," *Studi Romani* 59 (2011): 74–109; Claudia Conforti, "Via Pia: Rus in urbe," *Paragone: Arte,* 3rd ser., 82 (November 2008): 21–31; Fagioli and Madonna, "La Roma di Pio IV"; Simoncini, *Roma: Le trasformazioni urbane nel Cinquecento,* 1:209–221; and Spagnesi, *Roma: La basilica,* 122–129.

25. Gualandi, "'Roma resurgens,'" 151–152; and Simoncini, *Roma: Le trasformazioni urbane nel Cinquecento,* 1:213–214, for Via Merulana.

26. See Bellini, "La *Civitas Pia*"; Giulia Petrucci, "La città Pia: Un'espansione urbana del Cinquecento," *Storia Urbana* 17 (July–September 1993): 19–48; and Simoncini, *Roma: Le trasformazioni urbane nel Cinquecento,* 1:188–190.

27. The document is in the Archivio di Stato di Firenze, Carte Strozzi, cod. 233, transcribed in Wolfgang Lotz, "Gli 883 cocchi della Roma del 1594," In *Studi offerti a Giovanni Incisa della Rocchetta* (Rome: Società Romana di Storia Patria, 1973), 247–266. See also Delumeau, *Vie économique et sociale*, 1:443–446; John M. Hunt, "The Ceremonial Possession of a City: Ambassadors and Their Carriages in Early Modern Rome," *Royal Studies Journal* 3 (2016): 69–89; "Carriages, Violence, and Masculinity in Early Modern Rome," *I Tatti: Studies in the Italian Renaissance* 17 (Spring 2013): 175–196.

28. ASV, Bandi Sciolti I, no. 89, "per ristoro in parte del grand danno che di continuo fanno in l'ammattonati, et strade di Roma."

29. An edict, issued on March 7, 1583, was published by Scaccia Scarafoni, "L'antico statuto dei 'magistri stratarum,'" 297–300, no. XXI. The edict confirmed a directive issued in 1574 by the masters of the streets—see ASC, Carte Boccapaduli, Arm. II, maz. IV, b. 104b, fasc. 63, L. 1029, which dictated a tax of two scudi per month per coach and required each coach to have a license, and L.1032, 1036, 1037, and 1039, all of which concern taxes on coaches. This edict on L. 1029 was also recorded in the minutes of the Congregation on Streets, Bridges, and Fountains—see ASR, *Congregazione super viis pontibus et fontibus*, Registro 1, fol. 75r; and Genovese and Sinisi, *Pro ornatu et publica utilitate*, May 31, 1574 (Congregation no. 59), 58, and passim for the granting of many appeals for tax relief by those claiming to be impoverished by using the coach tax. See also Pecchiai, *Roma nel Cinquecento*, 382, who cites the 1574 edict, which was promulgated by the president (Andrea Spinola) and masters of the streets (Propero Boccapaduli and Marcello Negri). For prohibitions of prostitutes from riding in coaches, see Monica Kurzel-Runscheiner, *Töchter der Venus: Die Kurtisanen Roms im 16. Jahrhundert* (Munich: C. H. Beck, 1995), esp. 83–84 which provides further examples.

30. See Genovese and Sinisi, *Pro ornatu et publica utilitate*, xxx–xxxi; and see Charles Burroughs, "Absolutism and the Rhetoric of Topography: Streets in the Rome of Sixtus V," in Çelik, Favro, and Ingersoll, *Streets: Critical Perspectives*, 189–202; Marder, "Sixtus V and the Quirinal"; and Schiffmann, *Roma felix*, 21–55, which provides a detailed chronology based on archival documents.

31. Maurizio Crocco, *Roma, Via Felice da Sisto V a Paolo V* (Rome: Edizioni Kappa, 2002), 58–67, provides a résumé of conflicting interpretations; and see Enrico Guidoni, *L'urbanistica di Roma tra miti e progetti* (Rome: Laterza, 1990), 131–153.

32. See Crocco, *Roma, Via Felice*; Gualandi, "'Roma resurgens,'" 156–158; Simoncini, *Roma: La trasformazioni urbane nel Cinquecento*, 1:309–349; Simoncini, "*Roma restaurata*"; and Spagnesi, *Roma: La basilica*, 132–142.

33. Tomassetti et al., *Bullarum diplomatum et privilegiorum*, vol. 8, no. CII, 914–916.

34. Giovanni Francesco Bordini, *De rebus praeclare gestis a Sixto V. Pon. Max.* (Rome: J. Tornerius, 1588), 51–52. For Bordini, see Nello Vian, "Bordini, Giovanni Francesco," *DBI*, 12:507–508.

35. Duperac, *I vestigi dell'antichità*, no. 34.

36. Paving was a major concern of the *Congregazione super viis, pontibus et fontibus*, dealt with in numerous meetings. See ASR, *Congregazione super viis pontibus et fontibus*, Registro 1, and a list of instances (that also includes other matters having to do with streets) in Genovese and Sinisi, *Pro ornatu et publica utilitate*, 173–177.

37. Ludovica Cibin, *Selciato romano: Il sampietrino; materiale, lavorazione, evoluzione storica, tipologie, apparecchiature, posa in opera* (Rome: Gangemi, 2003), 48, for shape, and 57–95, for the historical development.

38. Tomassetti et al., *Bullarum diplomatum et privilegiorum*, vol. 7, no. CXX, 386–389 (clause 31) on 395–396. For the discussion at the time of Sixtus IV, see Cibin, *Selciato romano*, 67; and more generally Rinne, *Waters of Rome*, 203–213.

39. ASR, Presidenza delle Strade, b. 42, fols. 10r–v, 40v–41r, 42v–43r, 56v–57v, and 62r–v (for stone), 43v–44r (for stone or gravel), and 13v–14v (for both stone and brick). The remaining fifty-three licenses are for paving with brick.

40. BNC, Gesuitico 713/6, Guido Baldo Foglietta, "Discorso del mattonato o selicato di Roma." A published transcription of the tract is Guido Baldo Foglietta, "Discorso del mattonato o selicato di Roma," ed. U. Balzani, *ASRSP* 1 (1877): 371–376. The tract was written at some time during the pontificate of Sixtus V (as is known from the reference to Sixtus on 376). I have used the page numbers of the printed version, the text of which is identical to the manuscript with the exception of minor orthographic variations.

41. Foglietta, "Discorso del mattonato o selicato," 372: "quella testa [di chiodo] come punta di scarpello che taglia et distacca le pietre violentemente."

42. Ibid., 373: "escludendo i mattoni per materia molto tenera"; "porosità molto frangibile." For "astraco," see Max Pfister, ed., *Lessico etimologico italiano*, vol. 3.2 (Wiesbaden: Ludwig Reichert, 1991), s.v. "astracum," cols. 1927–1954.

43. Foglietta, *Discorso del mattonato o selicato*, 374: "moto violento di carozze, con Tevertini, colonne et altro, ha bisogna di fondamento gagliardo"; "la materia di bona compositione"; "bisogna usarci molta diligenza, cosa che non si è fatto sino a hoggi."

44. Ibid., 376. ASR, *Congregazione super viis pontibus et fontibus*, Registro 1, fols. 155r, 158v, 159v, 161r, and 163r; and Genovese and Sinisi, *Pro ornatu et publica utilitate*, 124 (Congregation no. 17), June 7, 1585; 127 (Congregation no. 120), September 28, 1585; 128 (Congregation no. 121), November 5, 1585; 129–130 (Congregation no. 123), February 6, 1586; and 131 (Congregation no. 124), May 18, 1586.

45. For the Capitoline contracts or licenses given to masons in 1586, see ASC, C.C., cred. IV, tom. 82, cat. 326 (1586–1589). The paving contracts for the year 1586 are, for stone, fols. 3r, 4r, 8r, 8v, 9r, 12r, 21r, 30r, and 31r; and for brick, 3v, 4r, 8v, 11r, 28r, 29r, 29v, and 36r. For this and related (later) documents, see Howard Hibbard, "Di alcune licenze rilasciate dai mastri di strade per opera di edificazione a Roma (1586–'89, 1602–'34)," *Bollettino d'Arte*, 5th ser, 52 (1967): 99–117.

46. ASV, Misc. Arm. XI, tome 86, fol. 228. The document is reproduced in F. Cerasoli, "Notizie circa la sistemazione di molte strade di Roma nel sec. XVI," *Bullettino della Commissione Archeologica Comunale di Roma* 28 (October–December 1901): 342–362; and Cibin, *Selciato romano*, 72–79. Another document records a series of mandati or pay orders issued between January and July 1587 totaling almost 4,000 scudi—ASR, Camerale I, Mandati Camerale, b. 936 (January 1587–December 1589), fols. 2r–v, 3r–v, 4r, 5v, 6r–v, 9v, 12r–v, and 14r.

47. BAV, Urb. lat. 1055, 30r–v, January 28, 1587: "Trattasi di mattonare Roma di pietre cotte, et levare i lastricati de selici per mille mali effetti, che causa questa opera."

48. ASR, *Congregazione super viis pontibus et fontibus*, Registro 1, fols. 178r and 179v; and Genovese and Sinisi, *Pro ornatu et publica utilitate*, 140 (Congregation nos. 131, 132), December 14, 1587, and February 3, 1588.

49. Elizabeth S. Cohen, "To Pray, To Work, To Hear, To Speak: Women in Roman Streets c. 1600," *Journal of Early Modern History* 12 (2008): 289–311; "Honor and Gender in the Streets of Early Modern Rome," *Journal of Interdisciplinary History* 22 (Spring 1992):

595–625; and Elizabeth S. Cohen and Thomas V. Cohen, "Open and Shut: Social Meanings of the Cinquecento Roman House," *Studies in the Decorative Arts* 9, no. 1 (Fall–Winter 2001–2002): 61–84. See also Laurie Nussdorfer, "The Politics of Space in Early Modern Rome," *Memoirs of the American Academy in Rome* 42 (1997): 161–181.

50. Stow, *Catholic Thought and Papal Jewry Policy*, which includes the Latin text and English translation of *Cum Nimis* as appendix 1, 291–298. The bull can also be found in Tomassetti et al., *Bullarum diplomatum et privilegiorum*, vol. 4, no. IV, 498–500.

51. Stow, *Catholic Thought and Papal Jewry Policy*, 291–298. For the actual conditions and construction of the ghetto, see Attilio Milano, *Il ghetto di Roma* (Rome: Staderini, 1964), esp. 71–82, for the general conditions under successive popes through Sixtus V, and 185–207.

52. Milano, *Il ghetto*, 74–82 and Kenneth R Stow, *Theater of Acculturation: The Roman Ghetto in the Sixteenth Century* (Seattle: University of Washington Press, 2001), 42–44, for the size of the ghetto.

53. Milano, *Il ghetto*, 74–82, emphasizes the great differences in enforcement among the popes, while Stow, *Catholic Thought and Papal Jewry Policy*, esp. 13–24, insists that the popes followed the same policies even if some made concessions because of temporary contingencies. Insightful studies that explore the complex ways in which the Jews dealt with the new regime include Bernard D. Cooperman, "Ethnicity and Institution Building among Jews in Early Modern Rome," *AJS Review: The Journal of the Association for Jewish Studies* 30 (April 2006): 119–145; "Licenses, Cartels, and Kehila: Jewish Moneylending and the Struggle against Restraint of Trade in Early Modern Rome," in *Purchasing Power: The Economics of Modern Jewish History*, ed. Rebecca Kobrin and Adam Teller (Philadelphia: University of Pennsylvania Press, 2015): 27–45; and Di Nepi, *Sopravvivere al ghetto*. For the history of the Roman Jews in an earlier period, see Anna Esposito, *Un'altra Roma: Minoranze nazionali e comunità ebraiche tra Medioevo e Rinascimento* (Rome: Il Calamo, 1995), and for an overview, see Foa and Stow, "Gli ebrei di Roma."

54. Lazar, *Working in the Vineyard*, 99–124.

55. Kurzel-Runtscheiner, *Töchter der Venus*, esp. 20–22 and 45–92; Cohen, "Seen and Known"; and see also Lazar, *Working in the Vineyard*, 37–70.

56. For a general overview, see Romano Canosa and Isabella Colonnello, *Storia della prostituzione in Italia dal Quattrocento alla fine del Settecento* (Rome: Sapere 1989), esp. 43–56; and see Cohen, "Seen and Known"; "Back Talk: Two Prostitutes' Voices from Rome c. 1600," *Early Modern Women* 2 (Fall 2007): 95–126; Kurzel-Runtscheiner, *Töchter der Venus*, esp. 1–24 and 64–92; and Storey, *Carnal Commerce*, esp. 115–138.

57. ASC, C.C., cred. I, cat. 1, tom. 37, fol. 237v (ad xiij Kal Junii MD lxvi), Consilium Publicum: "desiderava sommamente"; "in ogni modo . . . di un luogho per le meretrici et donne dishoneste, perche non voleva in modo alcuno che stessero et habitassero fra le donne honeste." Also in ASC, C.C., cred. I, cat. 1, tom. 23, fol. 7v. For an account of Pius V's policies against the prostitutes, see also Pastor, *History of the Popes*, 17:89–92.

58. BAV, Urb. lat. 1040, fol. 240v, May 25, 1566: "le cortegiane stiano in una, ò due parte della Città"; "in trastevere ò verso il Popolo."

59. ASC, C.C., cred. I, cat. 1, tom. 37, fol. 239r (ad iij Kal Junij MD lxvi), Consilium Ordinarium: "ad reclaustrum Meretricium"; "aliquantulum difficiles resolutiones." Also in ASC, C.C., cred. I, cat. 1, tom. 23, fol. 9v.

60. BAV, Urb. lat. 1040, June 29, 1566, fol. 257r: "Hieri sera a forza li sbiri, le meretrici sgombrorno Borgo et li Conservatori s'affaticano per trovar luoco da serrarle."

61. The *avviso* was written by Camillo Luzzara and sent from Rome to the Duke of Mantua. It is published by Antonino Bertolotti, "Repressioni straordinarie alla prostituzione in Roma nel secolo XVI," *Rivista di Discipline Carcerarie* 16 (1886): 505–517, on 510–511, which includes the transcription of an *avviso* in the Archivio di Stato di Mantova. See also Cohen, "Seen and Known," 401, and Storey, *Carnal Commerce*, 71–72. For the history of homes and/or convents for (reformed) prostitutes focusing on the Jesuits, see Lazar, *Working in the Vineyard*, 49–70.

62. BAV, Urb. lat. 1040, 27 Luglio 1566, fol. 269v–270r: "nelle altre strade men publiche di Roma."

63. BAV, Urb. lat. 1040, August 3, 1566, fol. 273r–v: "Tutta questa matina havete sonato il vostro campanello in Congregarii a far che? Per conservarli l'infamia [...] che dalle meretici siano habitate le piu belle strade di Roma sancta ové è sparso il sangue di tanti santi Martiri, ove sono tanto relique tanto devotioni ove e la sede Apostolica et tanta Religione"; and ASC, C.C., cred. I, tom. 37, cat. 37 (iij Nonas Augusti MD lxvi), fols. 249v–250r. Also in C.C., Cred. I, tom. 23, cat. 23, fol. 25r–v; and Bertolotti, "Repressione," 511 (document IV), where an *avviso* sent to Mantua also reports the pope's reaction; and see Cohen, "Seen and Known," 400.

64. See Storey, *Carnal Commerce*, 79–88; and for reports of the letter having been written and delivered, ASC, C.C., cred. I, tom. 37, cat. 37 (III nonas Augusti MD lxvi), fols. 250r–v, where the author of the letter, Girolamo Picchi, is mentioned (250r), "come affettionato cittadino." (Also in ASC C.C., cred. I, tom 23, cat. 23, fol. 25v); and BAV, Urb. lat. 1040, August 10, 1566, fol 269r. For the letter (which may be a seventeenth-century copy) see ASR, Fondo Santa Croce, Filza 188: "Lettera scritta da incerto a Papa Pio Quinto accioche gli ebrei e le meretrici non si scacciano da Roma con le ragioni allegate per il medemo effetto da Romano."

65. For example, BAV, Urb. lat. 1040 (March 15, 1567), fol. 383v; Bertolotti, "Repressioni straordinarie," 513 (document X), an *avviso* sent to Mantua from Rome; and see Storey, *Carnal Commerce*, 74; and Canosa and Colonnello, *Storia della prostituzione*, 53.

66. . See Bertolotti, "Repressioni straordinarie," 512–513 (September 7, 1566, document VII), and 513 (July 19, 1567, document XI), *avvisi* sent from Rome to the court of Mantua, and BAV, Urb lat. 1040, fol. 409v (May 24, 1567); and see Storey, *Carnal Commerce*, 74.

67. For example, July 19, 1567, BAV, Urb. lat. 1040, fol. 434r, where it is reported that those not leaving would be "enclosed [...] like the Jews" (serarano [...] come sono li hebrei); October 5, 1569, BAV, Urb. lat. 1041, pt. 1, fol. 158v: "Si levano due strade principali alle Meretrici, et si restringono in vicoli povertissimi" (two main streets are taken from the prostitutes, and they are restricted to very poor alleys); October 17, 1569, BAV, Urb. lat. 1041, pt. 1, fol. 180r, "si ha dato principio à fare li Portoni, et gli sono cominciati li fondamenti all'hortacio per restringare le Cortegiane" (making the gates has started and the foundations of the Ortaccio in order to restrict the prostitutes). See also Pastor, *History of the Popes*, 17:408; and Lanciani, *Storia degli scavi*, 4:25; February 4, 1570, BAV, Urb. lat. 1041, pt. 1, fol. 231v: "vi si metterano guardiani per concludere affatto queste povere cortegiane" (and guards will be put there to close in entirely these poor courtesans); February 18, 1570, Urb lat. 1041, pt. 1, fols. 224v–225r (for closure during Lent with daily rations given out).

68. Cohen, "Seen and Known," 402–403, 404–409.

69. BAV, Urb. lat. 1043 (September 12, 1573), fol. 325v; and BAV, Urb. lat. 1045 (October 20, 1576), fol. 176. This edict also ordered vagabonds to leave Rome because of the plague

and the threat of them spreading it. And see BAV, Urb. lat. 1050 (January 27, 1582), fol. 24r, which reports that the prostitutes have been evicted from the Borgo.

70. Martin, *Roma sancta*, 191. For poverty in Rome, see Luigi Fiorani, "Religione e povertà: Il dibattito sul pauperismo a Roma tra Cinque e Seicento," *Ricerche per la storia religiosa di Roma: Studi, documenti, inventari* 3 (1979): 43–131; "'Charità et pietate'"; Lazar, *Working in the Vineyard*, 12–28, 71–98; Maria T. B. Russo, "Problemi e istituti dell'assistenza romana nel Cinque e Seicento," *Studi Romani* 34 (1986): 230–252; and Paolo Simoncelli, "Note sul sistema assistenziale a Roma nel XVI secolo," in *Timore e carità: I poveri nell'Italia moderna*, ed. Giorgio Politi, Mario Rosa, and Franco della Peruta, Atti del Convegno, Cremona, March 28–30, 1980 (Cremona: Biblioteca Statale, 1982), 137–156.

71. See Nicholas Terpstra, *Cultures of Charity: Women, Politics, and the Reform of Poor Relief in Renaissance Italy* (Cambridge, MA: Harvard University Press, 2013), esp. 14–20 and 30–31, for diverse and evolving types of charity; and Terpstra, *Religious Refugees in the Early Modern World: An Alternative History of the Reformation* (New York: Cambridge University Press, 2015), 218–240. See also Brian Pullan, "Poveri, mendicanti e vagabondi (secoli XIV–XVII)," in *Storia d'Italia: Annali*, vol. 1, *Dal feudalesimo al capitalismo*, ed. Corrado Vivanti and Ruggiero Romano (Turin: Einaudi, 1978): 981–1047; "New Approaches to Poverty and New Forms of Institutional Charity in Late Medieval and Renaissance Italy"; Alessandro Pastore, "Il problema dei poveri agli inizi dell'età moderna: Linee generali," and Angela Groppi, "Birbanti e poveri benestanti: Attitudini e pratiche assistenziali nei confronti della vecchiaia nella Roma pontificia (secc. XVI–XVIII)," in Zamagni, *Povertà e innovazioni istituzionali*, 17–43, 185–205, and 259–277, respectively.

72. Simoncelli, "Note sul sistema assistenziale," 140–141 (on midcentury expulsions); and ASV, Misc. Arm. IV, tome 60, fol. 207r: "Bando contra le Cortigiane e altre persone scandalose"; "qual si voglia persona vagabonda e senza esercitio, o partito"; "tutti li mendicanti che sono sani e gagliardi."

73. Tomassetti et al., *Bullarum diplomatum et privilegiorum*, vol. 7, no. IX, 434–438: "pauperes quoque mendicantes seu eleemosynas petentes."

74. For the early development of Roman hospitals in their various functions, see Carla L. Keyvanian, *Hospitals and Urbanism in Rome, 1200–1500* (Leiden: Brill, 2015), and "Charity, Architecture and Urban Development in Post-Tridentine Rome: The Hospital of the SS.ma Trinità dei Pellegrini e Convalescenti (1548–1680)" (PhD diss., Massachusetts Institute of Technology, 2000), esp. 124–140. For hospitals as lodging for pilgrims, see Romani, *Pellegrini e viaggiatori*, 198–234.

75. See Camillo Fanucci, *Trattato di tutte l'opere pie dell'alma citta di Roma* (Rome: Lepido Facij and Stefano Paolini ad instanza di Bastiano de' Franceschi Senese, 1601), 60–65, esp. 63–64; and see Simoncelli, "Note sul sistema assistenziale," 143–146; and Terpstra, *Cultures of Charity*, 39–40.

76. Fanucci, *Trattato di tutte l'opere*, 65–66: "sacco rosso [i.e., rozzo] con maze rosse [i.e., rozze] in mano"; "spettacolo veramente pietoso, maraviglioso, e forse non mai piu visto il simile"; "con maggiore trionfo che non feccero mai gli antichi Romani."

77. See Annarosa Cerutti Fusco, "Il progetto di Domenico Fontana 'per ridurre il Coliseo di Roma ad habitatione' e le opera sistine di 'pubblica utilità,'" *Quaderni dell'Istituto di Storia dell'Architettura*, n.s., 12 (1988): 65–84.

78. Tomassetti et al., *Bullarum diplomatum et privilegiorum*, vol. 8, no. LXXXV, 847–853: "fieret ne pauperes ipsi per alienas civitates vagarentur"; "in hac alma Urbe pietatis et cari-

tatis altrice"; "plurimi pauperes fame, frigore, nuditate, rerumque omnium inopia confecti, ac variis morbis e incommodis conflictati (849)"; and see Anna Maria Affanni, "L'Ospedale dei Mendicanti a Ponte Sisto: L'istituzione dell'Ospedale dei Mendicanti disposto da Sisto V nel 1587; Studio della conformazione originaria attraverso i documenti di archivio e le illustrazioni coeve," in Porzio, *Impronte sistine*, 101–117; and Simoncelli, "Note sul sistema assistenziale," 147–148.

79. Tomassetti et al., *Bullarum diplomatum et privilegiorum*, vol. 8, no. LXXXV, 847–853 (clause 1, 849): "ne . . . mendici per vias et plateas tota Urbe dispersi ac vagantes, victum quaeritando defatigentur"; "ne, non modo publica loca aut privatarum aedium vestibula, sed ipsa quoque templa gemitibus et eiulatibus compleant"; "ne in ecclesiis in quibus debet esse quieta conversatio, quaerimoniis, lamentis ac vociferationibus tumultus excitent"; "ne [. . .] incertis sedibus Urbem peragrant, proprium parochum non agnoscentes, absque alicuius divini praecepti aut bonorum morum notitita, tanquam bruta animalia rationis expertia."

80. Tomassetti et al., *Bullarum diplomatum, et privilegiorum*, vol. 8, no. LXXXV, 847–853 (clause 1, 849): "eorumque ignaviae et nequitiae via praecludatur." See also Keyvanian, "Charity, Architecture and Urban Development," esp. 150–155.

81. Fanucci, *Trattato di tutte l'opere*, 67: "poveri mendicanti, e in tanto numero, che non si può stare ne andare per le strade, che continuamente l'huomo non sia attorniato da questi." And see Lazar, *Working in the Vinyard*, 22–24.

82. Cited in Fiorani, "Religione e povertà," esp. 87–91 (citation in note 93): "si può dire ch'a guise de santi martiri con inusitato tormento [. . .] poiché furono da infinite puncture di pidocchi morti e trafitti, non essendo mai stato possible il difendersi da loro." See also Henderson, "'Mal Francese' in Sixteenth-Century Rome," 506–507.

83. Blastenbrei, *Kriminalität in Rom*; "I Romani tra violenza e giustizia nel tardo Cinquecento," *Roma Moderna e Contemporanea* 5 (January–April 1997): 67–79; "Violence, Arms and Criminal Justice in Papal Rome, 1560–1600," *Renaissance Studies* 20 (February 2006): 68–87; Hunt, *Vacant See*; Laurie Nussdorfer, "The Vacant See: Ritual and Protest in Early Modern Rome," *Sixteenth Century Journal* 18 (Summer 1987): 173–189; and Nussdorfer, "Priestly Rulers, Male Subjects: Swords and Courts in Papal Rome," in *Violent Masculinities: Male Aggression in Early Modern Texts and Culture*, ed. Jennifer Feather and Catherine E. Thomas (New York: Palgrave MacMillan, 2013), 109–128. On the more detailed level of individual cases that came before the Roman Tribunal, see Thomas V. Cohen and Elizabeth S. Cohen, *Words and Deeds in Renaissance Rome* (Toronto: University of Toronto Press, 1993), esp. 3–32; and for the riveting story of a single violent episode, Thomas V. Cohen, *Love and Death in Renaissance Italy* (Chicago: University of Chicago Press, 2004).

84. For bandits and "soldier-bandits," see Delumeau, *Vie économique et sociale*, esp. 1:85–89 and 2:542–566; Michele Di Sivo, "Le costituzioni e i bandi di Sisto V: L'amministrazione della giustizia tra accentramento e crisi dello Stato Pontificio," *Archivi per la Storia* 4 (November 1991): 137–147; Fosi, *La società violenta*; "La giustizia e la sua immagine: Propaganda politica e realtà nel pontificato sistino," in Fagiolo and Madonna, *Sisto V*, 1:59–82; Hunt, *Vacant See*, esp. 115–131; and Pastor, *History of the Popes*, 20:523–540, and 21:72–89.

85. For Pius IV, see Tomassetti et al., *Bullarum diplomatum et privilegiorum*, vol. VII, no. XXXII, 102–106, and no. LXIV, 186–193; ASR Bandi, Coll. I, b. 2, no. 194, *Motus proprius* [. . .] *per quem declaratur quod in breve nuper contra homicidas editio* [. . .] (Rome: Antonio Blado, 1564); ASR, Bandi, Coll. I, b. 3, no. 33, *Motus proprius . . . quae gratiae, seu remissiones, aut compositiones ex causa, vel occasione homicidii* [. . .] (Rome: Antonio Blado, 1564); and a

collection of previous edicts, ASR, Bandi, Coll. I, b. 3, no. 61, *Bullae* [. . .] *contra impios, ac facinorosos homicidas* (Bologna: Johannes Rubrius, 1565). For Pius V, see Tomassetti et al., *Bullarum diplomatum et privilegiorum*, tom. VII, no. XIV, 452–458. For Gregory XIII, see ASR, Bandi, Coll. I, b. 5, no. 39, *Sommario di quello* [. . .] *publicate contra gl'homicidi, banditi* [. . .] (Rome: Haeredes Antonij Bladij, 1578), and Tomassetti et al., *Bullarum diplomatum et privilegiorum*, vol. VIII, no. CXXII, 355–357. For Sixtus V, see Tomassetti et al., *Bullarum diplomatum et privilegiorum*, vol. VIII, no. IX, 585–591; ASR, Bandi, Coll. I, b. 6, no. 164, *Bando contra banditi*, [. . .] (Rome: Heredi d'Antonio Blado, 1587); and ASR, Bandi, Coll. II, b. 294, no. 140, *Constitutione* [. . .] *sopra le pene e confiscationi criminali* [. . .] (Rome: Paolo Blado, 1589).

86. For the decree of the Council of Trent, see Norman P. Tanner, ed. *Decrees of the Ecumenical Councils* (London: Sheed and Ward / Washington, DC: Georgetown University Press, 1990), 2: 795 (Session 25, chap. 19). For Pius IV's bull, see Tomassetti et al., *Bullarum diplomatum et privilegiorum*, vol. VII, no. XXV, 83–86. For one of Gregory XIII's edicts, see ASR, Bandi, Coll. II, b. 294, no. 15, "*Constitutio qua Concilii Tridentini poenis subiiciuntur duellum* [. . .] (Rome: Haeredes Antonij Bladij, 1583); and for his bulls, Tomassetti et al., *Bullarium Romanum et privilegiorum*, vol. VIII, no. CXL, 399–400 (November 5, 1582) and no. CXLI, 400–401. For the duel in Italy, see Marco Cavina, *Il sangue dell'onore: Storia del duello* (Rome: Laterza, 2005), and Uwe Israel and Gherardo Ortalli, eds., *Il duello fra Medioevo ed età moderna: Prospettive storico-culturali* (Rome: Viella, 2009).

87. Tomassetti et al., *Bullarum diplomatum et privilegiorum*, vol. VII, no. LVIII, 171–172: "experientia cognovimus quamplura ex illis tormentulis igneis, archibusettis vulgariter nuncupatis (quae occulte deferri possunt et ad necem inferendam aptiora et fere inevitabilia esse videntur)"; "ob sicariorum copiam, audaciam."

88. Tomassetti et al., *Bullarum diplomatum et privilegiorum*, vol. VII, no. CCXVI, 965–967.

89. Blastenbrei, "Violence, Arms," esp. 74–83.

90. For Sixtus's public executions (and those of the popes before him), see Blastenbrei, *Kriminalität in Rom*, 301. For the effectiveness of Sixtus's measures, see Fosi, *La società violenta*, esp. 133–163, and "La giustizia e la sua imagine." See also Paglia, *La morte confortata*, a classic study of what might be called the moral economy of public executions in Rome.

CHAPTER 8

1. This account is derived primarily from Fontana, *Della trasportatione dell'Obelisco*, fol. 33r, "guadagnorno assai denari." See also Fontana, *Della trasportatione dell'Obelisco*, trans. Sullivan, 38–39.

2. Fontana, *Della trasportatione dell'Obelisco*, fol. 33r, "tutta la Città si rallegrò assai." Detailed accounts of the complex process of moving the Vatican Obelisk include Bern Dibner, *Moving the Obelisks* (1950; repr., Norwalk, CT: Burndy Library, 1991), 20–43; Brian A. Curran, Anthony Grafton, Pamela O. Long, and Benjamin Weiss, *Obelisk: A History* (Cambridge, MA: Burndy Library, 2009), esp. 102–139; Cesare D'Onofrio, *Gli obelischi di Roma: Storia e urbanistica di una città dall'età antica al XX secolo*, 3rd ed. (Rome: Romana Società Editrice, 1992), 143–184; Maria Luisa Riccardi et al., "Gli obelischi sistini," in Piera Sette and Benedetti, *Architetture per la città*, 13–38; William Barclay Parsons, *Engineers and Engineering in the Renaissance* (Cambridge, MA: MIT Press, 1968), 155–167. For Fontana's house in

the Borgo (now the Hotel Bramante in Vicolo delle Palline), see Giuseppe Bonaccorso and Michela Lucci, "Le case dei Fontana a Roma," in Faggiolo and Bonaccorso, *Studi sui Fontana*, 463–465; and Margherita Fratarcangeli, "Il cavaliere Domenico Fontana: La 'robba' per la nobilità," in Curcio, Navone, and Villari, *Studi su Domenico Fontana*, 56–57.

3. For Domenico Fontana and his large family workshop in Rome, see esp. Curcio, Navone, and Villari, *Studi su Domenico Fontana*; De Cavi, *Architecture and Royal Presence*; Fagiolo and Bonaccorso, *Studi sui Fontana*; Ippoliti, "Fontana, Domenico"; and Quast, "Domenico Fontana."

4. For the topography and history of the area in antiquity, see Paolo Carafa and Paola Pacchiarotti, "Regione XIV: *Transtiberim*," and Francesco De Stefano, "Appendice: La Basilica e la memoria di San Pietro," both in Carandini and Carafa, *Atlante di Roma antica*, 1:549–582 and 583–587, respectively; Ferdinando Castagnoli, *Il Vaticano nell'antichità classica* (Vatican City: Biblioteca Apostolica Vaticana, 1992); and Paolo Liverani, *La topografia antica del Vaticano* (Vatican City: Monumenti, Musei e Gallerie Pontificie, 1999), 21–28, for the circus. For the Vatican Obelisk in antiquity, see Géza Alföldy, *Der Obelisk auf dem Petersplatz in Rom: Ein historisches Monument der Antike* (Heidelberg: Carl Winter, 1990).

5. For the area in the medieval period, see Lorenzo Bianchi, *Ad Limina Petri: Spazio e memoria della Roma cristiana* (Rome: Donzelli editore, 1999). And see Curran et al., *Obelisk: A History*, esp. 35–83; D'Onofrio, *Gli obelischi di Roma*, 97–185; Erik Iversen, *Obelisks in Exile*, vol. 1, *The Obelisks of Rome* (Copenhagen: GEC Gad, 1968), 19–46, for the Vatican Obelisk. For interest in moving the Vatican Obelisk in the fifteenth century, see Brian A. Curran and Anthony Grafton, "A Fifteenth-Century Site Report on the Vatican Obelisk," *Journal of the Warburg and Courtauld Institutes* 58 (1995): 234–248.

6. Michele Mercati, *De gli obelischi di Roma*. [...] (Rome: Domenico Basa, 1589), 343–344, "E se si rompesse?" For a reprint edition with useful introduction, see Michele Mercati, *Gli obelischi di Roma*, ed. Gianfranco Cantelli (Bologna: Cappelli Editore, 1981); and see Elisa Andretta, "Mercati, Michele," *DBI*, 73:606–611. For his work on natural history, see esp. Bruno Accordi, "Michele Mercati (1541–1593) e la Metallotheca," *Geologica Romana* 19 (1980): 1–50; and Alix Cooper, "The Museum and the Book: The Metallotheca and the History of an Encyclopaedic Natural History in Early Modern Italy," *Journal of the History of Collections* 7 (1995): 1–23. For Michelangelo as an engineer, see Wallace, "Michelangelo Engineer," 96–107.

7. Dibner, *Moving the Obelisks*, 20–43; and Iversen, *Obelisks in Exile*, 1:19–46.

8. D'Onofrio, *Gli obelischi di Roma*, esp. 97–141. For the citation, see Pastor, *History of the Popes*, 20:598 and 598n2, "per maggior commodità della vista delle persone che veranno l'anno santo a Roma," citing an *avviso* of July 27, 1574, from the State Archive of Vienna.

9. Camillo was the older brother of the better-known Gabriele Paleotti (1522–1597), the latter known for his tract on sacred and profane images. See Paolo Prodi, *Il cardinal Gabriele Paleotti* (Rome: Edizioni di Storia e Letteratura, 1959–1967), 1:41–50 for Camillo. For the letter, see Lothar Sickel, "Un progetto per il trasporto dell'Obelisco Vaticano descritto da Camillo Paleotti nel febbraio 1581," *Strenna dei Romanisti* 68 (April 2007): 653–662, 659 for a transcription of the relevant part of the letter (which I cite here), the original of which is in Parma, Archivio di Stato, Carteggio farnesiano estero: Roma, b. 489: "un giovane siciliano"; "il quale mostra d'haver grandissimo giuditio con tutto che sia senza lettere, di cose di machine"; "non solo il grand'ingegno d'esso artifice ma ancora la sua buona opinione à chi lo mira che la cosa sia riuscibile"; "troppo longa sarebbe la scrittura." The recipient of the letter,

Antonio Anselmi (b. 1512) was the ex-secretary of Cardinal Pietro Bembo and had an active interest in the technical problems of architecture. See also Christof Thoenes, "Perché studiare Domenico Fontana," in Curcio, Navone, and Villari, *Studi su Domenico Fontana*, esp. 14–15. I warmly thank Lothar Sickel for kindly sending me a copy of his article. As Sickel points out, the Sicilian mentioned in the letter could not be the one mentioned in the Congregazione that submitted a proposal, since he (Giacomo Del Duca) had been in Rome for a long time, while the Sicilian mentioned in the letter was described as having just arrived—see Genovese and Sinisi, *Pro ornatu et publica utilitate*, 126. For the calculation of the weight of the obelisk, I have used Dibner, *Moving the Obelisks*, 59, and Parsons, *Engineers and Engineering*, 637, table 11, where Parsons gives one Roman *libbra* as .748 lbs.

10. Evelyn Lincoln, *Brilliant Discourse: Pictures and Readers in Early Modern Rome* (New Haven, CT: Yale University Press, 2014), 60–113. See also George Sarton, "Agrippa, Fontana and Pigafetta: The Erection of the Vatican Obelisk in 1586," *Archives Internationales d'Histoire des Sciences*, no. 8 (July 1949): 832–836; and Gian Luigi Barni, "Agrippa, Camillo," *DBI*, 1:503.

11. See D'Onofrio, *Le fontane di Roma*, 230; and Leonardo Lombardi, "Camillo Agrippa's Hydraulic Inventions on the Pincian Hill (1574–1578)," *Waters of Rome*, no. 5 (April 2008), www3.iath.virginia.edu/waters/first.html.

12. Camillo Agrippa, *Trattato* [. . .] *Di trasportar la guglia* [. . .] (Rome: Francesco Zanetti, 1583), 5, "et infiniti altri," and sig. Ai, "impiegato a inventioni non meno utili al ben publico, che honorevole." For Michelangelo and Antonio da Sangallo's reactions to moving the obelisk, see also Iversen, *Obelisks in Exile*, 1:28.

13. Agrippa, *Trattato* [. . .] *Di trasportar la guglia* [. . .], 3–5; and Michele Mercati, *De gli obelischi di Roma* (1589), 344–346. For a discussion of the meeting in the "Metallotheca Vaticana," see also D'Onofrio, *Gli obelischi*, 140–141.

14. Agrippa, *Trattato* [. . .] *Di trasportar la guglia* [. . .], 5–27; and Parsons, *Engineers and Engineering*, 155–156.

15. Agrippa, *Trattato* [. . .] *Di trasportar la guglia* [. . .], 29–47, and for Trajan's column, 37–46. See also Elio Nenci, "Camillo Agrippa: Un ingegnere rinascimentale di fronte ai problemi della filosofia naturale," *Physis*, n.s., 29 (1992): 102–109. For the Fossani brothers, see Lincoln, *Brilliant Discourse*, 74 and 251, note 30.

16. Fontana, *Della trasportatione dell'Obelisco*, fol. 4v–5r.

17. A succinct biography, including Domenico Fontana's arrival in Rome, is Paola Carla Verde, "Domenico Fontana," in Fagiolo and Bonaccorso, *Studi sui Fontana*, 421–427. For Villa Montalto, see Fontana, *Della trasportatione dell'Obelisco*, fols. 37r–38r, in which Fontana stresses the beauty of the gardens and fountains as well as the villa. See also Quast, *Die Villa Montalto*, and for Sixtus's plans for the Quirinal as a whole, Marder, "Sixtus V and the Quirinal." The vast scope of Fontana's work for Sixtus V can be appreciated by a perusal of his extant account books—Enrico Guidoni, Angela Marino, and Angela Lanconelli, "I 'Libri dei conti' di Domenico Fontana: Riepilogo generale delle spese e Libro I," *Storia della Città* 40 (1987): 45–77, and "I 'Libri dei conti' di Domenico Fontana: I monumenti antichi: Il Settizonio, le colonne coclidi, i 'massicci' di Termini, i cavalli del Quirinale: i Libri XIII, XIV, XIX, XX," *Storia della Città* 43 (1988): 86–104.

18. Domenico Fontana, *Della trasportatione dell'Obelisco*, fol. 8r, for the illustration of models. For Bonifacio, see Fabia Borroni Salvadori, "Bonifacio, Natale (Bonifatio, Bonifazio) detto Bonifacio da Sebenico o Natale Dalmatino," *DBI*, 12:201–204; Witcombe, *Copy-*

right in the Renaissance, esp. 143–147 and 278–283; and Witcombe, *Print Publishing*, esp. 286, 344.

19. Fontana, *Della trasportatione dell'Obelisco*, fol 4v: "per intelligenza, e per esperimento in simili affari." For the *avviso*, see BAV, Urb. lat. 1053, fol. 410r.

20. Ibid., fol. 5r, "li litterati, Mattematici, Architetti, Ingegnieri, e altri valent'uomini [. . .] acciò ch'ogniuno dicesse il parer suo intorno all'essecutione di tanta impresa"; "per dar tempo à molti valent'huomini forestieri che di varij luoghi concorrevano à Roma per mostrar le forze dell'ingegno loro intorno à cosa tanto desiderata da nostro Signore"; "la sua inventione."

21. Ibid., fol. 8r.

22. ASR, *Congregazione super viis pontibus et fontibus*, Registro 1, fols. 157r–158r: "parva impensa"; and Genovese and Sinisi, *Pro ornatu et publica utilitate*, 126–128 (Congregation no. 119). See also D'Onofrio, *Gli obelischi di Roma*, 148–149n5, 153n8, which two notes combined consist of a complete transcription of the Latin minutes of the congregation (the writing on the original document as it now exists is not completely legible).

23. Fontana, *Della trasportatione dell'Obelisco*, fol. 5r: "levai quella Guglia, e l'abbassai ordinatamente mostrando con parole a cosa per cosa la ragione, e il fondamento di ciascuno di quei movimenti, si come seguì poi apunto in effetto." See D'Onofrio, *Gli obeliscbi di Roma*, 475, for a transcription of part of Fontana's account book, which includes a payment to Colantonio Liante for 25 scudi for a model made on September 5, 1585.

24. Fontana, *Della trasportatione dell'Obelisco*, fol. 5r–v: "Hora considerati esquisitamente, e ponderati li discorsi, disegni, e edifitij di ciascheduno di noi, e disputato assai, all'ultimo si venne in questa conclusione, che il modo del muovere, e trasportar la Guglia da me ritrovato fusse il più facile sicuro, e meglio inteso per sortire prospero fine di tutti gli altri, ch'ivi furono proferiti: e di commune consentimento di tutta la congregatione fu eletto, e approvato per servirsene nel trasportar la Guglia lasciando da banda tutti gli altri"; "e assegnata a due valenti Architetti."

25. ASR, *Congregazione super viis pontibus et fontibus*, Registro 1, fols. 157r–158r; and Genovese and Sinisi, *Pro ornatu et publica utilitate*, 126–127 (Congregation no. 119).

26. Nicola Navone, "Alle origini dell'impresa Fontana," in Curcio, Navone, and Villari, *Studi su Domenico Fontana*, 61–73.

27. Fontana, *Della trasportatione dell'Obelisco*, fol. 5v: "Nostro Signore ordinò. ch'io solo dovesi dar principio all'opera, e eseguire la mia intentione." For other conflicts between Fontana and Ammannati, see Maurizia Cicconi, "'Bisognava un huomo valente': Ammannati versus Fontana nel carteggio di Guglielmo Sangalletti," in Curcio, Navone, and Villari, *Studi su Domenico Fontana*, 229–239.

28. Fontana, *Della trasportatione dell'Obelisco*, fols. 5v–6r.

29. Ibid., fol. 6r–v: "l'aiutino, obedischino, favorischino, e assistino"; "contrarijs non obstantibus quibuscumque."

30. Ibid., fols. 6v–7r; and Fontana, *Della trasportatione dell'Obelisco*, trans. Sullivan and ed. Rowland, "Translator's Note," where Sullivan provides the units of measurement used by Fontana as a *palmo* = 8.66 inches and a *canna* = 5.5 feet. See also Marconi, *Edificando Roma barocca*, 231–241, esp. 235, for the collection of materials; "*Genitor urbis ad usum fabricae*: Il trasporto fluviale dei materiali per l'edilizia nella Roma del Cinquecento," in Bonaccorso, "Le acque e la città," 143–166, which treats the transport of material for the Saint Peter's workshop (the transport networks would have been similar or the same for materials needed for the transport of the Vatican Obelisk); and *Castelli e ponti: Apparati per il restauro nell'opera*

di maestro Nicola Zabaglia per la fabbrica di San Pietro in Vaticano (Foligno: Il Formichiere, 2015).

31. Filippo Pigafetta, *Discorso* [. . .] *d'intorno all'historia della aguglia* [. . .] (Rome: Bartolomeo Grassi, 1586). For Pigafetta's life, see Daria Perocco, "Pigafetta, Filippo," *DBI*, 83: 578–582. And see Antonio Becchi, Domenico Bertoloni Meli, and Enrico Gamba, eds., *Guidobaldo del Monte (1545–1607): Theory and Practice of the Mathematical Disciplines from Urbino to Europe*, Max Planck Research Library for the History and Development of Knowledge, Proceedings 4 (Berlin: Edition Open Access, 2013); and Mary Henninger-Voss, "Working Machines and Noble Mechanics: Guidobaldo del Monte and the Translation of Knowledge," *Isis* 91 (June 2000): 233–259.

32. Perocco, "Pigafetta, Filippo"; and Pigafetta, *Discorso*, sig. A3v: "queste pietre smisurate"; "dalla cava naturalmente tratte, si come affermo io d'havere co'miei proprij occhi veduto."

33. Pigafetta, *Discorso*, sig. C3v–D3r, "co'palmi Romani, e co'piedi communi di geometria" (C3v).

34. Pigafetta, *Discorso*, sig. D4r–v. And for his theory of pulleys, see Guidobaldo del Monte, *Mechanicorum liber* (Pisa: H. Concordia, 1577), fols. 62r–105v (De trochlea), and *Le mechaniche* [. . .] *tradotte in volgare dal Sig. Filippo Pigafetta* (Venice: F. di Franceschi, 1581), fols. 56r–101v (Della Taglia). And see Domenico Bertoloni Meli, *Thinking with Objects: The Transformation of Mechanics in the Seventeenth Century* (Baltimore: Johns Hopkins University Press, 2006), 32–35.

35. Pigafetta, *Discorso*, sig. D6r.

36. Fontana, *Della trasportatione dell'Obelisco*, fols. 9r–10v.

37. Ibid., fols. 10v–11r; and see Antonio Becchi, "Cantieri d'inchiostro. Meccanica teorica e meccanica chirurgica nella seconda metà del Cinquecento," in Curcio, Navone, and Villari, *Studi su Domenico Fontana*, 91–103. For an introduction to machine books, see Marcus Popplow, "Why Draw Pictures of Machines? The Social Contexts of Early Modern Machine Drawings," in *Picturing Machines, 1400–1700*, ed. Wolfgang Lefèvre (Cambridge, MA: MIT Press, 2004), 17–52.

38. Fontana, *Della trasportatione dell'Obelisco*, fols. 10v–13r, 10r for the houses.

39. Ibid., fols. 13r–14r: "e moltissimi foristieri concorsi da tutte le parti dell'Italia per vedere spettacolo così nuovo, e maraviglioso"; "poi che quest'opera era drizzata à Gloria di Dio."

40. Ibid., fol. 14r–v: "Che ne mostrò grandissima allegrezza." For astragals, see Diane Favro, "Obelisk Bones," *Seleucia* 8 (2018): 13–28.

41. Ibid., fol. 16r.

42. Ibid.

43. Ibid., fols. 16v–17r: "tutto il popolo ne sentì infinito giubilo, e in segno l'Architetto fu accompagnato da tamburi e trombe a casa sua."

44. Ibid., 23r–v.

45. Ibid., fols. 25r, 33r–v: "fu bellissimo spettaculo per molti rispetti."

46. Ibid., fols. 33v–34v.

47. Vincenzo Scamozzi, *L'idea della architettura universale* (Venice: Presso L'Autore, 1616), 335–336: "in varij tempi si affaticavano molti elevati ingegni della Città, e altrove per ritrovare il modo più sicuro, e artificioso per doverlo condurre, de' quali nella nostra gioventù, mentre eravamo à Roma vedemmo parte d'essi"; "chi poteva far le gratie"; "trattò per dir cosi la sostanza delle inventioni raccontate."

48. See Marconi, *Edificando Roma barocca*, 231–241, esp. 234.

49. Veronica Biermann, "Ortswechsel: Überlegungen zur Bedeutung der Bewegung schwerer Lasten für die Wirkung und Rezeption monumentaler Architektur am Beispiel des Vatikanischen Obelisken," in *Perspektiven der Spolienforschung*, vol. 1, *Spoliierung und Transposition*, ed. Stefan Altekamp, Carmen Marcks-Jacobs, and Peter Seiler (Berlin: Walter de Gruyter, 2013), esp. 135–139, for Guerra. On the fame of the Vatican Obelisk transport (and of Fontana) into the next century and beyond, see Costanza Caraffa, "Domenico Fontana e gli obelischi: Fortuna critica del "Cavaliere della Guglia," in Curcio, Navone, and Villari, *Studi su Domenico Fontana*, 21–47.

50. Fontana, *Della trasportatione dell'Obelisco*, fols. 70v–74r. The discovery of the ancient *cloaca* was reported by Flaminio Vacca (1538–1603) in an account of 1594, published in the eighteenth century—Flamino Vacca, *Memorie di varie antichità* [. . .] in *Roma antica distinta per rioni* [. . .] (Rome: Gio. Lorenzo Barbiellini Libraro, 1741), 1:214–266, on 216. For the Lateran Obelisk, see also Iversen, *Obelisks in Exile*, 55–64; and Riccardi, "Gli obelischi sis tini," 53–70. For the Circus Maximus, see Chiara Bariviera, "Regione XI: *Circus Maximus*," in Carandini and Carafa *Atlante di Roma antica*, 1: 421–445; and Marialetizia Buonfiglio, "La *cloaca circi* e le acque della valle del Circo Massimo," in Bianchi, *La Cloaca Maxima*, 166–183.

51. Fontana, *Della trasportatione dell'Obelisco*, fols. 75r–77v (Flamino Obelisk at Piazza del Popolo) and 76r–77r (obelisk at Santa Maria Maggiore); and see Iversen, *Obelisks in Exile*, 1:65–75 (Piazza del Popolo) and 47–54 (Santa Maria Maggiore); and Riccardi et al., "Gli obelischi sistini," 13–18, 71–89. For the project of erecting the obelisk at Santa Maria Maggiore, see also the illuminating account book, ASV, Archivium Arcis, Arm. B. 13, which itemizes the costs of the project, thereby providing many details.

52. Fontana, *Della trasportatione dell'Obelisco*, fol. 71r.

53. Michael W. Cole, "Perpetual Exorcism in Sistine Rome," in *The Idol in the Age of Art: Objects, Devotions and the Early Modern World*, ed. Michael W. Cole and Rebecca Zorach (Farnham, Surrey: Ashgate, 2009), 57–76, 65 for indulgences.

54. Curran, *Egyptian Renaissance*, esp. 51–63, 227–243; and see also Giovanni Cipriani, *Gli obelischi egizi: Politica e cultura nella Roma barocca* (Florence: Leo S. Olschki, 1993); and Curran et al., *Obelisk: A History*, esp. 140–159.

55. For the Hermetic corpus, see Brian P. Copenhaver, ed. and trans., *Hermetica: The Greek Corpus Hermeticum and the Latin Asclepius in a New English Translation* (Cambridge: Cambridge University Press, 1992). The classic study of hieroglyphs and their relationship to Renaissance humanism, now in English translation with a useful commentary, is Karl Giehlow, *The Humanist Interpretation of Hieroglyphs in the Allegorical Studies of the Renaissance*, ed. and trans. Robin Raybould (Leiden: Brill, 2015). See also Cipriani, *Gli obelischi egizi*; Brian A. Curran, "'De sacrarum litterarum aegyptiorum interpretatione': Reticence and Hubris in Hieroglyphic Studies of the Renaissance; Piero Valeriano and Annius of Viterbo," *Memoirs of the American Academy in Rome* 43–44 (1998/1999): 139–182; Curran, *Egyptian Renaissance*, esp. 51–65, 89–105, 227–243; Curran et al., *Obelisk: A History*, esp. 141–177; and Corinne Mandel, "Simbolismo ermetico negli obelischi e colonne della Roma sistina," in Fagiolo and Madonna, *Sisto V*. 1:659–692.

56. Mercati, *De gli obelischi* (1589), 25–44, 82–130.

57. Fontana, *Della trasportatione dell'Obelisco*, fols. 75r–77v, on the Flamina Obelisk at Piazza del Popolo with illustration; and see Curran et al., *Obelisk: A History*, esp. 156–159.

58. For Trajan's column, see F. Cerasoli, "La Colonna Traiana e le sue adiacenze nei secoli

XV, XVI e XVII," *Bullettino della Commissione Archeologica Comunale di Roma* 29 (January–March 1901): 300–308; Filippo Coarelli, *La Colonna Traiana* (Rome: Colombo, 1999); Lynne C. Lancaster, "Building Trajan's Column," *American Journal of Archaeology* 103 (July 1999): 419–439; Frank Lepper and Sheppard Frere, *Trajan's Column: A New Edition of the Cichorius Plates* (Gloucester, UK: Alan Sutton, 1988); Pietro Petraroia and Sergio Lombardi. "Colonna Traiana," in Madonna, *Roma di Sisto V*, 406–407; Salvatore Settis, "La colonne Trajane: Invention, composition, disposition," *Annales Histoire, Sciences Sociales* 40 (September–October 1985): 1151–1194; and Mark Wilson Jones, *Principles of Roman Architecture* (New Haven, CT: Yale University Press, 2000), 160–175.

59. For the Column of Marcus Aurelius, see Martin Beckmann, *The Column of Marcus Aurelius: The Genesis and Meaning of a Roman Imperial Monument* (Chapel Hill: University of North Carolina Press, 2011), esp. 35–36, 45–47, for the confusion of the Column of Antoninus with that of Marcus Aurelius; C. Caprino, A. M. Colini, G. Gatti, M. Pallottino, and P. Romanelli, *La Colonna di Marco Aurelio* (Rome: L'Erma di Bretschneider, 1955); Filippo Coarelli, *La Colonna di Marco Aurelio / The Column of Marcus Aurelius*, trans. Helen Patterson (Rome: Editore Colombo, 2008); and Giangiacomo Martines, ed., "La Colonna Antonina "in *Via del Corso: Una strada lunga 2000 anni*, ed. Cesare D'Onofrio (Rome: Edizioni de Luca, 1999), 111–149. See also Guidoni, Marino, and Lanconelli, "I 'Libri dei Conti,'" 98–104.

60. Karmon, *Ruin of the Eternal City*, 39–41, discusses earlier (and not very successful) attempts to maintain them.

61. Fontana, *Della trasportatione dell'Obelisco*, fol. 99r–v: "ben maggior difficultà"; "era ridotta a tal termine, che pareva impossibile, non che difficile a ristorarla"; "a tale che spaventava chi la rimirava"; "con grandissima diligenza"; "con grande arte, e spesa." See also Simoncini, *Roma: Le trasformazioni urbane nel Cinquecento*, 1:364–365.

62. See Lanciani, *Storia degli scavi*, 4:176–178, and Jennifer Montagu, *Roman Baroque Sculpture: The Industry of Art* (New Haven, CT: Yale University Press, 1989), 56.

63. Fontana, *Della trasportatione dell'Obelisco*, fols. 33v–34v; and see Curran et al., *Obelisk: A History*, esp. 141–143; and Mandel, "Simbolismo ermetico."

64. ASC, C.C., cred. I, tom. 29, cat. 29, fol. 192r–v (XI Kal Maij 1589).

65. ASC, C.C., cred. I, tom. 29, cat. 29, fol. 193r–v (VI Nonas Maij 1589; and fols. 194r–195v (IV Nonas Maij 1589): "damnum pauperibus vidimus Pupillis, orphanis et alijs pauperibus exinde resultans"; "in constructione aliarum stratarum, erectione Piramidum, restauratione Columnae Antonine, aedificatione tot et tantorum aedificiorum ad divinum cultum ac privatum et publicum usum destinatum et in conductione aquae Felicis ad publicam urbis et particularium personarum utilitatum"; "multas deprecationes."

66. ASC, C.C., cred. I, tom. 29, cat. 29, fols. 197v–198r.

67. ASV, Misc. Arm. IV-V, tom. 58, fol. 3: "Bando contro quelli che menano Bufale sciolte per Roma"; "Vedendosi giornalmente per esperientia, quanto pericolo apporti à diverse persone con morte ben spesso"; "le Bufale sciolte e senza ligame."

68. ASV, Misc. Arm. IV-V, tom. 58, fol. 153r: "Visum est Sanctiss. D. N. reformare Statuta, et Bannimenta Magistrorum Iustitiariorum contra Caprarios, et alios damnum dantes cum Capris, et alijs animalibus in Vineis, Cannetis, et aliis locis, ut infra, videlicet."

69. For a thorough study, including relevant documentation, see Domenico Chiari, *Il territorio pontino in epoca sistina: Immagini di riforma e vita nello Stato della Chiesa, 1585–1590* (Terracina: Comune di Terracina, 1990); and see ASR, Camerale II, Paludi Pontine, b. 1.

70. Chiari, *Il territorio pontino*, 88; and for the *avviso*, BAV, Urb. lat. 1058 (April 4, 1590),

fol. 158v: "È vero che, le paludi Pontine disecate l'anno passato [. . .] sono ritornate per le continue pioggie nello stato primiero, et si fà conto, che sono più di ducento, et tanti giorni, che non fà altro che piovere"; and (April 7, 1590), fol. 166r: "per le continue piogge di durata di 4 mesi."

71. Chiari, *Il territorio pontino,* 90. Citation of Cicatelli in Henderson, "'Mal Francese' in Sixteenth-Century Rome," 506–507 (translation by Henderson).

72. BAV, Vat. lat. 7484, fols. 74r–140v: "Memorie di Tre anni del pontificato di Sisto V, cioè 1585, 1586, e 1587"; fol. 80r: "Di nessuna cosa fu quest'anno più fecondo, che di gabelle, impositioni, nuovi offitii, ed inventioni di cavar denari"; "odiosissime"; "più d'ogn'altro per non esser avvezzo a portar quelle gravezze"; "E quindi nacque l'odio, che acerbissimo da questo tempo in poi i suoi popoli le portarono." And see Butzek, *Die kommunalen Repräsentationsstatuen der Päpste,* 306–307, for the new taxes.

73. BAV, Urb. lat. 1058, fol. 442v: "dimandando la statua di Sisto per strascinarla"; "questi arcigogolanti." See Butzek, *Die kommunalen Repräsentationsstatuen der Päpste,* 308–309, and 501–502, for a transcription of the *avvisi* passage with useful notes; and Hunt, *Vacant See,* 189–190, which includes further references. For Fontana's refuge in the Palazzo Sforza, see Fontana, *Della trasportatione dell'Obelisco Vaticano, 1590,* ed. Carugo with an introduction by Portoghesi, xii.

74. See Simona Benedetti, "Il Ponte Felice," in Piera Sette and Benedetti, *Architetture per la città,* 223–239; C. Paola Scavizzi, "Il Ponte Felice al Borghetto nel quadro della viabilità territoriale," in Fagiolo and Madonna, *Sisto V,* 1:623–638, esp. 627–628; Verde, "Domenico Fontana," 425; and Fontana's own description in Domenico Fontana, *Della trasportatione dell'Obelisco Vaticano* [. . .] (Libro II) (Naples: Constantino Vitale, 1604), fols. 20v–22r.

75. The document is in ASV, Borghese, IV, 280, fols. 146r–160v, and is published in full by J. A. F. Orbaan, "Die Selbstverteidigung des Domenico Fontana, 1592–1593," *Repertorium für Kunstwissenschaft* 46 (1925): 174–189. For further details concerning fines and payments, see J. A. F. Orbaan, "Il caso Fontana," *Bollettino d'Arte,* 1st ser., 9 (1915): 165–168. For Fontana's career in Naples, see De Cavi, *Architecture and Royal Presence;* "La committenza spagnola di Domenico e Giulio Cesare Fontana (1592–1627)"; Paolo Mascilli Migliorini, "Il Palazzo Reale di Napoli e la città vicereale"; Maria Raffaela Pessolano, "Idraulica e modernità: Il porto di Napoli nell'immagine della città vicereale"; and Grete Stefani and Giovanni Di Maio, "Idraulica e modernità: Il canale Conte di Sarno," all in Curcio, Navone, and Villari, *Studi su Domenico Fontana,* 161–183, 185–195, 197–211, and 213–227, respectively. For his 1604 treatise, Fontana, *Della trasportatione dell'Obelisco Vaticano* [. . .] (Libro II); and see Fulvio Lenzo, "'Che cosa è architetto': La polemica con gli ingegneri napoletani e l'edizione del *Libro secondo,*" in Curcio, Navone, and Villari, *Studi su Domenico Fontana,* 265–287.

Bibliography

ARCHIVAL DOCUMENTS AND MAPS

Italy

Florence

Galleria degli Uffizi, Gabinetto dei Disegni e delle Stampe

Uffizi, Architettura No. 4236A [map of the *Acqua Vergine* by Pirro Ligorio]
Uffizi, 288A and 289A [drawings of canalization of the Tiber River attributed to Bartolomeo de Rocchi]

Naples

Biblioteca Nazionale di Napoli (BNN)

Mss. XIII.B.1–10 [Pirro Ligorio, *I libri delle antichità,* 10 vols.]
Ms. XIII.B.9 [Pirro Ligorio on ancient rivers, fountains, and lakes]
Mss. XII.G.59/60 [On fountains and hydrology by Giovanni Antonio Nigrone]

Rome

Archivio di Stato di Roma (ASR)

Archivio

Archivio delle soldatesche e galere

Busta 15 [account books for Roman fortifications under Paul III, 1537–1549]

Busta 16 [account books for fortifications from 1546–1549; and under Paul IV, 1556–1558]
Busta 17 [account books for the fortification of the Borgo, 1561–1571]

Camerale I
Mandati Camerale
Buste 907, 912, 914, 916, 918, 921 [payments to Antonio Trevisi for work on the Acqua Vergine]
Mandati Camerale
Busta 936 (January 1587–December 1589) [contracts for street paving] Notai RCA [Reverenda Camera Apostolica]

Camerale II
Paludi Pontine
Busta 1 [documents on draining of Pontine marshes under Sixtus V]
Tevere
Busta 12, fasc. 106 [copy of concession of mill to Matteo da Castello]

Congregazione super viis, pontibus et fontibus
Congregazione super viis pontibus et fontibus, Registro 1, 1567–1588

Fondo Santa Croce
Filza 188 [letter of Girolomo Picchi concerning the prostitutes of Rome]

Notai RCA [Reverenda Camera Apostolica]
Busta 453. Notary Hieron de [Girolamo di] Tarano [includes contract for Antonio Trevisi for reconstruction of the Acqua Vergine and contracts to others for same]

Notai del Tribunale delle Acque e Strade. Notary Octavius Gracchus [Ottavio Gracco]
Busta 6, prot. II [contracts for Acqua Vergine conduits in the Campo Marzio]

Presidenza degli acquedotti urbani
Busta 2 [records of Acqua Vergine water gifted or sold]
Busta 5, fols. 1–37 [records of Acqua Felice water sold during papacy of Sixtus V]
Buste 82, 83, 84 [account books for construction of the Acqua Felice, 1585–1588]

Presidenza delle Strade
Busta 2 [book of fines for violations of clean-street regulations, 1467]
Busta 42 [contracts for street paving]
Busta 445 [Tax collection record for taxes collected by Antonio Trevisi as submaster of the streets for repair of Ponte Quattro Capi (Ponte Fabricio) and taxes collected for other infrastructure projects]

Biblioteca

Bandi
Coll. 1, b. 2, no. 5 [Julius III edict about tax collection for street cleaning]
Coll. I, b. 2, no. 194, *Motus proprius* [. . .] *per quem declaratur quod in breve nuper contra homicidas edito* [. . .]. Rome: Antonio Blado, 1564. [Against homicides]
Coll. I, b. 3, no. 33, *Motus proprius* [. . .] *quae gratiae, seu remissiones, aut compositiones ex causa, vel occasione homicidii* [. . .]. Rome: Antonio Blado, 1564. [Against homicides]

Coll. I, b. 3, no. 61, *Bullae* [. . .] *contra impios, ac facinorosos homicidas*. Bologna: Johannes Rubrius, 1565. [Against homicides]

Coll. I, b. 5, no. 39, *Sommario di quello* [. . .] *publicate contra gl'homicidi, banditi* [. . .]. Rome: Haeredes Antonij Bladij, 1578

Coll. I, b. 6, no. 10, *Erectio magistratus officii quatuordecim magistrorum viarum urbis* (Rome: Haeredes Antonio Bladij, 1588)." [March 7, 1587, edict of Sixtus V creating fourteen masters of the streets]

Coll. I, b. 6, no. 164, *Bando contra banditi,* [. . .]. Rome: Heredi d'Antonio Blado, 1587. [Against bandits]

Coll. II, b. 294, no. 15, *Constitutio qua Concilii Tridentini poenis subiicuntur duellum* [. . .]. Rome: Haeredes Antonij Bladij, 1583

Coll. II, b. 294, no. 140, *Constitutione* [. . .] *sopra le pene e confiscationi criminali* [. . .]. Rome: Paolo Blado, 1589

Archivio Storico Capitolino (ASC)

Carte Boccapaduli, Arm. II, maz. IV, b. 104b, fasc. 63, L. 1029v–1030r

Carte Boccapaduli, Arm. II, maz. IV, b. 104b, fasc. 63, L. 1042, 1043 [regulations for Acqua Virgine distribution]

Carte Boccapaduli, Arm. II, maz. IV. b. 104b, fasc. 66, L. 1096 [*oncie* sizes for the Acqua Vergine]

Fondo Camera Capitolina (C.C.)

Decreti di consigli, magistrati e cittadini romani

C.C. cred. I, tom. 1, cat. 1

C.C. cred. I, tom. 17, cat. 17

C.C., cred. I, tom. 20, cat. 20

C.C. cred. I, tom. 21, cat. 21

C.C., cred. I, tom. 22, cat. 22

C.C., cred. I, tom. 23, cat. 23

C.C., cred. I, tom. 24, cat. 24

C.C., cred. I, tom. 25, cat. 25

C.C., cred. I, tom. 26, cat. 26

C.C., cred. I, tom. 27, cat. 27

C.C., cred. I, tom. 29, cat. 29

C.C., cred. I, tom. 36, cat. 36

C.C., cred. I, tom. 37, cat. 37

C.C., cred. I, tom. 38, cat. 38

C.C., cred. IV, tom. 82, cat. 326 [notarial book of licenses and contracts given by the masters of the streets, including contracts for street paving, 1586–1589]

C.C., cred. IV, tom. 95, cat. 339 [purchase of lime and lumber transport for Ponte Santa Maria; contracts relevant to the Capitoline Fountain]

C.C., cred. VI, tom. 50, cat. 451, stragr. 23, *Congregazione super Aqua Salonis*

C.C., cred. VI, tom. 53, cat. 454, stragr. 25 [grants of Acqua Vergine water under Gregory XIII]

C.C., cred. VI, tom. 59, cat. 460 [report of Giacomo Della Porta on Guglielmo Della Porta's pipe disaster; Letter of Matteo Bartolino da Castello on necessity of further shoring up a pier of Ponte Santa Maria]

Biblioteca Angelica (BA)

Anonimo Ms. 153, fols. 30r–31v, "Relatione: dell'inundatione del Tevere in Roma [...] nel 1557." [Notes on Tiber River flooding]

GG. 11.22 [Printed copy of Ludovico Gómez, *De prodigiosis Tyberis inundationibus* [...] *ad annum M. D. XXXI* [...] (Rome: Francesco Minizio Calvo, 1531), with marginal annotations in same hand as Anonimo Ms. 153]

Ms. 247 [Bartolomeo Marliani's studies and translations of the plays of Aristophanes]

Biblioteca dell'Istituto di Archeologia e Storia dell'Arte (BIASA)

Roma IX.16.3.10 [Stefano Duperac, *Specimen, seu perfecta urbis antiquae imago* (Rome: Antonio Lafreri, 1573)]

Roma X.648, inv. 47738 [Mario Cartaro, *Novissimae urbis Romae accuratissima descriptio* (Rome: Mario Cartaro, 1576)]

Biblioteca Nazionale Centrale (BNC)

Gesuitico 713/6, Guido Baldo Foglietta, "Discorso del mattonato o selicato di Roma."

P.A.1. [Mario Cartaro, *Novissimae urbis Romae accuratissima descriptio* (Rome, Mario Cartaro, 1576)]

Turin

Archivio di Stato di Torino (AST)

Ms. J.a.III.3.–Ms. J.a.II.5 [Pirro Ligorio, *I libri delle antichità*, an alphabetical dictionary of antiquities, 18 vols.]

Ms. J.a.II.6. (XIX)–Ms. J.a.II.17 (XXX) [Pirro Ligorio, *I libri delle antichità*, 12 vols.]

United Kingdom

London

British Library (BL)

Maps C.25.d.9 (1.) [Pirro Ligorio, *Anteiquae urbis imago accuratissime ex vetusteis monumenteis formata* (Rome: M. and Fr. Tramezini, 1561)]

Maps* 155. (5.) [Pirro Ligorio, *Urbis Romae situs cum iis quae adhuc conspiciuntur veter. monument. reliquiis Pyrrho Ligorio Neap. invent.* (Venice: Michaelis Tramezini, 1552)]

Maps* 155. (7.) [Stefano Duperac, *Urbis Romae sciographia ex antiquis monumentis accuratiss. delineata* (Rome: Lorenzo Vaccari 1574; repr., Francesco Villamena, 1590s)]

Maps* 155. (10.) [Mario Cartaro, *Celeberrimae urbis antiquae fidelissima topographia post omnes alias aeditiones accuratissime delineata* (Rome: Mario Cartaro, 1579)]

Maps* 155. (35.) [Pirro Ligorio, *Urbis Romae* [...] (Venice: Michele Tramezzino, 1553)]

Maps* 23805. (7.) [Mario Cartaro, *Urbis Romae descriptio* (Rome: Mario Cartaro, 1575)]

Maps* 23805. (8.) [Stefano Duperac, *Nova urbis Romae descriptio* (Rome: Antonio Lafreri, 1577)]

Maps* 23810. (4.) [Stefano Duperac, *Urbis Romae sciographia ex antiquis monumentis accuratiss. delineata* (Rome: Lorenzo Vaccari 1574)—fragment containing privilege from Gregory XIII]

Maps S. T. R. 175. [Leonardo Bufalini, *Roma* (1551; repr. Rome: Antonio Blado, 1560)]

United States

Austin, Texas

Harry Ransom Center, University of Texas at Austin

Ranuzzi Family Collection Ph 2827.11, fol. 194r–195r, "Scrittura atenente a rimedi per l'inondazione del Tevere che sieguono in Roma fatta nel pontificato di Pio IV."

Ranuzzi Family Collection Ph 12827.12, fol. 196r–v, "Avvertimenti dell'Ill. Sig.r Gabrio Serbeloni sopra la fortificazione de i Borghi di Roma da osservarsi per conservazione di quella"

Chicago, Illinois

Newberry Library

Novacco 4F 250 [Giovanni Antonio Dosio, *Roma* (Rome: Bartolomeo Faletti, 1561]

Novacco 4F 256 (PrCtO) [Antonio Tempesta, *Recens prout hodie iacet almae urbis Romae cum omnibus viis aedificiisque prospectus accuratissime delineatus* (Rome, 1593)]

Vatican City

Archivio Segreto Vaticano (ASV)

Arch. Concist., Acta Vicecanc. 9 [record of consistory meetings, 1559–1564]

Archivium Arcis (A.A.), Arm. B. 13 [account book for transporting and raising obelisk at Santa Maria Maggiore]

Arm LII, tom. III, fol. 223r–v [grant to Cardinal Montepulciano of an *oncia* of Acqua Vergine water]

Bandi Sciolti I, no. 89 [tax for coaches]

Borghese IV, 280, fols. 145r–160v [Domenico Fontana's defense against charges concerning the bridge at Borgetto (Ponte Sisto)]

Misc. Arm. IV–V. tom. 58, fol. 3 [edict against leading untied buffalo through Rome]

Misc. Arm. IV-V, tom. 58, fol. 153r [edict against uncontrolled goats]

Misc. Arm. IV-V, tom. 60, fol. 207r [edict against courtesans and other scandalous people]

Misc. Arm. IV-V, tom. 74, no. 120 [general edict of the masters of the streets]

Misc. Arm. XI, tom. 86, fol. 228 [list of streets paved under Sixtus V]

Misc. Arm. LII, tom. III, fol. 223r–v [grant of an *oncia* of water to Cardinal Montepulciano]

Rubiccellone della Famiglia Naro. Indice 726 [index for papers of the Naro family]

Biblioteca Apostolica Vaticana (BAV)

Manoscritti

Avvisi di Roma

Urb. lat. 1038 (1554–1559)
Urb. lat. 1039 (1559–1562)
Urb. lat. 1040 (1565–1568)
Urb. lat. 1041 (1569–1570)
Urb. lat. 1043 (1572–1574)
Urb. lat. 1045 (1576–1577)
Urb. lat. 1050 (1582)
Urb. lat. 1053 (1585)
Urb. lat. 1054 (1586)
Urb. lat 1055 (1587)
Urb. lat. 1058 (1590)

Barb. lat. 2800. Firmano, Giovanni Francesco. *Caeremoniale Io. Francisci Firmani Maceratensis. Capellae S.mi. D. N. Papae Caerimoniarus Clerici* [. . .]

Barb. lat. 4391 (A) [collection of drawings of fortified walls surrounding the Borgo and diverse cities of Italy]

Barb. lat. 4391 (B) [collection of drawings of fortified walls surrounding Rome and the Borgo]

Barb. lat. 4432 [partial copy of Leonardo Bufalini, *Roma* (1551; repr. Rome: Antonio Blado, 1560)]

Codex Chigiano H II 43 [documents, some printed, most handwritten, concerning acqueducts, fountains, water distribution, and flood prevention measures]

Ms. Ross 1204 [copy of Bartolomeo Marliani, *Urbis Romae topographia* [. . .]. Rome: Valerij, Dorici, et Aloisij fratris, Academiae Romanae Impressorum, 1544, annotated by Benedetto Egio]

Ottob. lat. 3364– Ottob. lat. 3377. Ligorio, Pirro. *I libri delle antichità*. [seventeenth-century copies made for Queen Christiana of 14 of the 18 volumes of the alphabetical dictionary in Turin]

Vat. lat. 3434 [contains transcriptions of no longer extant fragments of the Forma Urbis]

Vat. lat. 3439 [writings by Onofrio Panvinio]

Vat. lat. 3587. "Agustinus Steucus de restituenda navigatione Tiberis"

Vat. Lat. 6783 [Ms version of Onofrio Panvinio's *Reipublicae Romanae commentariorum libri tres*]

Vat. lat 7246, 17r–25v [writing concerning navigation of the Tiber and restoration of the Acqua Vergine, author identified as Agostino Steuco]

Vat. lat. 7484, fols. 74r–140v. "Memorie di tre anni del pontificato di Sisto V, cioè 1585, 1586, e 1587"

Stampati

Bacci, Andrea, Del Tevere [. . .] libri tre [. . .]. Venice: [Aldo Manuzio], 1576. Aldine II 98 [letter on flooding to Clement VIII, "Alla santità di nostro sig. Clemente VIII Pontefice Massimo," in ten unnumbered MS fols. bound in front of the printed volume]

I. G. BAV, R.G. Scienze IV 199, int. 10, fols. 174r–176v [description of 1589 Tiber River flood]

Stampe Geogr. I. 620 Riserva [Leonardo Bufalini, *Roma* (1551; repr., Rome: Antonio Blado, 1560)]

Stampe Geogr.1.621. Riserva [Antonio Tempesta, *Recens prout hodie iacet almae urbis Romae cum omnibus viis aedificiisque prospectus accuratissime delineatus* (Rome: Antonio Tempesta, 1606)]

PRIMARY PRINTED SOURCES

Agrippa, Camillo. *Trattato* [. . .] *Di trasportar la guglia* [. . .]. Rome: Francesco Zanetti, 1583.

Albèri, Eugenio, ed. *Le relazioni degli ambasciatori veneti al senato durante il secolo Decimosesto.* Vol. 10 (ser. 2, vol. 4). Florence: Società Editrice Fiorentina, 1857.

Bacci, Andrea. *Del Tevere: Della natura et bonta dell'acque & delle inondationi Libri II.* Rome: Vincenzo Luchino, 1558.

———. *Del Tevere* [. . .] *libri tre* [. . .]. Venice: [Aldo Manuzio], 1576.

———. *Del Tevere dell'eccell. dottore medico e filosofo Andrea Baccio libro quarto.* Rome: Stampatori Camerali, 1599.

Balzani, U., ed. "Discorso del mattonato o selicato di Roma" [by Guido Baldo Foglietta]. *Archivio della Società Romana di Storia Patria* 1 (1877): 371–376.

Bardo, Marc Antonio. *Facultates magistratus curatorum viarum* [. . .]. Rome: Antonio Blado, 1565.

———. *Tractatus de tempore* [. . .]. 1565. Reprint, Cologne: Johannus Gymnicus, 1581.

———. *Tractatus iuris protomiseos sive congrui* [. . .] *Ad interpretationem bullarum* [. . .] *super aedificiis construendis, ad decorum Almae Urbis.* Rome: Antonio Blado, 1565.

Bariviera, Chiara, and Pamela O. Long. "An English Translation of Agostino Steuco's *De Aqua Virgine in Urbem revocanda.* (Lyon: Gryphius, 1547) [*On Bringing Back the Acqua Vergine to Rome*]." *Waters of Rome* 8 (August 2015). http://www3.iath.virginia.edu/waters/Journal8BarivieraLong.pdf.

Biondo Flavio. *Roma instaurata / Rome restaurée.* Edited and translated by Anne Raffarin-Dupuis. 2 vols. Paris: Société d'édition Les Belles Lettres, 2005–2012.

Bonelli, Renato, ed. "Lettera a Leone X." In *Scritti rinascimentali di architettura,* edited by Arnaldo Bruschi et al., 461–484. Milan: Il Polifilo, 1978.

Bordini, Giovanni Francesco. *De rebus praeclare gestis a Sixto V. Pon. Max.* Rome: J. Tornerius, 1588.

Bosio, Giacomo. *Dell'istoria della sacra religione et illma. militia de San Giovanni gierosolimitano.* 3 vols. Rome: Stamperia Apostolica Vaticana, 1594–1602.

Botero, Giovanni. *On the Causes of the Greatness and Magnificence of Cities.* Translated and with an introduction by Geoffrey Symcox. Toronto: University of Toronto Press, 2012.

Buffa, Sebastian, ed. *Antonio Tempesta: Italian Masters of the Sixteenth Century.* Vols. 35–37 of *The Illustrated Bartsch,* edited by Walter L. Strauss. New York: Abaris Books, 1983–1984.

Cancellieri, Francesco. *Storia de' solenni possessi de' sommi pontefici* [. . .]. Rome: L. Lazzarini, 1802.

Carpo, Mario, and Francesco Furlan, eds. *Leon Battista Alberti's Delineation of the City of Rome (Descriptio urbis Romae).* Critical edition by Jean-Yves Boriaud and Francesco Furlan,

English translation by Peter Hicks. Tempe: Arizona Center for Medieval and Renaissance Studies, 2007.

Center for the Advanced Study in the Visual Arts, National Gallery of Art. "History of the Accademia di San Luca, c. 1590–1635: Documents from the Archivio di Stato di Roma." https://www.nga.gov/accademia/en/intro.html.

Cherubini, P., A. Modigliani, D. Sinisi, and O. Verdi. "Un libro di multe per la pulizia delle strade sotto Paolo II (21 luglio–12 ottobre 1467)." *ASRSP* 107 (1984): 52–274.

Clarante, Paolo. *Della inondatione del Tevere, e della nuova foce del medesimo.* Perugia: Pietroiacomo Petrucci, 1577. (BAV, R. G. Scienze IV.199 [12]).

———. "Ricordi di m. Paolo Clarante da Terni medico in Roma." (BAV, Miscellana R.I. IV. 1551 [int. 99].)

Copenhaver, Brian P., ed. and trans. *Hermetica: The Greek Corpus Hermeticum and the Latin Asclepius in a New English Translation.* Cambridge: Cambridge University Press, 1992.

del Monte, Guidobaldo. *Mechanicorum liber.* Pisa: H. Concordia, 1577.

———. *Le mechaniche* [...] *tradotte in volgare dal Sig. Filippo Pigafetta.* Venice: F. di Franceschi, 1581.

De Marchi, Francesco. *Della architettura militare libri tre.* Brescia: Comino Presegni, 1599.

Dosio, Giovanni Antonio. *Urbis Romae aedificiorum illustriumquae supersunt reliquiae* [...] *ut hodie cernuntur descriptae* [...]. Rome, 1569.

———. *Le antichità di Roma: Aedificiorum illustriumquae supersunt reliquiae* [...]. Edited by Franco Borsi. 1569. Reprint, Rome: Edizioni Colombo Ristampe, 1970.

Duperac, Stefano [Étienne Dupérac]. *I vestigi dell'antichità di Roma* [...]. Rome: Lorenzo della Vaccheria, 1575.

Ehrle, Francesco. *Roma al tempo di Clemente VIII: La pianta di Roma di Antonio Tempesta del 1593 riprodotta da una copia vaticana del 1606.* Vatican City: Biblioteca Apostolica Vaticana, 1932.

———. *Roma al tempo di Giulio III: La pianta di Roma di Leonardo Bufalini del 1551 riprodotta dall'esemplare esistente nella Biblioteca Vaticana.* 1911. Reprint, Vatican City: Biblioteca Apostolica Vaticana, 2011.

———. *Roma prima di Sisto V: La pianta di Roma du Pérac-Lafréry del 1577 riprodotta dall'esemplare esistente nel Museo Britannico.* Vatican City: Biblioteca Apostolica Vaticana, 1908.

Evans, Harry B. *Aqueduct Hunting in the Seventeenth Century: Raffaello Fabretti's* De aquis et aquaeductibus veteris Romae. Ann Arbor: University of Michigan Press, 2002.

Fabricius, Georg. *Roma: Antiquitatum libri due* [...]. Basel: Johannes Oporinus, 1550.

Fanucci, Camillo. *Trattato di tutte l'opere pie dell'alma citta di Roma.* Rome: Lepido Facij e Stefano Paolini ad instanza di Bastiano de' Franceschi Senese, 1601.

Fauno, Lucio. "Alli lettori Lucio Fauno." In *Delle antichità della citta di Roma* [...]. Venezia: Michele Tramezzino, 1548.

———. *Delle antichità della citta di Roma* [...]. Venice: Michele Tramezzino, 1548. Edited by Margaret Daly Davis and Charles Davis. *Fontes* 13, September 15, 2008. http://archiv.ub.uni-heidelberg.de/artdok/580/1/Daly_Davis_Davis_Fontes13.pdf.

Fontana, Domenico. *Della trasportatione dell'Obelisco Vaticano* [...]. Roma: Domenico Basa, 1590.

———. *Della trasportatione dell'Obelisco Vaticano* [...] (Libro II) Naples: Costantino Vitale, 1604.

———. *Della trasportatione dell'Obelisco Vaticano, 1590*. Edited by Adriano Carugo with an introduction by Paolo Portoghesi. Milan: Il Polifilo, 1978.

———. *Della trasportatione dell'Obelisco Vaticano, Rome 1590*. Translated by David Sullivan with a commentary by Ingrid D. Rowland. Oakland, CA: Octavio, 2002.

Fontana, Francesco. *Relazione dello stato vecchio, e nuovo dell'Acqua Felice* [. . .]. Rome: Giovanni Francesco Buagni, 1696.

Frutaz, Amato Pietro. *Le piante di Roma*. 3 vols. Rome: Istituto di Studi Romani, 1962.

Fulvio, Andrea. *L'antichità di Roma di Andrea Fulvio antiquario* [. . .] *con le aggiuntioni e annotationi di Girolamo Ferrucci* [. . .]. Venice: Girolamo Francini, 1588.

Gamucci, Bernardo. *Le antichità della città di Roma* [. . .]. 1569. Reprint, Rome: Centro Editoriale Internazionale, 1993.

———. *Libri quattro dell'antichità della città di Roma* [. . .]. Venice: G. Varisco, 1565.

Genovese, Carmen, and Daniela Sinisi. Pro ornatu et publica utilitate: *L'attività della Congregazione cardinalizia* super viis, pontibus et fontibus *nella Roma di fine '500*. Rome: Gangemi Editore, 2010.

Gómez, Ludovico. *De prodigiosis Tyberis inundationibus* [. . .] *ad annum M. D. XXXI* [. . .]. Rome: Francesco Minizio Calvo, 1531.

Guicciardini, Luigi. *The Sack of Rome*. Translated and edited by James H. McGregor. New York: Italica Press, 1993.

Hülsen, Christian, ed. *Das Skizzenbuch des Giovannantonio Dosio im staatlichen Kupferstichkabinett zu Berlin*. Berlin: Heinrich Keller, 1933.

Lansford, Tyler. *The Latin Inscriptions of Rome: A Walking Guide*. Baltimore: Johns Hopkins University Press, 2009.

Lemerle, Frédérique, ed. and trans. *Les* Annotations *de Guillaume Philandrier sur le* De architectura *de Vitruve, Livres I à IV*. Paris: Picard, 2000.

———, ed. and trans. *Les* Annotations *de Guillaume Philandrier sur le* De architectura *de Vitruve, Livres V à VII*. Paris: Classiques Garnier, 2011.

Ligorio, Pirro. *Libro dei fiumi, dei fonti e dei laghi antichi*. Edited by Robert W. Gaston Rome: De Luca Editori d'Arte, 2015.

———. *Libro di M. Pyrrho Ligori napolitano delle antichità di Roma* [. . .] *con le Paradosse del medesimo autore* [. . .]. Venice: Michele Tramezino, 1553. Facsimile of the first edition, with an introduction by Margaret Daly Davis, *Fontes* 9, July 16, 2003. http://archiv.ub.uni-heidelberg.de/artdok/volltexte/2008/562.

Marliani, Bartolomeo. *Annales consulum, dictatorum, censorumque Romanorum* [. . .]. Rome: Antonio Blado, 1560.

———. *Antiquae Romae topographia libri septem*. Rome: Antonio Blado in Aedibus D. Ioan Bap. De Maximis, 1534.

———. "B. Marliani topographiae urbis Romae haec nuper adiecta." In *Urbis Romae topographia: Nuper ab ipso auctore nonnullis erroribus* [1544]. http://archive.org/details/vrbisromaetopogr00marl.

———. *Consulum, dictatorum censorumque series* [. . .]. Rome: R. Ian, 1549.

———. *Urbis Romae topographia* [. . .]. Rome: Valerij, Dorici, et Aloisij fratris, Academiae Romanae Impressorum, 1544.

———. *Urbis Romae topographia: Nuper ab ipso auctore nonnullis erroribus sublatis emendate, Addita etiam interpretatione nominum* [. . .]. Rome: Valerij, Dorici, et Aloisij, fratris, Academia Romanae Impressorum, 1544. [Version at Rubenstein Library of Duke University, A-17 fm348U c. 1; and at Rome, Biblioteca Casanatense, Vol. misc. 1001 4.]

Martin, Gregory. *Roma sancta (1581)*. Edited by George Bruner Parks. Rome: Edizioni di Storia e Letteratura, 1969.

Mercati, Michele. *De gli obelischi di Roma* [. . .]. Rome: Domenico Basa, 1589.

———. *Gli obelischi di Roma*. Edited by Gianfranco Cantelli. Bologna: Cappelli Editore, 1981.

Modio, Giovanni Battista. *Il Tevere* [. . .]. Rome: Vincenzo Luchino, 1556.

Montaigne, Michel de. *Montaigne's Travel Journal*. Translated by Donald M. Frame. Repr. ed. San Francisco: North Point Press, 1983.

Oldradi, [Angelo]. *Aviso della pace* [. . .] *del diluvio* [. . .]. Rome: Antonio Blado, 1557. [Copy in Biblioteca Casanatense, Rome, Vol. Misc. 1927/18.]

Palladio, Andrea. *Palladio's Rome: A Translation of Palladio's Two Guidebooks to Rome*. Edited and translated by Vaughan Hart and Peter Hicks. New Haven, CT: Yale University Press, 2006.

Panvinio, Onofrio. *Antiquae urbis imago* [. . .]. Paris: Egidius et Nicolas Gilles, 1588.

———. *Fasti et triumphi romani* [. . .]. Venice: Jacopo Strada, 1557.

———. *Fastorum libri V* [. . .]. Venice: Vincenzo Valgrisio, 1558.

———. *Reipublicae Romanae commentariorum libri tres*. Venice: Vincenzo Valgrisio, 1558.

———. *XXVII Ponitificum Maximorum elogia et imagines* [. . .] *delineatae*. Rome: Antonio Lafreri, 1568.

Parigioli, Lorenzo. *Nuovo discorso sopra il diluvio di Roma*. Rome: Heredi d'Antonio Blado, 1579. [BNCF, Coll. 64.2]

Peto, Luca. *De mensuris et ponderibus romanis et graecis* [. . .] *Libri quinque*. Venice: [P. Manutius], 1573.

———. *De restitutione ductus Aquae Virginis*. Rome: B. Tosius, 1570.

———. *Discorso* [. . .] *intorno alla cagione della eccessiva inondatione del Tevere* [. . .]. Rome: Giuseppe degl'Angeli, 1573.

———. "Liber sextus varias lectiones continens." In Peto, *De mensuris et ponderibus*, 73–80.

Petroni, Alessandro Traiano. *De Aqua Tiberina*. Rome: Valerius et Aloisius Doricus, 1552.

———. *De victu romanorum et de sanitate tuenda libri quinque*. Rome: In Aedibus Populi Romani, 1581.

Pigafetta, Filippo. *Discorso* [. . .] *d'intorno all'historia della aguglia* [. . .]. Rome: Bartolomeo Grassi, 1586.

Scamozzi, Vincenzo. *L'idea della architettura universale*. Venice: Presso L'Autore, 1615.

Stanford Digital *Forma Urbis Romae* Project. http://formaurbis.stanford.edu/.

Statuta almae urbis Romae [. . .]. Rome: In Aedibus Populi Romani, 1580.

Steuco, Agostino. *De Aqua Virgine in urbem revocanda*. Lyon: Sebastianus Gryphius, 1547.

———. *De restituenda navigatione Tiberis*. Lyon: Sebastianus Gryphius, 1547.

———. *De via Pauli et de fontibus inducendis in eam*. Rome: Balthasar Cartularius [n.d., not after 1543].

Tanner, Norman P., ed. *Decrees of the Ecumenical Councils*. 2 vols. London: Sheed and Ward / Washington, DC: Georgetown University Press, 1990.

Tolomei, Claudio. "Lettera al Conte Agostino de' Landi," edited by Sandro Benedetti and Tommaso Scalesse. In *Trattati con l'aggiunta degli scritti di architettura di Alvise Cornaro,*

Francesco di Giorgi, Claudio Tolomei, Giangiorgio Trissino, Giorgio Vasari, edited by Elena Bassi et al., 32–61. Milan: Il Polifilo, 1985.

Tomassetti, Luigi, et al., eds. *Bullarum diplomatum et privilegiorum Sanctorum Romanorum Pontificum Taurinensis editio*. 25 vols. Turin: Seb. Franco and Henrico Dalmazzo, 1857–1872). http://www.icar.beniculturali.it/biblio/_view_volume.asp?ID_VOLUME=2120.

Trevisi, Antonio. *Fondamento del edifitio* [...] *sopra la innondatione del fiume* [...]. Rome: Antonio Blado, 1560.

University of Chicago Library. The *Speculum Romanae Magnificentae* Digital Collection. http://speculum.lib.uchicago.edu.

Vacca, Flaminio. *Memorie di varie antichità* [...]. In *Roma antica distinta per rioni* [...], 1:214–266. Rome: Gio. Lorenzo Barbiellini Libraro, 1741.

Vagnetti, Luigi. "La 'Descriptio Urbis Romae': Uno scritto poco noto di Leon Battista Alberti (contributo alla storia del rilevamento architettonico e topografico)." *Quaderni dell'Istituto degli Elementi di Architettura e Rilievo dei Monumenti* 1 (October 1968): 25–80.

Valentini, Roberto, and Giuseppe Zucchetti, eds. *Codice topografico della città di Roma*. 4 vols. Rome: Tipografia del Senato, 1940–1953.

Williams, Kim, Lionel March, and Stephen Wassell. *The Mathematical Works of Leon Battista Alberti*. Basel: Birkhäuser, 2010.

SECONDARY SOURCES

Accordi, Bruno. "Michele Mercati (1541–1593) e la Metallotheca." *Geologica Romana* 19 (1980): 1–50.

Acidini [Luchinat], Cristina. "Dosi, Giovanni Antonio (Giovan Antonio, Giovannantonio), detto Dosio." *DBI*, 41:516–523.

———. "Roma antica." In Borsi er al., *Giovanni Antonio Dosio*, 27–166.

Adam, Jean-Pierre. *Roman Building: Materials and Techniques*. Translated by Anthony Mathews. London: Routledge, 1999.

Adams, Nicholas, and Laurie Nussdorfer. "The Italian City, 1400–1600." In *The Renaissance from Brunelleschi to Michelangelo: The Representation of Architecture*, edited by Henry A. Millon and Vittorio Magnago Lampugnani, 205–230. New York: Rizzoli, 1997.

Adams, Nicholas, and Simon Pepper. "The Fortification Drawings." In Frommel and Adams, *Architectural Drawings of Antonio da Sangallo the Younger*, 60–74.

Affani, Anna Maria. "L'Ospedale dei Mendicanti a Ponte Sisto: L'istituzione dell'Ospedale dei Mendicanti disposto da Sisto V nel 1587; Studio della conformazione originaria attraverso i documenti di archivio e le illustrazioni coeve." In Porzio, *Impronte sistine*, 101–117.

Ago, Renata. *Economia barocca: Mercato e istituzioni nella Roma del Seicento*. Rome: Donzelli, 1998.

Aicher, Peter J. *Guide to the Aqueducts of Ancient Rome*, Wauconda, IL: Bolchazy-Carducci, 1995.

Ait, Ivana. "Uno spazio produttivo: Il Tevere nel basso Medioevo." *Rivista Storica del Lazio* 10 (2002): 3–15.

———. "Il Tevere e le attività produttive a Roma nel basso Medioevo." In Travaglini, *La città e il fiume*, 81–94.

Albanese, Massimiliano. "Marliani (Marliano), Bartolomeo (Giovanni Bartolomeo)." *DBI*, 70:597–600.

Aldrete, Gregory S. *Floods of the Tiber in Ancient Rome*. Baltimore: Johns Hopkins University Press, 2007.

Alexander, John. *From Renaissance to Counter-Reformation: The Architectural Patronage of Carlo Borromeo during the Reign of Pius IV*. Rome: Bulzoni, 2007.

Alföldy, Géza. *Der Obelisk auf dem Petersplatz in Rom: Ein historisches Monument der Antike*. Heidelberg: Carl Winter, 1990.

Almagià, Roberto. "Intorno a un cartografo italiano del secolo XVI." *Rivista Geografica Italiana* 20 (February–March 1913): 99–112.

———. "Pirro Ligorio cartografo." *Rendiconti delle Sedute dell'Accademia Nazionale dei Lincei: Classe di Scienze, Morali, Storiche e Filologiche*, 8th ser., 11 (1956): 49–61.

Altieri, Piero. "L'attuazione della riforma tridentina." In *Storia della Chiesa di Cesena*, edited by Marino Mengozzi, vol. 1, 261–287. Cesena: Editrice Stilgraf, 1998.

Amadio, Sonia. "Famiglie in carriera ed artisti nella Roma barocca. I Naro." *Studi Romani* 45 (July–December 1997): 314–330.

Anderson, Paul A. "Master Carpenters in Renaissance and Baroque Rome: The Collaboration of Artists, Architects and Artisans on Monumental Commissions in the Cinquecento and Seicento." 2 vols. PhD diss. University of California, Santa Barbara, 2008.

Andres, Glen M. "Le jardin du cardinal Ricci." In Chastel and Morel, *La Villa Médicis*, 2:342–350.

———. *The Villa Medici in Rome*. 2 vols. New York: Garland, 1976.

Andretta, Elisa. "Mercati, Michele." *DBI*, 73:606–611.

Andretta, Stefano. "Le istituzioni e l'esercizio del potere." In Pinelli, *Roma del Rinascimento*, 93–121.

Annoscia, Giorgia Maria. *Fonti e strutture per la conoscenza del sistema idrico di Roma nel Medioevo*. Rome: ARACNE, 2007.

Anselmi, Alessandra. "Porta, Giacomo della." *DA*, 25:258–261.

Antinori, Aloisio. "Le basiliche romane meridionali da Sisto V a Paolo V." In Fagiolo and Madonna, *Sisto V*, 1:497–518.

———, ed. "Le reti dell'acqua dal tardo Cinquecento al Settecento." Special Issue, *Roma Moderna e Contemporanea* 16 (July–December 2008).

———. "L'uso privato dell'Acqua Vergine dalla realizzazione della nuova rete idrica alla revisione generale del 1631." In Antinori, "Le reti dell'acqua," 231–272.

Antonucci, Micaela. "Le porte di Roma nei progetti di Antonio da Sangallo il Giovane." *Roma Moderna e Contemporanea* 22 (January–June 2014): 17–35.

Aragozzini, Giovanna, and Marco Nocca. *Le piante di Roma dal Cinquecento all'Ottocento*. Rome: Dino Audino, 1993.

Arce, J. "Arcus Titi (Via Sacra)." In Steinby, *Lexicon topographicum*, 1:109–111.

Ash, Eric H., ed. *Expertise: Practical Knowledge and the Early Modern State*. Osiris, 2nd ser., vol. 25. Chicago: University of Chicago Press, 2010.

———. *Power, Knowledge, and Expertise in Elizabethan England*. Baltimore: Johns Hopkins University Press, 2004.

Ashby, Thomas. *The Aqueducts of Ancient Rome*. Edited by I. A. Richmond. Oxford: Clarendon Press, 1935.

Asor Rosa, Laura. "Fauno, Lucio." *DBI*, 45:377–378.

———. "Gamucci, Bernardo." *DBI*, 52:132–133.

Baernstein, P. Renée. "Regional Intermarriage among the Italian Nobility in the Sixteenth Century." In *Marriage in Premodern Europe: Italy and Beyond*, edited by Jacqueline Murray, 201–219. Toronto: Centre for Reformation and Renaissance Studies, 2012.

Ballon, Hillary, and David Friedman. "Portraying the City in Early Modern Europe: Measurement, Representation, and Planning." In Woodward, *History of Cartography*, pt. 1, 680–704.

Baratta, Mario. "La pianta d'Imola di Leonardo da Vinci." *Bollettino della Società Geografica Italiana* 12 (1911): 945–967.

Barberi, Francesco. "Blado, Antonio." *DBI*, 10:753–757.

Bariviera, Chiara. "Regione XI: *Circus Maximus*." In Carandini and Carafa, *Atlante di Roma antica*, 1:421–445.

———. "Regione XII: *Piscina Publica*." In Carandini and Carafa, *Atlante di Roma antica*, 1:375–387.

Barkan, Leonard. *Unearthing the Past: Archaeology and Aesthetics in the Making of Renaissance Culture*. New Haven, CT: Yale University Press, 1999.

Barletti, Emanuele, ed. *Giovan Antonio Dosio da San Gimignano: Architetto e scultor fiorentino tra Roma, Firenze e Napoli*. Florence: Edizione Firenze, 2011.

Barni, Gian Luigi. "Agrippa, Camillo." *DBI*, 1:503.

Battaglini Di Stasio, Renata. "Bartolini, Matteo, detto—Matteo da Castello." *DBI*, 6:616–617.

Bauer, Clemens. "Die Epochen der Papstfinanz." *Historische Zeitscrift* 138 (1928): 473–503.

Bauer, Heinrich. "Cloaca, Cloaca Maxima." In Steinby, *Lexicon topographicum*, 1:288–290.

Bauer, Stefan. "Panvinio, Onofrio." *DBI*, 81:36–39.

Beard, Mary. *The Roman Triumph*. Cambridge, MA: Harvard University Press, 2007.

Becchi, Antonio. "Cantieri d'inchiostro: Meccanica teorica e meccanica chirurgica nella seconda metà del Cinquecento." In Curcio, Navone, and Villari, *Studi su Domenico Fontana*, 91–103.

Becchi, Antonio, Domenico Bertoloni Meli, and Enrico Gamba, eds. *Guidobaldo del Monte (1545–1607): Theory and Practice of the Mathematical Disciplines from Urbino to Europe*. Max Planck Research Library for the History and Development of Knowledge, Proceedings 4. Berlin: Edition Open Access, 2013.

Becker, Frank. "Intorno all'acquedotto Vergine in Roma." In Conforti and Hopkins, *Architettura e tecnologia*, 158–177.

Beckmann, Martin. *The Column of Marcus Aurelius: The Genesis and Meaning of a Roman Imperial Monument*. Chapel Hill: University of North Carolina Press, 2011.

Bedon, Anna. "Della Porta, Giacomo." *DBI*, 37:160–170.

———. "L'orgoglio e la borsa: I palazzi romani della seconda metà del XVI secolo nelle strategie finanziarie delle famiglie municipali." In Curcio, Navone, and Villari, *Studi su Domenico Fontana*, 143–159.

Bellini, Federico. *La basilica di San Pietro da Michelangelo a Della Porta*. 2 Vols. Rome: Argos, 2011.

———. "La *Civitas Pia* e le fortificazione vaticane di Pio IV." *Studi Romani* 61 (January–December 2013): 42–76.

———. "I collegi e gli insediamenti nazionali nella Roma di Gregorio XIII (con una nota su Sant'Atanasio dei Greci e la Trinità dei Monti)." *Città e Storia* 2 (2007): 111–130.

———. "I grandi cantieri: Campidoglio, San Pietro, *Studium Urbis*." In Conforti and Tuttle, *Storia*, 4:66–93.

———. "Michelangelo, La Strada e la Porta Pia." *Studi Romani* 59 (2011): 74–109.

———. "Le porte romane di Pio IV (1559–1565)." In "Entrare in città: Le porte di Roma," edited by Giuseppe Bonaccorso and Claudia Conforti. Special Issue, *Roma Moderna e Contemporanea* 22 (January–June 2014): 37–61.

Bellini, Paolo. "Stampatori e mercanti di stampe in Italia nei secoli XVI e XVII." *I Quaderni del Conoscitore di Stampe* 26 (1975): 19–45. Translated as "Printmakers and Dealers in Italy." *Print Collector* 13 (May–June 1975): 17–45.

Beltrani, Giovanni. "Leonardo Bufalini e la sua pianta topografica di Roma." *Rivista Europea* 22 (1880): 5–28, 361–387.

Benedetti, Sandro, and Giuseppe Zander. *L'arte in Roma nel secolo XVI*. Vol. 1, *L'architettura*. Bologna: Cappelli, 1990.

Benedetti, Simona. "L'acquedotto Felice da Porta Furba alla Mostra del Mosè." In Piera Sette and Benedetti, *Architetture per la città*, 90–129.

———. "Il Ponte Felice." In Piera Sette and Benedetti, *Architetture per la città*, 223–239.

Benocci, Carla. "Giovanni e Domenico Fontana ed i 'sistemi di acque e fontane' nei giardini romani in età sistina." In Fagiolo and Madonna, *Sisto V*, 1:545–557.

Bertoloni Meli, Domenico. *Thinking with Objects: The Transformation of Mechanics in the Seventeenth Century*. Baltimore: Johns Hopkins University Press, 2006.

Bertolotti, Antonino. *Artisti bolognesi, ferraresi ed alcuni altri del già Stato Pontificio in Roma nei secoli XV, XVI e XVII*. 1885. Reprint, Bologna: Arnaldo Forni, 1962.

———. *Artisti francesi in Roma nei secoli XV, XVI e XVII*. 1886. Reprint, Bologna: Arnaldo Forni, 1975.

———. *Artisti lombardi a Roma nei secoli XV, XVI, XVII*. 2 vols. 1881. Reprint, Bologna: Arnaldo Forni, 1969–1970.

———. *Artisti subalpini in Roma nei secoli XV, XVI e XVII*. 1884. Reprint, Bologna: Arnaldo Forni, 1965.

———. *Artisti veneti in Roma nei secoli XV, XVI e XVII*. 1884. Reprint, Bologna: Arnaldo Forni, 1965.

———. "Bartolomeo Marliano, archeologo nel secolo XVI." *Atti e memorie delle RR Deputazioni di Storia Patria per le Provincie dell'Emilia*, 4th ser., pt. 2 (1880): 107–138.

———. "Nuovi documenti intorno all'architetto Antonio da Sangallo (il Giovane) ed alla sua famiglia: III." *Buonarroti*, 3rd ser., 4 (1892): 278–287.

———. "Repressioni straordinarie alla prostituzione in Roma nel secolo XVI." *Rivista di Discipline Carcerarie* 16 (1886): 505–517.

Bertoni, Luisa. "Crispi, Tiberio." *DBI*, 30:801–803.

Besse, Jean-Marc, and Pascal Dubourg Glatigny. "Cartographier Rome au XVIe siècle (1544–1599)." In *Rome et la science moderne entre Renaissance et Lumière*, edited by Antonella Romano, 369–414. Rome: École française de Rome, 2008.

Bevilacqua, Irene. "Acque e mulini nella Roma del Seicento." *Città e Storia* 5 (2010): 99–140.

Bevilacqua, Mario. "L'immagine di Roma moderna da Bufalini a Nolli: Un modello europeo." In Bevilacqua and Fagiolo, *Piante di Roma*, 62–95.

———. *Il monte dei Cenci: Una famiglia romana e il suo insediamento urbano tra Medioevo ed età barocca*. Rome: Gangemi, 1988.

Bevilacqua, Mario, and Marcello Fagiolo, eds. *Piante di Roma dal Rinascimento ai Catasti.* Rome: Artemide Edizioni, 2012.

Bianchi, Elisabetta, ed. *La Cloaca Maxima e i sistemi fognari dall'antichità ad oggi.* Rome: Palombi, 2014.

———. "Il primo tratto della Cloaca Maxima e gli adattamenti realizzati per la costruzione dei Fori Imperiali." In Bianchi, *La Cloaca Maxima,* 82–107.

Bianchi, Elisabetta, and Luca Antognoli. "La Cloaca Massima dal Foro Romano al Velabro dagli studi di Heinrich Bauer alle nuove indagini." In Bianchi, *La Cloaca Maxima,* 108–153.

Bianchi, Lorenzo. *Ad Limina Petri: Spazio e memoria della Roma cristiana.* Rome: Donzelli, 1999.

Biermann, Veronica. "Ortswechsel: Überlegungen zur Bedeutung der Bewegung schwerer Lasten für die Wirkung und Rezeption monumentaler Architektur am Beispiel des Vatikanischen Obelisken." In *Perspektiven der Spolienforschung.* Vol. 1, *Spoliierung und Transposition,* edited by Stefan Altekamp, Carmen Marcks-Jacobs, and Peter Seiler, 123–156. Berlin: Walter de Gruyter, 2013.

Biow, Douglas. *The Culture of Cleanliness in Renaissance Italy.* Ithaca, NY: Cornell University Press, 2006.

Blastenbrei, Peter. "Glücksritter und Heilige: Motivstruktur und Täterprofile bei der Accoltiverschwörung gegen Pius IV. im Jahre 1564." *Quellen und Forschungen aus italienischen Archiven und Bibliotheken* 70 (1990): 441–490.

———. *Kriminalität in Rom, 1560–1585.* Tübingen: Max Niemeyer, 1995.

———. "I Romani tra violenza e giustizia nel tardo Cinquecento." *Roma Moderna e Contemporanea* 5 (January–April 1997): 67–79.

———. "Violence, Arms and Criminal Justice in Papal Rome, 1560–1600." *Renaissance Studies* 20 (February 2006): 68–87.

Bocquet, Denis. "Storia urbana e storia della decisione: L'arginamento del Tevere a Roma (1870–1880)." In Travaglini, *La città e il fiume,* 323–342.

Boiteux, Martine. "Fêtes et cérémonies romaines au temps des Carrache." In *Les Carrache et les décors profanes: Actes du colloque,* 183–213. Collection de L'École française de Rome, 106. Rome: L'École française de Rome, 1988.

———. "La noblesse romaine sur la scène: Contribution à l'étude de l'évergétisme romain à l'époque moderne." In *La nobilità delle città capitali,* edited by Martine Boiteux, Catherine Brice, and Carlo M. Travaglini, 131–180. Rome: Università degli Studi Roma Tre, CROMA, 2009.

———. "Parcours rituels romains à l'époque moderne." In Visceglia and Brice, *Cérémonial et ritual à Rome,* 26–87.

———. "Rivalità festive: Rituali pubblici romani al tempo di Sisto V." In Fagiolo and Madonna, *Sisto V,* 1:357–392.

Bonacini, Claudio. *Bibliografia delle arti scrittorie e della calligrafia.* Florence: Sansoni, 1953.

Bonaccorso, Giuseppe, ed. "Le acque e la città (XV–XVI secolo)." Special issue, *Roma Moderna e Contemporanea* 17 (January–December 2009).

———. "Il 'ponte dinamico': I traghetti e l'attraversamento del Tevere a Roma nei secoli XVI–XVIII." In Calabi and Conforti, *I ponti delle capitali d'Europa,* 89–103.

———. "Roma e le sue acque potabili nel Cinquecento: La competizione con il Tevere." In Bonaccorso, "Le acque e la città," 73–90.

Bonaccorso, Giuseppe, and Michela Lucci. "Le case dei Fontana a Roma." In Fagiolo and Bonaccorso, *Studi sui Fontana*, 463–472.

Bonaventura, Maria Antonietta. "L'industria e il commercio delle incisioni nella Roma del '500." *Studi Romani* 8 (July–August 1960): 430–436.

Bonora, Elena. *Roma 1564: La congiura contro il papa*. Rome: Laterza, 2011.

Borghese, Daria, ed. *La Casina di Pio IV in Vaticano*. Turin: Umberto Allemandi, 2010.

Borromeo, Agostino. "Cesi, Pier Donato." *DBI*, 24:261–266.

———. "Gregorio XIII, papa." *DBI*, 59:204–219.

Borroni Salvadori, Fabia. "Bonifacio, Natale (Bonifatio, Bonifazio), detto Bonifacio da Sebenico o Natale Dalmatino." *DBI*, 12:201–204.

———. "Cartaro, Mario." *DBI*, 20:796–799.

———. *Carte, piante e stampe storiche delle raccolte lafreriane della Biblioteca Nazionale di Firenze*. Rome: Istituto Poligrafico e Zecca dello Stato, 1980.

Borsi, Franco, Cristina Acidini, Fiammetta Mannu Pisani, and Gabriele Morolli, eds. *Giovanni Antonio Dosio: Roma antica e i disegni di architettura agli Uffizi*. Rome: Officina Edizioni, 1976.

Borsi, Stefano. *Roma di Sisto V: La pianta di Antonio Tempesta, 1593*. Rome: Officina Edizioni, 1986.

Bradley, Mark, and Kenneth Stow, eds. *Rome, Pollution and Propriety: Dirt, Disease and Hygiene in the Eternal City from Antiquity to Modernity*. Cambridge: Cambridge University Press, 2012.

Bragaglia Venuti, Cristina. "Etienne Dupérac and Pirro Ligorio." *Print Quarterly* 23 (December 2006): 408–413.

———. "Etienne Dupérac e i paesaggi della Loggia di Pio IV." *Rivista dell'Istituto Nazionale d'Archeologia e Storia dell'Arte*, 3rd ser., 57 (2002): 279–309.

Brentano, Carrol. "Della Porta, Guglielmo." *DBI*, 37:192–199.

Broise, Henri, and Vincent Jolivet. "Villa Médicis: L'Antiquité." In Chastel and Morel, *La Villa Médicis*, 2:9–40.

Brothers, Cammy. "Architecture, History, Archaeology: Drawing Ancient Rome in the Letter to Leo X and in Sixteenth-Century Practice." In *Coming About: A Festschrift for John Shearman*, edited by Lars R. Jones and Louisa C. Mathew, 135–140. Cambridge, MA: Harvard Art Museums, 2001.

Brotton, Jerry. *Trading Territories: Mapping the Early Modern World*. Ithaca, NY: Cornell University Press, 1998.

Bultrini, Emiliano. "L'Acqua Crabra: Un fiume scomparso vicende del confine naturale tra Roma e la Civitas Tusculana." *ASRSP* 135 (2012): 63–83.

Buonara, Paolo. "Gli acquedotti del papa: Restitutio o innovazione?" In Antinori, "Le reti dell'acqua," 323–334.

Buonfiglio, Marialetizia. "La *cloaca circi* e le acque della valle del Circo Massimo." In Bianchi, *La Cloaca Maxima*, 166–183.

Buonora, Paolo, and Manuel Vaquero Piñeiro. "Il sistema idraulico di Roma in età moderna: Assetti di potere e dinamiche produttive." In Travaglini, *La città e il fiume*, 147–168.

Burns, Howard. "Pirro Ligorio's Reconstruction of Ancient Rome: The *Anteiquae Urbis Imago* of 1561." In Gaston, *Pirro Ligorio*, 19–91.

Burroughs, Charles. "Absolutism and the Rhetoric of Topography: Streets in the Rome of Sixtus V." In Çelik, Favro, and Ingersoll, *Streets: Critical Perspectives*, 189–202.

———. *From Signs to Design: Environmental Process and Reform in Early Renaissance Rome.* Cambridge, MA: MIT Press, 1990.

Bury, Michael. *The Print in Italy, 1550–1620.* London: British Museum Press, 2001.

———. "The Taste for Prints in Italy to c. 1600." *Print Quarterly* 2 (March 1985): 12–26.

Butters, Suzanne B. "Ferdiand et le jardin du Pincio." In Chastel and Morel, *La Villa Médicis,* 2:351–410.

Butters, Suzanne B., and Pier Nicola Pagliara. "Il Palazzo dei Tribunali e via Giulia a Roma / The Palazzo dei Tribunali and Via Giulia in Rome." *Zodiac* 14 (1995–1996): 14–29.

Butzek, Monika. *Die kommunalen Repräsentationsstatuen der Päpste des 16. Jahrhunderts in Bologna, Perugia und Rom.* Bad Honnef: Bock und Herchen, 1978.

Cafà, Valeria. "The Via Papalis in Early Cinquecento Rome: A Contested Space between Roman Families and Curials." *Urban History* 37 (2010): 434–451.

Cajano, Elvira. "L'acquedotto Felice dalla campagna alla città di Roma." In Porzio, *Impronte sistine,* 17–25.

Calabi, Donatella, and Claudia Conforti, eds. *I ponti delle capitali d'Europa dal Corno d'Oro alla Senna.* Milan: Electa, 2002.

Camerota, Filippo. "Dal compasso alla lente: Il rinnovamento delle tecniche di cartografia urbana tra Rinascimento e Barocco." In Bevilacqua and Fagiolo, *Piante di Roma,* 108–115.

Campbell, Brian. *Rivers and the Power of Ancient Rome.* Chapel Hill: University of North Carolina, 2012.

Canepari, Eleonora. *La construction du pouvoir local: Élites municipales, liens sociaux et transactions économiques dans l'espace urbain, Rome, 1550–1650.* Rome: École française de Rome, 2017.

———. "Mestiere e spazio urbano nella costruzione dei legami sociali degli immigrati a Roma in età moderna." In *L'Italia delle migrazioni interne: Donne, uomini, mobilità in età moderna e contemporanea,* edited by Angiolina Arru and Franco Ramella, 33–76. Rome: Donzelli, 2003.

Canosa, Romano, and Isabella Colonnello. *Storia della prostituzione in Italia dal Quattrocento alla fine del Settecento.* Rome: Sapere 1989.

Caperna, Maurizio. *La Lungara.* Vol. 1, *Storia e vicende edilizie dell'area tra il Gianicolo e il Tevere.* Rome: Edizioni Quasar, 2013.

Caprino, C., A. M. Colini, G. Gatti, M. Pallottino, and P. Romanelli. *La Colonna di Marco Aurelio.* Rome: L'Erma di Bretschneider, 1955.

Caracciolo, Alberto. "Sovrano pontefice e sovrani assoluti." *Quaderni storici* 18 (April 1983): 279–286.

Carafa, Paolo. "Le Mura Aureliane." In Carandini and Carafa, *Atlante di Roma antica,* 1:85–89.

Carafa, Paolo, and Paola Pacchiarotti. "Regione XIV: *Transtiberim.*" In Carandini and Carafa, *Atlante di Roma antica,* 1:549–582.

Caraffa, Costanza. "Domenico Fontana e gli obelischi: Fortuna critica del 'Cavaliere della Guglia.'" In Curcio, Navone, and Villari, *Studi su Domenico Fontana,* 21–47.

Carandini, Andrea, and Paolo Carafa, eds. *Atlante di Roma antica: Biografia e ritratti della città.* 2 vols. Milan: Mondadori Electa, 2012.

Carandini, Silvia. "L'effimero spirituale: Feste e manifestazioni religiose nella Roma dei papi in età moderna." In Fiorani and Prosperi, *Roma, la città del papa,* 519–553.

Caravale, Mario. "Le istituzioni temporali della Chiesa sotto Sisto V." In Fagiolo and Madonna, *Sisto V*, 1:39–58.

Caravale, Mario, and Alberto Caracciolo, eds. *Lo Stato Pontificio da Martino V a Pio IX*. Storia d'Italia, vol. 14. Turin: Unione Tipografico-Editrice Torinese, 1978.

Carbonetti Vendittelli, Cristina. "Documentazione inedita riguardante i *magistri edificiorum urbis* e l'attività della loro curia nel secoli XIII e XIV." *ASRSP* 113 (1990): 169–188.

Carettoni, Gianfilippo, Antonio M. Colini, Lucos Cozza, and Guglielmo Gatti. *La pianta marmorea di Roma antica: Forma Urbis Romae*. 2 vols. Rome: Comune di Roma, 1960.

Carlino, Andrea. *Books of the Body: Anatomical Ritual and Renaissance Learning*. Translated by John Tedeschi and Anne C. Tedeschi. Chicago: University of Chicago Press, 1999.

Carocci, Sandro. "The Papal State." In *The Italian Renaissance State*, edited by Andrea Gamberini and Isabella Lazzarini, 69–89. Cambridge: Cambridge University Press, 2012.

Cassiani, Chiara, and Myriam Chiabò, eds. *Pomponio Leto e la prima Accademia Romana*. Rome: Roma nel Rinascimento, 2007.

Cassiani, Gennaro. "Modio, Giovanni Battista." *DBI*, 75:238–241.

———. "Patrigno Tevere: Le obiezioni sperimentali di Giovanni Battista Modio al 'dogma' della potabilità dell'acqua del Tevere a metà Cinquecento." *Roma nel Rinascimento* (2014): 357–372.

Castagnoli, Ferdinando. *Il Vaticano nell'antichità classica*. Vatican City: Bibllioteca Apostolica Vaticana, 1992.

Cattaneo, Annalisa. "Mario Cartaro, incisore viterbese del XVI secolo." *Grafica d'Arte* 9, no. 35 (1998): 2–9; 11, no. 41 (2000): 6–14; and 11, no. 42 (2000): 3–11.

Cavazzini, Patrizia. *Painting as a Business in Early Seventeenth-Century Rome*. University Park: Pennsylvania State University Press, 2008.

Cavina, Marco. *Il sangue dell'onore: Storia del duello*. Rome: Laterza, 2005.

Cazzato, Vincenzo. "Momenti romani di Antonio Trevisi." In Vetrugno, *Antonio Trevisi*, 131–150.

Ceen, Allan. "Bufalini 1551: Distortion and Rectification." In Bevelacqua and Fagiolo, *Piante di Roma*, 128–133.

Celani, Enrico. "Alcune iscrizioni sulle inondazioni del Tevere." *Bollettino della Commissione Archeologica Communale* 23 (1895): 283–295.

Çelik, Zeynep, Diane Favro, and Richard Ingersoll, eds. *Streets: Critical Perspectives on Public Space*. Berkeley: University of California Press, 1994.

Cerasoli, Francesco. "Censimento della popolazione di Roma dall'anno 1600 al 1739." *Studi e Documenti di Storia e Diritto* 12 (1891): 169–199.

———. "La Colonna Traiana e le sue adiacencze nei secoli XV, XVI e XVII." *Bullettino della Commissione Archeologica Comunale di Roma* 29 (January–March 1901): 300–308.

———. "Notizie circa la sistemazione di molte strade di Roma nel sec. XVI." *Bullettino della Commissione Archeologica Comunale di Roma* 28 (October–December 1901): 342–362.

Ceresa, Massimo. "Fulvio, Andrea." *DBI*, 50:709–712.

Cerutti Fusco, Annarosa. "Il progetto di Domenico Fontana 'per ridurre il Coliseo di Roma ad habitatione' e le opera sistine di 'pubblica utilità.'" *Quaderni dell'Istituto di Storia dell'Architettura*, n.s., 12 (1988): 65–84.

Chambers, D. S. *Renaissance Cardinals and Their Worldly Problems*. Aldershot, Hampshire: Variorum, Ashgate, 1997.

Chastel, André. *The Sack of Rome, 1527*. Translated by Beth Archer. Princeton, NJ: Princeton University Press, 1983.

Chastel, André, and Philippe Morel, eds. *La Villa Médicis*. Vol. 2, *Études*. Rome: Académie de France á Rome, 1991.

Chiabò, Maria, Giusi D'Alessandro, Paola Piacentini, and Concetta Ranieri, eds. *Alle origini della nuova Roma: Martino V (1417–1431)*. Atti del Convegno, Roma, March 2–5, 1992. Rome: Istituto Storico Italiano per il Medio Evo, 1992.

Chiari, Domenico. *Il territorio pontino in epoca sistina: Immagini di riforma e vita nello Stato della Chiesa, 1585–1590*. Terracina: Comune di Terracina, 1990.

Chiomenti Vassalli, Donata. *Paolo IV e il processo Carafa: Un caso d'ingiusta giustizia nel Cinquecento*. Milan: Gruppo Ugo Mursia, 1993.

Cibin, Ludovica. *Selciato romano: Il sampietrino; materiale, lavorazione, evoluzione storica, tipologie, apparecchiature, posa in opera*. Rome: Gangemi, 2003.

Cicconi, Maurizia. "'Bisognava un huomo valente': Ammannati versus Fontana nel carteggio di Guglielmo Sangalletti." In Curcio, Navone, and Villari, *Studi su Domenico Fontana*, 229–239.

Cipriani, Giovanni. *Gli obelischi egizi: Politica e cultura nella Roma barocca*. Florence: Leo S. Olschki, 1993.

Coarelli, Filippo. *La Colonna di Marco Aurelio / The Column of Marcus Aurelius*. Translated by Helen Patterson. Rome: Editore Colombo, 2008.

———. *La Colonna Traiana*. Rome: Colombo, 1999.

———. *Il Foro Romano*. 2 vols. Rome: Edizioni Quasar, 1983–1985.

Coffin, David R. *Gardens and Gardening in Papal Rome*. Princeton, NJ: Princeton University Press, 1991.

———. *Pirro Ligorio: The Renaissance Artist, Architect, and Antiquarian*. University Park: Pennsylvania State University, 2004.

———. *The Villa in the Life of Renaissance Rome*. Princeton, NJ: Princeton University Press, 1979.

Cohen, Elizabeth S. "Back Talk: Two Prostitutes' Voices from Rome c. 1600." *Early Modern Women* 2 (Fall 2007): 95–126.

———. "'Courtesans' and 'Whores': Words and Behavior in Roman Streets." In "Women in the Renaissance: An Interdisciplinary Forum (MLA 1981)," edited by Ann Rosiland Jones and Betty S. Travitsky. Special issue, *Women's Studies* 19 (1991): 201–208.

———. "The Early Accademia di San Luca and Artists in Rome: A Historian's Observations." In Lukehart, *Accademia Seminars*, 324–345.

———. "Honor and Gender in the Streets of Early Modern Rome." *Journal of Interdisciplinary History* 22 (Spring 1992): 595–625.

———. "Seen and Known: Prostitutes in the Cityscape of Late-Sixteenth-Century Rome." *Renaissance Studies* 12 (September 1998): 392–409.

———. "To Pray, To Work, To Hear, To Speak: Women in Roman Streets c. 1600." *Journal of Early Modern History* 12 (2008): 289–311.

Cohen, Elizabeth S., and Thomas V. Cohen. "Open and Shut: Social Meanings of the Cinquecento Roman House." *Studies in the Decorative Arts* 9 (Fall–Winter 2001–2002): 61–84.

Cohen, Thomas V. *Love and Death in Renaissance Italy*. Chicago: University of Chicago Press, 2004.

Cohen Thomas V., and Elizabeth S. Cohen. *Words and Deeds in Renaissance Rome*. Toronto: University of Toronto Press, 1993.

Cole, Michael W. "Perpetual Exorcism in Sistine Rome." In *The Idol in the Age of Art: Objects, Devotions and the Early Modern World*, edited by Michael W. Cole and Rebecca Zorach, 57–76. Farnham, Surrey: Ashgate, 2009.

Coldagelli, Umberto. "Boncompagni, Giacomo." *DBI*, 11:689–692.

"Colocci, Angelo." *DBI*, 27:105–111.

Colzi, Francesco. *Il debito pubblico del Campidoglio: Finanza comunale e circolazione dei titoli a Roma fra Cinque e Seicento*. Naples: Edizioni Scientifiche Italiane, 1999.

Conforti, Claudia. "Il cantiere di Michelangelo al Ponte Santa Maria a Roma (1548–49)." In Calabi and Conforti, *I ponti delle capitali d'Europa*, 75–87.

———. *La città del tardo Rinascimento*. Rome: Laterza, 2005.

———. "Modes and Techniques of Building on Water in 16th-Century Rome." In *Practice and Science in Early Modern Italian Building: Towards an Epistemic History of Architecture*, edited by Hermann Schlimme, 31–42. Milan: Electa, 2006.

———. "Roma: Architettura e città." In Conforti and Tuttle, *Storia*, 4:26–65.

———. "Via Pia: Rus in urbe." *Paragone: Arte*, 3rd ser., 82 (November 2008): 21–31.

Conforti, Claudia, and Andrew Hopkins, eds. *Architettura e tecnologia: Acque, tecniche e cantieri nell'architettura rinascimentale e barocca*. Rome: Nuova Argos, 2002.

Conforti, Claudia, and Richard J. Tuttle, eds. *Storia dell'architettura italiana*. Vol. 4, *Il secondo Cinquecento*. Milan: Electa, 2001.

Connors, Joseph. "Alliance and Enmity in Roman Baroque Urbanism." *Römisches Jahrbuch der Bibliotheca Hertziana* 25 (1989): 207–294. Translated by Marco Cupellaro as *Alleanze e inimicizie: L'urbanistica di Roma barocca* (Rome: Laterza, 2005).

———. *Borromini and the Roman Oratory: Style and Society*. New York: Architectural History Foundation; Cambridge, MA: MIT Press, 1980.

Cooper, Alix. "The Museum and the Book: The Metallotheca and the History of an Encyclopaedic Natural History in Early Modern Italy." *Journal of the History of Collections* 7 (1995): 1–23.

Cooperman, Bernard D. "Ethnicity and Institution Building among Jews in Early Modern Rome." *AJS Review: The Journal of the Association for Jewish Studies* 30 (April 2006): 119–145.

———. "Licenses, Cartels, and Kehila: Jewish Moneylending and the Struggle against Restraint of Trade in Early Modern Rome." In *Purchasing Power: The Economics of Modern Jewish History*, edited by Rebecca Kobrin and Adam Teller, 27–45. Philadelphia: University of Pennsylvania Press, 2015.

Corbo, Anna Maria. *Artisti e artigiani in Roma al tempo di Martino V e Eugenio IV*. Rome: De Luca, 1969.

Corsi, Stefano, and Pina Ragionieri, eds. *Speculum Romanae Magnificentiae: Roma nell'incisione del Cinquecento*. Florence: Mandragora, 2004.

Courtright, Nicola. *The Papacy and the Art of Reform in Sixteenth-Century Rome: Gregory XIII's Tower of the Winds in the Vatican*. Cambridge: Cambridge University Press, 2003.

Crawford, M. H. "Benedetto Egio and the Development of Greek Epigraphy." In *Antonio Agustin between Renaissance and Counter-Reform*, edited by M. H. Crawford, 133–153. London: Warburg Institute, 1993.

Crescentini, Claudio, and Antonio Martini, eds. *Le confraternite romane: Arte storia committenza.* Rome: Fondazione Marco Besso, 2000.

Crespi, Mario. "Bacci, Andrea." *DBI*, 5:29–30.

Crocco, Maurizio. *Roma, Via Felice da Sisto V a Paolo V.* Rome: Edizioni Kappa, 2002.

Culatti, Marcella. *Villa Montalto Negroni: Fortuna iconografica di un luogo perduto di Roma.* Venice: Istituto Veneto di Scienze, Lettere ed Arti, 2009.

Curcio, Giovanna, Nicola Navone, and Sergio Villari, eds. *Studi su Domenico Fontana.* Mendrisio: Fondazione Archivio del Moderno, 2011.

Curran, Brian A. "'De sacrarum litterarum aegyptiorum interpretatione': Reticence and Hubris in Hieroglyphic Studies of the Renaissance; Pierio Valeriano and Annius of Viterbo." *Memoirs of the American Academy in Rome* 43/44 (1998/1999): 139–182.

———. *The Egyptian Renaissance: The Afterlife of Ancient Egypt in Early Modern Italy.* Chicago: University of Chicago Press, 2007.

Curran, Brian A., and Anthony Grafton. "A Fifteenth-Century Site Report on the Vatican Obelisk." *Journal of the Warburg and Courtauld Institutes* 58 (1995): 234–248.

Curran, Brian A., Anthony Grafton, Pamela O. Long, and Benjamin Weiss. *Obelisk: A History.* Cambridge, MA: Burndy Library, 2009. Distributed by MIT Press.

Daly Davis, Margaret. "Andrea Palladio's 'L'Antichita di Roma' of 1554." *Pegasus: Berliner Beiträge zum Nachleben der Antike* 9 (2007): 151–192.

———. "Introduction." In *Libro di M. Pyrrho Ligori napolitano delle antichità di Roma.* Venice: Michele Tramezino, 1553. Facsimile of the first edition, with an introduction by Margaret Daly Davis. *Fontes* 9, July 16, 2003. http://archiv.ub.uni-heidelberg.de/artdok/volltexte/2008/562.

———. "Two Early 'Fundberichte': Lucio Fauno and the Study of Antiquities in Farnese Rome." In *Opere e giorni: Studi su mille anni di arte europea dedicati a Max Seidel*, edited by Klaus Bergdolt and Giorgio Bonsanti, 525–532. Venice: Marsilio, 2001.

D'Amico, John F. *Renaissance Humanism in Papal Rome: Humanists and Churchmen on the Eve of the Reformation.* Baltimore: Johns Hopkins University Press, 1983.

Dandelet, Thomas J. *Spanish Rome, 1500–1700.* New Haven, CT: Yale University Press, 2001.

Davies, Surekha. *Renaissance Ethnography and the Invention of the Human: New Worlds, Maps and Monsters.* Cambridge: Cambridge University Press, 2016.

De Caro, Gaspare. "Bosio, Giacomo." *DBI*, 13:261–264.

De Cavi, Sabina. *Architecture and Royal Presence: Domenico and Giulio Cesare Fontana in Spanish Naples (1592–1627).* Newcastle upon Tyne: Cambridge Scholars, 2009.

———. "La committenza spagnola di Domenico e Giulio Cesare Fontana (1592–1627)." In Curcio, Navone, and Villari, *Studi su Domenico Fontana*, 161–183.

De Certau, Michel. "Carlo Borromeo, santo." *DBI*, 20:260–269.

Delph, Ronald K. "From Venetian Visitor to Curial Humanist: The Development of Agostino Steuco's 'Counter'-Reformation Thought." *Renaissance Quarterly* 47(Spring 1994): 102–139.

———. "Polishing the Papal Image in the Counter-Reformation: The Case of Agostino Steuco." *Sixteenth Century Journal* 23 (Spring 1992): 35–47.

———. "*Renovatio, Reformatio*, and Humanist Ambition in Rome." In *Heresy, Culture, and Religion in Early Modern Italy: Contexts and Contestations*, edited by Ronald K. Delph, Michelle M. Fontaine, and John J. Martin, 73–92. Kirksville, MO: Truman State University Press, 2006.

Del Re, Niccolò. *La Curia Capitolina e tre altri antichi organi giudiziari romani.* Rome: Fondazione Marco Besso, 1993.

———. *La Curia Romana: Lineamenti storico-giuridici.* 4th ed. Vatican City: Libreria Editrice Vaticana, 1998.

———. "Luca Peto giureconsulto e magistrato capitolino (1512–1581)." In *Scritti in onore di Filippo Caraffa,* 309–337. Anagni: Istituto di Storia e di Arte del Lazio Meridionale, Centro di Anagni, 1986.

———. *Monsignor Governatore di Roma.* Rome: Istituto di Studi Romani Editore, 1972.

Delumeau, Jean. *Vie économique et sociale de Rome dans la seconde moitié du XVIe siècle.* 2 vols. Bibliothèque des Écoles française d'Athène et de Rome, 184. Paris: E. de Boccard, 1957–1959.

Dennis, Kimberly L. "Camilla Peretti, Sixtus V, and the Construction of Peretti Family Identity in Counter-Reformation Rome." *Sixteenth Century Journal* 43 (2012): 71–101.

———. "Rediscovering the Villa Montalto and the Patronage of Camilla Peretti." In *Wives, Widows, Mistresses, and Nuns in Early Modern Italy: Making the Invisible Visible through Art and Patronage,* edited by Katherine A. McIver, 55–73. Farnham, Surrey: Ashgate, 2012.

Denunzio, Antonio Ernesto. "Brevi considerazioni intorno a due ambiti della committenza dosiana: La corte farnesiana di Roma e quella vicereale di Napoli." In Barletti, *Giovan Antonio Dosio,* 139–145.

DeSilva, Jennifer Mara. "Taking Possession: Rituals, Space, and Authority." *Royal Studies Journal* 3 (2016): 1–17.

De Stefano, Francesco. "Appendice: La Basilica e la memoria di San Pietro." In Carandini and Carafa, *Atlante di Roma antica,* 1:583–587.

Deswarte-Rosa, Sylvie. "Le cardinal Giovanni Ricci de Montepulciano." In Chastel and Morel, *La Villa Médicis,* 2:110–169.

———. "Les gravures de monuments antiques d'Antonio Salamanca à l'origine du *Speculum Romanae Magnificentiae.*" *Annali di Architettura* 1 (1989): 47–62.

De Vincentiis, Amedeo. "Papato, stato e curia nel XV secolo: Il problema della discontinuità." *Storica* 8 (2002): 91–115.

Dey, Henrik W. *The Aurelian Wall and the Refashioning of Imperial Rome, AD 271–855.* Cambridge: Cambridge University Press, 2011.

Dibner, Bern. *Moving the Obelisks.* 1950. Reprint: Norwalk, CT: Burndy Library, 1991.

Di Carlantonio, Lorena. "I percorsi degli acquedotti e la trasformazione del paesaggio urbano." *Roma Moderna e Contemporanea* 17 (January–December 2009): 31–41.

Di Martino, Vittorio, Roswitha Di Martino, and Massimo Belati. Huc Tiber ascendit: *Le memorie delle inondazioni del Tevere a Roma.* Rome: Arbor Sapientiae, 2017.

Di Nepi, Serena. *Sopravvivere al ghetto: Per una storia sociale della comunità ebraica nella Roma del Cinquecento.* Rome: Viella, 2013.

Di Sivo, Michele. "Le costituzioni e i bandi di Sisto V: L'amministrazione della giustizia tra accentramento e crisi dello Stato Pontificio." *Archivi per la Storia* 4 (November 1991): 137–147.

Ditchfield, Simon. *Liturgy, Sanctity and History in Tridentine Italy: Pietro Maria Campi and the Preservation of the Particular.* Cambridge: Cambridge University Press, 1995.

———. "Reading Rome as a Sacred Landscape, c. 1586–1635." In *Sacred Space in Early Modern Europe,* edited by Will Coster and Andrew Spicer, 167–192. Cambridge: Cambridge University Press, 2005.

Di Teodoro, Francesco P. *Raffaello, Baldassar Castiglione e la lettera a Leone X*. Rev. ed. Bologna: Minerva, 2003.

Donato, Maria Pia, and Jill Kraye, eds. *Conflicting Duties: Science, Medicine and Religion in Rome, 1550–1750*. London: Warburg Institute, 2009.

D'Onofrio, Cesare. *Castel Sant'Angelo e Borgo tra Roma e papato*. Rome: Romana Società Editrice, 1978.

———. *Le fontane di Roma*. 3rd ed. Rome: Romana Società Editrice, 1986.

———. *Gli obelischi di Roma: Storia e urbanistica di una città dall'età antica al XX secolo*. 3rd ed. Rome: Romana Società Editrice, 1992.

———. *Il Tevere: L'isola tiberina, le inondazioni, i molini, i porti, le rive, i muraglioni, i ponti di Roma*. 4th ed. Rome: Romana Società Editrice, 1982.

Dooley, Brendan. *The Social History of Skepticism: Experience and Doubt in Early Modern Culture*. Baltimore: Johns Hopkins University Press, 1999.

Dorez, Léon. *La cour du Pape Paul III d'après le registres de la trésorerie secrete*. 2 vols. Paris: Librairie Ernest Leroux, 1932.

Dos Reis Souza Lima, Gaspar. "Repertorio delle fabbriche dei nuovi ordini religiosi." In Simoncini, *Roma: Le trasformazioni urbane nel Cinquecento*, 2:57–88.

Doti, Gerardo. "Laparelli, Francesco." *DBI*, 63:699–701.

Edgerton, David. *The Shock of the Old: Technology and Global History since 1900*. Oxford: Oxford University Press, 2007.

Enzi, Silvia. "Le inondazioni del Tevere a Roma tra il XVI e XVIII secolo nelle fonti bibliotecarie del tempo." *Mélanges de l'École française de Rome, Italie et Méditerranée* 118 (2006): 13–20.

Esposito, Anna. *Un'altra Roma: Minoranze nazionali e comunità ebraiche tra Medioevo e Rinascimento*. Rome: Il Calamo, 1995.

———. "La città e i suoi abitanti." In Pinelli, *Roma del Rinascimento*, 3–47.

———. "I 'diluvi' del Tevere tra '400 e '500." *Rivista Storica del Lazio* 10 (2002): 17–26.

———. "Estouteville (Tuttavilla), Guillaume (Guglielmo)." *DBI*, 43:456–460.

———. "Le inondazioni del Tevere tra tardo Medioevo e prima età moderna: Leggende, racconti, testimonianze." *Mélanges de l'École française de Rome, Italie et Méditerranée* 118 (2006): 7–12.

———. "Il Tevere e Roma." In *Le calamità ambientali nel tardo Medioevo europeo: Realtà, percezioni, reazioni*, edited by Michael Matheus, Gabriella Piccinni, Giuliano Pinto, and Gian Maria Varanini, 257–275. Atti del XII convegno del Centro di Studi sulla civiltà del tardo Medioevo, S. Miniato, May 21–June 2, 2008. Florence: Firenze University Press, 2010.

Evans, Harry B. *Water Distribution in Ancient Rome: The Evidence of Frontinus*. Ann Arbor: University of Michigan Press, 1994.

Fagan, Garrett G. *Bathing in Public in the Roman World*. Ann Arbor: University of Michigan Press, 1999.

Fagiolo, Marcello. "L'effimero di Stato: Dal conclave al possesso." In Fagiolo, *La festa a Roma*, 2:8–25.

———, ed. *La festa a Roma dal Rinascimento al 1870*. 2 vols. Rome: Umberto Allemandi, 1997.

———. "Intorno a S. Pietro in Montorio e a Giovanni Fontana: Opere viarie, panorami e fontane." In Fagiolo and Bonaccorso, *Studi sui Fontana*, 121–140.

———, ed. *Natura e artificio: L'ordine rustico, le fontane, gli automi nella cultura del Manierismo europeo*. Rome: Officina Edizioni, 1981.

———. "Piante di Roma antica e moderna: L'ideologia e i metodi di rappresentazione." In Bevilacqua and Fagiolo, *Piante di Roma*, 22–61.

———. "Il significato dell'acqua e la dialettica del giardino: Pirro Ligorio e la 'filosofia' della villa cinquecentesca." In Fagiolo, *Natura e artificio*, 176–189.

Fagiolo, Marcello, and Giuseppe Bonaccorso, eds. *Studi sui Fontana: Una dinastia di architetti ticinesi a Roma tra Manierismo e Barocco*. Rome: Gangemi Editore, 2008.

Fagiolo, Marcello, and Maria Luisa Madonna. "Il revival del trionfo classico: Da Alessandro VI alla sfilata dei Rioni." In Fagiolo, *La festa a Roma*, 1:34–49.

———. "La Roma di Pio IV. 2. Il sistema dei centri direzionali e la rifondazione della città." *Arte Illustrata* 6 (August 1973): 186–212.

———, eds. *Sisto V*. 2 vols. Rome: Istituto Poligrafico e Zecca dello Stato, Libreria dello Stato, 1992.

Fasano Guarini, Elena. "Ferdinando I de' Medici, granduca di Toscana." *DBI*, 46:258–278.

———. "'Rome, Workshop of All the Practices of the World': From the Letters of Cardinal Ferdinando de' Medici to Cosimo I and Francesco I." In Signorotto and Visceglia, *Court and Politics in Papal Rome*, 53–77.

Favro, Diane. "Obelisk Bones." *Seleucia* 8 (2018): 13–28.

Fea, Carlo. *Storia I: Delle acque antiche sorgenti in Roma perdute, e modo di ristabilirle; II: Dei condotti antico-moderni delle acque, Vergine, Felice, e Paola e loro autori*. Roma: Stamperia della Reverenda Camera Apostolica, 1832.

Feci, Simona. "Pio V, papa, santo." *DBI*, 83:814–825.

Federici, V. "Di Mario Cartaro, incisore viterbese del secolo XVI." *ASRSP* 21 (1898): 535–552.

Ferraro, Richard J. "The Nobility of Rome, 1500–1700: A Study of Its Composition, Wealth, and Investment." PhD diss. University of Wisconsin–Madison, 1994.

Ferrary, Jean-Louis. *Onofrio Panvinio et les antiquité romaines*. Rome: École française di Rome, 1996.

Filippi, Dunia. "Regione VIII: *Forum Romanum Magnum*." In Carandini and Carafa *Atlante di Roma antica*, 1:143–214.

Finocchi Ghersi, Lorenzo. "Francesco Fontana." In Fagiolo and Bonaccorso, *Studi sui Fontana*, 438–439.

Fiorani, Francesca. *The Marvel of Maps: Art, Cartography and Politics in Renaissance Italy*. New Haven, CT: Yale University Press, 2005.

Fiorani, Luigi. "'Charità et pietate': Confraternite e gruppi devoti nella città rinascimentale e barocca." In Fiorani and Prosperi, *Roma, la città del papa*, 429–476.

———. "Religione e povertà: Il dibattito sul pauperismo a Roma tra Cinque e Seicento." *Ricerche per la storia religiosa di Roma: Studi, documenti, inventari* 3 (1979): 43–131.

Fiorani, Luigi, and Adriano Prosperi, eds. *Roma, la città del papa: Vita civile e religiosa dal giubileo di Bonifacio VIII al giubileo di papa Wojtyla*. Storia d'Italia, 16. Turin: Giulio Einaudi, 2000.

Fiore, Francesco Paolo. "Rilievo topografico e architettura a grande scala nei disegni di Antonio da Sangallo il Giovane per le fortificazioni di Roma al tempo di papa Paolo III." In *Il disegno di architettura*, edited by Paolo Carpeggiani and Luciano Patetta, 175–180. Atti del Convegno Milano 15–18 febbraio 1988. Milan: Guerini and Associates, 1989.

Firpo, Luigi. "Botero, Giovanni." *DBI*, 13:352–362.

Foa, Anna, and Kenneth Stow. "Gli ebrei di Roma: Potere rituale e società in età moderna." In Fiorani and Prosperi, *Roma, la città del papa,* 555–581.

Fosi, Irene. "Conti, Torquato." *DBI,* 28:479–480.

———. "Court and City in the Ceremony of the *Possesso* in the Sixteenth Century." In Signorotto and Visceglia, *Court and Politics in Papal Rome,* 31–52.

———. *La giustizia del papa: Sudditi e tribunali nello Stato Pontificio in età moderna.* Rome: Laterza, 2007. Translated by Thomas V. Cohen as *Papal Justice: Subjects and Courts in the Papal State, 1500–1750* (Washington, DC: Catholic University of America Press, 2011).

———. "La Giustizia e la sua imagine: Propaganda politica e realtà nel pontificato sistino." In Fagiolo and Madonna, *Sisto V,* 1:59–82.

———. "'Parcere subiectis, debellare superbos': L'immagine della giustizia nelle cerimonie di possesso a Roma e nelle legazioni dello Stato Pontificio nel Cinquecento." In Visceglia and Brice, *Cérémonial et ritual à Rome,* 89–115.

———. *La società violenta: Il banditismo nello Stato Pontificio nella seconda metà del Cinquecento.* Rome: Edizioni dell'Ateneo, 1985.

Fragnito, Gigliola. "Le corti cardinalizie nella Roma del Cinquecento." *Rivista Storica Italiana* 106 (1994): 5–41.

———. "Ricci, Giovanni." *DBI,* 87:246–249.

Franzoni, Claudio. "'Urbe Roma in pristinam formam renascente': Le antichità di Roma durante il Rinascimento." In Pinelli, *Roma del Rinascimento,* 291–336.

Fratarcangeli, Margherita. "Il cavaliere Domenico Fontana: La 'robba' per la nobilità." In Curcio, Navone, and Villari, *Studi su Domenico Fontana,* 49–59.

———. "Giovanni Fontana e la sua stripe: Edifici d'acque e inondazioni del Tevere." In Fagiolo and Bonaccorso, *Studi sui Fontana,* 339–354.

Freiberg, Jack. "Pope Gregory XIII, Jurist." In "Art and Science in the Rome of Gregory XIII Boncompagni (1572–1585)." Special issue, *Memoirs of the American Academy in Rome* 54 (2009): 41–60.

Freudenberger, Theobald. *Augustinus Steuchus aus Gubbio, Augustinerchorherr und päpstlicher Bibliothekar (1497–1548) und sein literarisches Lebenswerk.* Münster in Westfalen: Aschendorffschen, 1935.

Friedman, David H. "La pianta di Imola di Leonardo, 1502." In *Rappresentare la città: Topografie urbane nell'Italia di antico regime,* edited by Marco Folin, 121–144. Reggio Emilia: Edizioni Diabasis, 2010.

Friedman, David H. with Paul Schlapobersky. "Leonardo Bufalini's Orthagonal *Roma* (1551)." *Thresholds* 28 (2003): 10–16.

Frommel, Christoph L. "Papal Policy: The Planning of Rome during the Renaissance." In *Art and History: Images and Their Meaning,* edited by Robert I. Rotberg and Theodore K. Rabb, 39–65. Cambridge: Cambridge University Press, 1988.

Frommel, Christoph L., and Nicholas Adams, eds. *The Architectural Drawings of Antonio da Sangallo the Younger and His Circle.* Vol. 1, *Fortification, Machines, and Festival Architecture.* New York: Architectural History Foundation; Cambridge, MA: MIT Press, 1994.

Frosini, Pietro. *Il Tevere: Le inondazioni di Roma e i provvedimenti presi dal governo italiano per evitarle.* Rome: Accademia Nazionale dei Lincei, 1977.

Fubini, Ricardo. "Biondo Flavio." *DBI,* 10:536–559.

Fumagalli, Elena. "La committenza cardinalizia a Roma." In Conforti and Tuttle, *Storia,* 94–107.

Funiciello, Renato, et al. ed. *La geologia di Roma: Il centro storico*. 2 vols. Rome: Istituto Poligrafico e Zecca dello Stato, 1995.

Gadol, Joan. *Leon Battista Alberti: Universal Man of the Early Renaissance*. Chicago: University of Chicago Press, 1969.

Galliazzo, Vittorio. *I ponti romani*. 2 vols. Treviso: Canova, 1994–1995.

Gamrath, Helge. Roma sancta renovata: *Studi sull'urbanistica di Roma nella seconda metà del sec. XVI con particolare riferimento al pontificato di Sisto V (1585–1590)*. Analecta Romana Instituti Danici, supp. 12. Rome: "L'Erma" di Bretschneider, 1987.

Gaston, Robert W. "'La mirabile natura dell'acque correnti': Pirro Ligorio's View of the Waters of the 'Old' World." In *La civiltà delle acque tra Medioevo e Rinascimento*, edited by Arturo Calzona and Daniela Lamberini, 1:75–87. Atti del Convegno internazionale, Mantova 1–4 Ottobre 2008. Florence: Leo S. Olschki, 2010.

———, ed. *Pirro Ligorio: Artist and Antiquarian*. Milan: Silvana, 1988.

———. "Pirro Ligorio, the Casino of Pius IV, and Antiques for the Medici: Some New Documents." *Journal of the Warburg and Courtauld Institutes* 47 (1984): 205–209.

Gautier Dalché, Patrick. *La géographie de Ptolémée en occident (IVᵉ–XVIᵉ siècle)*. Turnhout, Belgium: Brepols, 2009.

———. "The Reception of Ptolemy's *Geography* (End of the Fourteenth to Beginning of the Sixteenth Century)." In Woodward, *History of Cartography*, pt. 1, 285–364.

Gerbino, Anthony. *François Blondel: Architecture, Erudition, and the Scientific Revolution*. London: Routledge, 2010.

Gerbino, Anthony, and Stephen Johnston. *Compass and Rule: Architecture as Mathematical Practice in England*. New Haven, CT: Yale University Press, 2009.

Gersbach, Karl A. "The Books and Personal Effects of Young Onorfrio Panvinio, OSA, in Vat. Lat. 7205." *Analecta Augustiniana* 52 (1989): 51–76.

———. "Onofrio Panvinio, OSA, and His Florentine Correspondents Vincenzio Borghini, OSB, Pietro Vettori, Francesco de' Medici." *Analecta Augustiniana* 40 (1997): 207–280.

———. "Onofrio Panvinio's Brother Paolo, and His Role in the Posthumous Edition of the 'De primatu Petri et apostolicae sedis potestate' and the Purchase of Onofrio's Manuscripts for the Vatican Library." *Analecta Augustiniana* 56 (1993): 241–264.

Giehlow, Karl. *The Humanist Interpretation of Hieroglyphs in the Allegorical Studies of the Renaissance*. Edited and Translated by Robin Raybould. Leiden: Brill, 2015.

Giuliani, C. F., and P. Verduchi. "Forum Romanum (età tarda)." In Steinby, *Lexicon topographicum*, 2:342–343.

Gnoli, Umberto. *Topografia e toponomastica di Roma medioevale e moderna*. 2nd ed. Edited by Livio Jannattoni. Foligno: Edizioni dell'Arquata, 1984.

Goubert, Jean-Pierre. "The Development of Water and Sewerage Systems in France, 1850–1950." In Tarr and Dupuy, *Technology and the Rise of the Networked City*, 116–136.

Gouwens, Kenneth. *Remembering the Renaissance: Humanist Narratives of the Sack of Rome*. Leiden: Brill, 1998.

Grafton, Anthony. *Leon Battista Alberti: Master Builder of the Italian Renaissance*. New York: Hill and Wang, 2000.

Grazia Ercolino, Maria. "Lippi, Giovanni di Bartolomeo, detto Nanni di Baccio Bigio." *DBI*, 65:209–212.

Groppi, Angela. "Birbanti e poveri benestanti: Attitudini e pratiche assistenziali nei confronti della vecchiaia nella Roma pontificia (secc. XVI–XVIII)." In Zamagni, *Povertà e innovazioni istituzionali*, 259–277.

Gualandi, Maria Letizia. "'Roma resurgens': Fervore edilizio, trasformazioni urbanistiche e realizzazioni monumentali da Martino V Colonna a Paolo V Borghese." In Pinelli, *Roma del Rinascimento*, 123–160.

Guasco, Maurilio, and Angelo Torre, eds. *Pio V nella società e nella politica del suo tempo*. Bologna: Il Mulino, 2005.

Guidoni, Enrico. *L'urbanistica di Roma tra miti e progetti*. Rome: Laterza, 1990.

Guidoni, Enrico, Angela Marino, and Angela Lanconelli. "I 'Libri dei conti' di Domenico Fontana: I monumenti antichi: Il Settizonio, le colonne coclidi, i 'massicci' di Termini, i cavalli del Quirinale: I Libri XIII, XIV, XIX, XX." *Storia della Città* 43 (1988): 86–104.

———. "I 'Libri dei conti' di Domenico Fontana: Riepilogo generale delle spese e Libro I." *Storia della Città* 40 (1987): 44–81.

Guidoni, Enrico, and Giulia Petrucci. *Roma, Via Alessandrina: Una strada "tra due fondali" nell'Italia delle corti (1492–1499)*. Rome: Edizioni Kappa, 1997.

Guillerme, André. "The Genesis of Water Supply, Distribution, and Sewerage Systems in France, 1800–1850." In Tarr and Dupuy, *Technology and the Rise of the Networked City*, 91–115.

Hager, Helmut. "Fontana, Carlo." *DBI*, 48:624–636.

Hallman, Barbara McClung. *Italian Cardinals, Reform, and the Church as Property, 1492–1563*. Berkeley: University of California Press, 1985.

Harley, J. B. *The New Nature of Maps: Essays in the History of Cartography*. Edited by Paul Laxton. Baltimore: Johns Hopkins University Press, 2001.

Hart, Vaughan, and Peter Hicks, eds. *Paper Palaces: The Rise of the Renaissance Architectural Treatise*. New Haven, CT: Yale University Press, 1998.

Harvey, P. D. A. *The History of Topographical Maps: Symbols, Pictures, and Surveys*. London: Thames and Hudson, 1980.

Headley, John M. "Geography and Empire in the Late Renaissance: Botero's Assignment, Western Universalism, and the Civilizing Process." *Renaissance Quarterly* 53 (Winter 2000): 1119–1155.

Heiken, Grant, Renato Funiciello, and Donatella De Rita. *The Seven Hills of Rome: A Geological Tour of the Eternal City*. Princeton, NJ: Princeton University Press, 2005.

Helas, Philine, and Gerhard Wolf. "'E fece uno granni bene alla città di Roma': Considerazioni sulle opere di Martino V per la città di Roma." In *Martino V: Genazzano, il pontefice, le idealità; Studi in onore di Walter Brandmüller*, edited by Pierantonio Piatti and Rocco Ronzani, 219–240. Rome: Roma nel Rinascimento / Centro Culturale Agostiniano, 2009.

Henderson, John. "'Mal francese' in Sixteenth-Century Rome: The Ospedale di San Giacomo in Augusta and the 'incurabili.'" In Sonnino, *Popolazione e società a Roma*, 483–523.

Henninger-Voss, Mary. "Working Machines and Noble Mechanics: Guidobaldo del Monte and the Translation of Knowledge." *Isis* 91 (June 2000): 233–259.

Hibbard, Howard. "Di alcune licenze rilasciate dai mastri di strade per opera di edificazione a Roma (1586-'89, 1602–'34)." *Bollettino d'Arte*, 5th ser., 52 (1967): 99–117.

Hook, Judith. *The Sack of Rome, 1527*. London: Macmillan, 1972.

Höper, C. "Tempesta, Antonio." *DA*, 30:428–429.

Hopkins, John N. N. "The Cloaca Maxima and the Monumental Manipulation of Water in Archaic Rome." *Waters of Rome*, no. 4 (March 2007). http://www3.iath.virginia.edu/waters/first.html.

———. "'The 'Sacred Sewer': Tradition and Religion in the Cloaca Maxima." In Bradley and Stow, *Rome, Pollution and Propriety*, 81–102.

Hsia, R. Po-Chia. *The World of Catholic Renewal, 1540–1770*. Cambridge: Cambridge University Press, 1998.

Hughes, Thomas P. *Networks of Power: Electrification in Western Society, 1880–1930*. Baltimore: Johns Hopkins University Press, 1983.

Hülsen, Christian. "Piante icnografiche incise in marmo." *Mittheilungen des kaiserlich deutschen archäologischen Instituts: Römische Abtheilung* 5 (1890): 46–63.

———. *Saggio di bibliografia ragionata delle piante icnografiche e prospettiche di Roma dal 1551 al 1748*. 1915. Reprint, Rome: Bardi, 1969.

———. "Das *Speculum Romanae Magnificentiae* des Antonio Lafreri." In *Collectanea variae doctrinae Leoni S. Olschki*, 121–170. Munich: Jacque Rosenthal, 1921.

Hunt, John M. "Carriages, Violence, and Masculinity in Early Modern Rome." *I Tatti: Studies in the Italian Renaissance* 17 (Spring 2013): 175–196.

———. "The Ceremonial Possession of a City: Ambassadors and Their Carriages in Early Modern Rome." *Royal Studies Journal* 3 (2016): 69–89.

———. *The Vacant See in Early Modern Rome: A Social History of the Papal Interregnum*. Leiden: Brill, 2016.

Huppert, Ann C. *Becoming an Architect in Renaissance Italy: Art, Science, and the Career of Baldassarre Peruzzi*. New Haven, CT: Yale University Press, 2015.

———. "Mapping Ancient Rome in Bufalini's Plan and in Sixteenth-Century Drawings." *Memoirs of the American Academy in Rome* 53 (2008): 79–98.

Infelise, Mario. "Roman *avvisi*: Information and Politics in the Seventeenth Century." In Signorotto and Visceglia, *Court and Politics in Papal Rome*, 212–228.

———. *Prima dei giornali: Alle origini della pubblica informazione (secoli XVI e XVII)*. Rome: Laterza, 2002.

Ingersoll, Richard. "Piazza di Ponte and the Military Origins of Panopticism." In Çelik, Favro, and Ingersoll, *Streets: Critical Perspectives*, 177–188.

———. "The Ritual Use of Public Space in Renaissance Rome." PhD diss., University of California, Berkeley, 1985.

Insolera, Italo. *Roma: Immagini e realtà dal X al XX secolo*. 3rd ed. Rome: Laterza, 1985.

Ippoliti, Alessandro. "Fontana, Domenico." *DBI*, 48:638–643.

———. "Fontana, Giovanni." *DBI*, 48:676–677.

Israel, Uwe, and Gherardo Ortalli, eds. *Il duello fra Medioevo ed età moderna: Prospettive storico-culturali*. Rome: Viella, 2009.

Iversen, Erik. *Obelisks in Exile*. Vol. 1, *The Obelisks of Rome*. Copenhagen: GEC Gad, 1968.

Jacks, Philip. *The Antiquarian and the Myth of Antiquity: The Origins of Rome in Renaissance Thought*. Cambridge: Cambridge University Press, 1993.

Jansen, Dirk J. "Strada, (1) Jacopo Strada." *DA*, 29:737–740.

Jedin, Hubert. "Kardinal Giovanni Ricci (1497–1574)." *Lateranum*, n.s., 15 (1949): 269–358.

Jolivet, Vincent. "La localization des toponymes de la Rome antique à partir des Régionnaires: Une étude de cas." In Leone, Palombi, and Walker, *Res bene gestae*, 103–135.

Julia, Dominique. "L'accoglienza dei pellegrini a Roma." In Fiorani and Prosperi, *Roma, la città del papa*, 823–861.

Karmon, David. "Restoring the Ancient Water Supply System in Renaissance Rome: The Popes, the Civic Administration, and the Acqua Vergine." *Waters of Rome*, no. 3 (August 2005). http://www3.iath.virginia.edu/waters/first.html.

———. *The Ruin of the Eternal City: Antiquity and Preservation in Renaissance Rome*. New York: Oxford University Press, 2011.

Karrow, Robert W., Jr. *Mapmakers of the Sixteenth Century and Their Maps*. Chicago: Speculum Orbis for the Newberry Library, 1993.

Keller, Fritz-Eugen. "Une villa de la Renaissance sur le site d'une villa antique." In Chastel and Morel, *La Villa Médicis*, 2:64–77.

Kelly, J. N. D. *The Oxford Dictionary of Popes*. Oxford: Oxford University Press, 1986.

Keyvanian, Carla L. "Charity, Architecture and Urban Development in Post-Tridentine Rome: The Hospital of the SS.ma Trinità dei Pellegrini e Convalescenti (1548–1680)." PhD diss., Massachusetts Institute of Technology, 2000.

———. *Hospitals and Urbanism in Rome, 1200–1500*. Leiden: Brill, 2015.

Koloski-Ostrow, Ann Olga. *The Archaeology of Sanitation in Roman Italy: Toilets, Sewers, and Water Systems*. Chapel Hill: University of North Carolina Press, 2015.

Knaebel, Georges. "Historical Origins and Development of a Sewerage System in a German City: Bielefeld, 1850–1904." In Tarr and Dupuy, *Technology and the Rise of the Networked City*, 186–206.

Krautheimer, Richard. *The Rome of Alexander VII, 1655–1667*. Princeton, NJ: Princeton University Press, 1985.

Kruft, Hanno-Walter. "Guglielmo della Porta." *DA*, 25:255–257.

Kurzel-Runtscheiner, Monica. *Töchter der Venus: Die Kurtisanen Roms in 16. Jahrhundert*. Munich: C. H. Beck, 1995.

Lamberini, Daniela. "Francesco De Marchi." *DBI*, 38:447–454.

———. "Francesco De Marchi: Ritratto di un cortigiano del Cinquecento, virtuoso e dilettante di architettura militare." *Storia Architettura* 10 (1987): 69–88.

Lancaster, Lynne C. "Building Trajan's Column." *American Journal of Archaeology* 103 (July 1999): 419–439.

———. *Concrete Vaulted Construction in Imperial Rome: Innovations in Context*. Cambridge: Cambridge University Press, 2005.

Lanciani, Rodolfo. *I comentarii di Frontino intorno le acque e gli aquedotti*. Rome: Salviucci, 1880.

———. *Storia degli scavi di Roma e notizie intorno le collezioni romane di antichità*. Edited by Leonello Malvezzi Campeggi. 7 vols. Rome: Quasar, 1980–2002.

Landau, David, and Peter Parshall. *The Renaissance Print, 1470–1550*. New Haven, CT: Yale University Press, 1994.

Laureys, Marc. "Bartolomeo Marliano (1488–1566): Ein Antiquar des 16. Jahrhunderts." In *Antiquarische Gelehrsamkeit und Bildende Kunst: Die Gegenwart der Antike in der Renaissance*, edited by Gunter Schweikhart, 151–167. Cologne: Walther König, 1996.

———. "Marliano (Bartolomeo) (1488–1566)." In *Centuriae Latinae*, vol. 2, *Cent une figures humanistes de la Renaissance aux Lumières*, edited by Colette Nativel, 499–503, Geneva: Librairie Droz, 2006.

Laureys, Marc, and Anna Schreurs. "Egio, Marliano, Ligorio, and the Forum Romanum in the 16th Century." *Humanistica Lovaniensia* 45 (1996): 385–405.

Lazar, Lance Gabriel. *Working in the Vineyard of the Lord: Jesuit Confraternities in Early Modern Italy*. Toronto: University of Toronto Press, 2005.

Lazzaro, Claudia. *The Italian Renaissance Garden: From the Conventions of Planting, Design, and Ornament to the Grand Gardens of Sixteenth-Century Central Italy*. New Haven, CT: Yale University Press, 1990.

Lee, Egmont, ed. Descriptio urbis: *The Roman Census of 1527*. Rome: Bulzoni, 1985.

———. *Sixtus IV and Men of Letters*. Rome: Edizioni di Storia e Letteratura, 1987.

Lefevre, Renato. "La costituzione edilizia romana del 1574: Contenuto, valore e applicazione." *Economia e Storia* 19 (1972): 20–39.

———. "Schede su Matteo Bartolini da Castello architetto in Roma nel tardo Cinquecento." *Rassegna degli Archivi di Stato* 27 (1967): 142–160.

Le Gall, Joël. *Le Tibre, fleuve de Rome dans l'antiquité*. Paris: Presses Universitaires de France, 1953.

Leicht, Pier Silverio. "L'editore veneziano Michele Tramezino ed i suoi privilegi." In *Miscellanea di scritti di bibliografia ed erudizione in memoria di Luigi Ferrari*, 357–367. Florence: Leo S. Olschki, 1952.

Lemaître, Nicole. *Saint Pie V*. [Paris]: Fayard, 1994.

Lenzo, Fulvio. "'Che cosa è architetto': La polemica con gli ingegneri napoletani e l'edizione del *Libro secondo*." In Curcio, Navone, and Villari, *Studi su Domenico Fontana*, 265–287.

Leone, Anna, Domenico Palombi, and Susan Walker, eds. Res bene gestae: *Ricerche di storia urbana su Roma antica in onore di Eva Margareta Steinby*. Lexicon Topographicum Urbis Romae, supp. 4. Rome: Edizioni Quasar di Severino Tognon Sil, 2007.

Lepper, Frank, and Sheppard Frere. *Trajan's Column: A New Edition of the Cichorius Plates*. Gloucester, UK: Alan Sutton, 1988.

Leuschner, Eckhard. *Antonio Tempesta: Ein Bahnbrecher des römischen Barock und seine europäische Wirkung*. Petersberg: Michael Imhof, 2005.

———. "Prolegomena to a Study of Antonio Tempesta's 'Map of Rome.'" In Bevilacqua and Fagiolo, *Piante di Roma*, 158–167.

"Ligorio, Pirro." *DBI*, 65:109–114.

Lincoln, Evelyn. *Brilliant Discourse: Pictures and Readers in Early Modern Rome*. New Haven, CT: Yale University Press, 2014.

———. *The Invention of the Italian Renaissance Printmaker*. New Haven, CT: Yale University Press, 2000.

Lindgren, Uta. "Land Surveys, Instruments, and Practitioners in the Renaissance." In Woodward, *History of Cartography*, pt. 1, 407–508.

Liverani, Paolo. *La topografia antica del Vaticano*. Vatican City: Monumenti, Musei e Gallerie Pontificie, 1999.

Loffredo, Fernando. "La villa di Pedro de Toledo a Pozzuoli e una sicura provenienza per il *Fiume* di Pierino da Vinci al Louvre." *Rinascimento Meridionale* 2 (2011): 93–113.

Lombardi, Leonardo. "Camillo Agrippa's Hydraulic Inventions on the Pincian Hill (1574–1578)." *Waters of Rome*, no. 5 (April 2008). www3.iath.virginia.edu/waters/first.html.

———. "Geologia, idrogeologia e idrologia del bacino dello Spinon." In Bianchi, *La Cloaca Maxima*, 68–73.

Lombardi, Sergio. "Matteo Castello (Castelli)." In Madonna, *Roma di Sisto V*, 554.

Lombardo, Maria Luisa. "Le gabelle della città di Roma nel quadro dell'attività amministrativo-finanziaria della Camera Urbis nel secolo XV." In *Economia e società a*

Roma tra Medioevo e Rinascimento: Studi dedicati ad Arnold Esch, edited by Anna Esposito and Luciano Palermo, 205–228. Rome: Vielli, 2005.

Long, Pamela O. *Artisan/Practitioners and the Rise of the New Sciences, 1400–1600*. Corvallis: Oregon State University Press, 2011.

———. "The Contribution of Architectural Writers to a 'Scientific Outlook' in the Fifteenth and Sixteenth Centuries." *Journal of Medieval and Renaissance Studies* 15 (Fall 1985): 265–298.

———. "The Craft of Premodern European History of Technology: Past and Future Practice." *Technology and Culture* 51 (July 2010): 698–714.

———. "Hydraulic Engineering and the Study of Antiquity: Rome, 1557–70." *Renaissance Quarterly* 61 (Winter 2008): 1098–1138.

———. "Multi-tasking 'Pre-professional' Architect/Engineers and Other Bricolagic Practitioners as Key Figures in the Elision of Boundaries between Practice and Learning in Sixteenth-Century Europe: Some Roman Examples." In *The Structures of Practical Knowledge*, edited by Matteo Valleriani, 223–246. Dordrecht: Springer, 2017.

———. *Openness, Secrecy, Authorship: Technical Arts and the Culture of Knowledge from Antiquity to the Renaissance*. Baltimore: Johns Hopkins University Press, 2001.

———. "Trading Zones in Early Modern Europe." *Isis* 106 (December 2015): 840–847.

Lorizzo, Loredana. "Nicolas van Aelst's Will and a List of His Plates." *Print Quarterly* 31 (2014): 3–20.

Losito, Maria. *Pirro Ligorio e il Casino di Paolo IV in Vaticano: L'"essempio" delle "cose passate."* Rome: Palombi, 2000.

Lotz, Wolfgang. "Gli 883 cocchi della Roma del 1594." In *Studi offerti a Giovanni Incisa della Rocchetta*, 247–266. Rome: Società Romana di Storia Patria, 1973.

Lowry, Bates. "Notes on the *Speculum Romanae Magnificentiae* and Related Publications." *Art Bulletin* 34 (March 1952): 46–50.

Luciani, Roberto, ed., *Roma sotterranea: Porta San Sebastiano, 15 ottobre 1984–14 gennaio 1985*. Rome: Palombi, 1984.

Lukehart, Peter M., ed. *The Accademia Seminars: The Accademia di San Luca in Rome, c. 1590–1635*. Washington, DC: National Gallery of Art, 2009.

Lurin, Emmanuel. "Etienne Dupérac, graveur, peintre et architecte (vers 1535?–1604): Une artiste-antiquaire entre L'Italie et la France." PhD diss., L'Université Paris IV, Sorbonne, 2006.

———. "Étienne Dupérac vedutista e cartografo: La costruzione della pianta di Roma del 1577." In *La città dei cartografi: Studi e ricerche di storia urbana*, edited by Cesare De Seta, Brigitte Marin, and Marco Iuliano, 49–59. Naples: Electa Napoli, 2008.

———. "Les restitutions de scènes antiques: Onofrio Panvinio iconographe et inventeur d'images." In *Programme et invention dans l'art de la Renaissance*, edited by Michele Hochmann, Julian Kliemann, Jérémie Koering, and Philippe Morel, 153–173. Rome: Académie de France à Rome, 2008.

Madonna, Maria Luisa. "Il Genius Loci di Villa d'Este: Miti e misteri nel sistema di Pirro Ligorio." In Fagiolo, *Natura e artificio*, 190–226.

———. "L'ingresso di Carlo V a Roma." In Fagiolo, *La festa a Roma*. 1:50–65.

———, ed. *Roma di Sisto V: Le arti e la cultura*. Rome: Edizioni De Luca, 1993.

Maffioli, Cesare S. *La via delle acque (1500–1700): Appropriazione delle arti e trasformazione delle matematiche*. Florence: Leo S. Olschki, 2010.

Maier, Jessica. "Leonardo Bufalini e la prima pianta a stampa di Roma, 'la più bella di tutte le cose.'" In Bevelacqua and Fagiolo, *Piante di Roma*, 117–127.
———. "Mapping Past and Present: Leonardo Bufalini's Plan of Rome (1551)." *Imago Mundi* 59 (April 2007): 1–23.
———. *Rome Measured and Imagined: Early Modern Maps of the Eternal City*. Chicago: University of Chicago Press, 2015.
———. "*Roma Renascens*: Sixteenth-Century Maps of the Eternal City." In *Rome: Continuing Encounters between Past and Present*, edited by Dorigen Caldwell and Lesley Caldwell, 35–55. Surrey: Ashgate, 2011.
———. "A 'True Likeness': The Renaissance City Portrait." *Renaissance Quarterly* 65 (Fall 2012): 711–752.
Mandel, Corinne. "Simbolismo ermetico negli obelischi e colonne della Roma sistina." In Fagiolo and Madonna, *Sisto V*, 1:659–692.
Mandowsky, Erna, and Charles Mitchell, eds. *Pirro Ligorio's Roman Antiquities: The Drawings in MS XIII, B.7 in the National Library in Naples*. London: Warburg Institute, University of London, 1963.
Manfredi, Tommaso. "L'architetto sottomaestro delle strade." In In urbe architectus: *Modelli, disegni, misure; La professione dell'architetto Roma, 1680–1750*, edited by Bruno Contardi and Giovanna Curcio, 281–289. Rome: Àrgos, 1991.
———. *La costruzione dell'architetto: Maderno, Borromini, i Fontana e la formazione degli architetti ticinesi a Roma*. Rome: Argos, 2008.
Mangiasciutto, Silvia. "Giovanni Fontana." In Fagiolo and Bonaccorso, *Studi sui Fontana*, 419–420.
Marciano, Antonella. "Disegno e rappresentazione tra antico e moderno." In Barletti, *Giovan Antonio Dosio*, 75–113.
———. *Giovanni Antonio Dosio: Fra disegno dell'antico e progetto*. Naples: La Scuola di Pitagora, 2008.
Marciari, John. "Artistic Practice in Late Cinquecento Rome and Girolamo Muziano's Accademia di San Luca." In Lukehart, *Accademia Seminars*, 196–223.
Marconi, Nicoletta. *Castelli e ponti: Apparati per il restauro nell'opera di maestro Nicola Zabaglia per la fabbrica di San Pietro in Vaticano*. Foligno: Il Formichiere, 2015.
———. *Edificando Roma barocca: Macchine, apparati, maestranze e cantieri tra XVI e XVIII secolo*. Città di Castello: Edimond, 2004.
———. "*Genitor urbis ad usum fabricae*: Il trasporto fluviale dei materiali per l'edilizia nella Roma del Cinquecento." In Bonaccorso, "Le acque e la città," 143–166.
———. "I ponti: Teoria e pratica nei trattati di architettura tra XVI e XVIII secolo." In Calabi and Conforti, *I ponti delle capitali d'Europa*, 38–57.
Marconi, Paolo. "Contributo alla storia delle fortificazioni di Roma nel Cinquecento e nel Seicento." *Quaderni dell'Istituto di Storia dell'Architettura*, 13th ser., fasc. 73–78 (1966): 109–130.
Marder, Tod A. "Sisto V e la Fontana del Mosè." In Fagiolo and Madonna, *Sisto V*, 1:519–543.
———. "Sixtus V and the Quirinal." *Journal of the Society of Architectural Historians* 37 (December 1978): 283–294.
Marino, John. "Administrative Mapping in the Italian States." In *Monarchs, Ministers, and Maps: The Emergence of Cartography as a Tool of Government in Early Modern Europe*, edited by David Buisseret, 5–25. Chicago: University of Chicago Press, 1992.

Marr, Alexander. *Between Raphael and Galileo: Mutio Oddi and the Mathematical Culture of Late Renaissance Italy*. Chicago: University of Chicago Press, 2011.

Marta, Roberto. *Tecnica costruttiva a Roma nel Medioevo/Construction Techniques of the Middle Ages in Rome*. Rome: Edizione Kappa, 1989.

Martin, Craig. *Renaissance Meteorology: Pomponazzi to Descartes*. Baltimore: Johns Hopkins University Press, 2011.

Martines, Giangiacomo, ed. "La Colonna Antonina." In *Via del Corso: Una strada lunga 2000 anni*, edited by Cesare D'Onofrio, 111–149. Rome: Edizioni de Luca, 1999.

Martini, Antonio. *Arti, mestieri e fede nella Roma dei papi*. Bologna: Casa Editrice Licinio Cappelli, 1965.

Mascilli Migliorini, Paolo. "Il Palazzo Reale di Napoli e la città vicereale." In Curcio, Navone, and Villari, *Studi su Domenico Fontana*, 185–195.

Matteini, Federica. "Orsini, Fulvio." *DBI*, 79:649–653.

Mazzetti, Ernesto, ed. *Cartografia generale del Mezzogiorno e della Sicilia*. 2 vols. Naples: Edizioni Scientifiche Italiane, 1972.

McCahill, Elizabeth. *Reviving the Eternal City: Rome and the Papal Court, 1420–1447*. Cambridge, MA: Harvard University Press, 2013.

McCuaig, William. "The Fasti Capitolini and the Study of Roman Chronology in the Sixteenth Century." *Athenaeum* 79 (1991): 141–159.

McGinniss, Lawrence R., and Herbert Mitchell. *Catalogue of the Earl of Crawford's "Speculum Romanae Magnificentiae" Now in the Avery Architectural Library*. New York: Avery Architectural Library, Columbia University, 1976.

Megna, Laura. "'Acque et immonditie del fiume': Inondazioni del Tevere e smaltimento dei rifiuti a Roma tra Cinque e Settecento." *Mélanges de l'École française de Rome, Italie et Méditerranée* 118 (2006): 21–34.

Melosi, Martin V. *The Sanitary City: Urban Infrastructure in America from Colonial Times to the Present*. Baltimore: Johns Hopkins University Press, 2000.

Meneghini, Roberto. "La cartografia antica e il catasto di Roma imperiale." In Leone, Palombi, and Walker, *Res bene gestae*, 205–218.

Menniti Ippolito, Antonio. *I papi al Quirinale: Il sovrano pontefice e la ricerca di una residenza*. Rome: Viella, 2004.

Merrill, Elizabeth. "The *Professione di Architetto* in Renaissance Italy." *Journal of the Society of Architectural Historians* 76 (March 2017): 13–35.

Milano, Attilio. *Il ghetto di Roma*. Rome: Staderini, 1964.

Miller, Naomi. *Mapping the City: The Language and Culture of Cartography in the Renaissance*. London: Continuum, 2003.

Miller, Peter N., ed. *Momigliano and Antiquarianism: Foundations of the Modern Cultural Sciences*. Toronto: University of Toronto Press, 2014.

Millon, Henry A., Craig Hugh Smyth, and Francesca Consagra. "The Project for the Castel Sant'Angelo in the Dyson Perrins Codex." In Striker, *Architectural Studies*, 111–117.

Miranda, Salvador. *The Cardinals of the Holy Roman Church*. Miami: Florida International University Library, 1998–2015. http://www2.fiu.edu/~mirandas/cardinals.htm.

———. "Cesi, *seniore*, Pierdonato." In Miranda, *Cardinals*.

———. "Cueva Y Toledo, Bartolomé de la." In Miranda, *Cardinals*.

———. "Serbelloni, Giovanni Antonio," In Miranda, *Cardinals*.

———. "Sforza, Alessandro." In Miranda, *Cardinals*.

———. "Sforza di Santa Fiora, Guido Ascanio." In Miranda, *Cardinals.*

Mirri, Edoardo, ed. *Francesco Laparelli: Cortonese Architect at Malta.* Cortona: Tiphys Edizioni, 2009.

Misiti, Maria Cristina. "Antonio Salamanca: Qualche chiarimento biografico alla luce di un'indagine sulla presenza spagnola a Roma nel '500." In *La stampa in Italia nel Cinquecento,* edited by Marco Santoro, 1:545–563. Atti del Convegno, Rome, October 17–21, 1989. Rome: Bulzoni, 1992.

Mocchegiani Carpano, Claudio. "Le cloache dell'antica Roma." In Luciani, *Roma sotterranea,* 164–178.

Modigliani, Anna. "Artigiani e botteghe nella città." In Chiabò et al., *Alle origini della nuova Roma,* 455–477.

———. *Disegni sulla città nel primo Rinascimento romano: Paolo II.* Rome: Roma nel Rinascimento, 2009.

———. "Giganti, Girolamo." *DBI,* 54:663–666.

Momigliano, Arnaldo. *The Classical Foundations of Modern Historiography.* Berkeley: University of California Press, 1990.

Montagu, Jennifer. *Roman Baroque Sculpture: The Industry of Art.* New Haven, CT: Yale University Press, 1989.

Motta, Rossella. "L'acquedotto Felice." In *Il trionfo dell'acqua,* 220–225.

———. "La decadenza degli acquedotti antichi e la conduzione dell'acqua Mariana." In *Il trionfo dell'acqua,* 203–205.

Moyer, Gordon. "Aloisius Lilius and the 'Compendium Novae Rationis Restituendi Kalendarium.'" In *Gregorian Reform of the Calendar: Proceedings of the Vatican Conference to Commemorate Its 400th Anniversary, 1582–1982,* edited by G. V. Coyne, M. A. Hoskin, and O. Pedersen, 171–188. Vatican City: Specola Vaticana, 1983.

Muecke, Frances. "Humanists in the Roman Forum." *Papers of the British School at Rome* 71 (2003): 207–233.

Nanni, Stefania, and Maria Antonietta Visceglia, eds. "La città del perdono: Pellegrinaggi e anni santi a Roma in età moderna, 1550–1750." Special issue, *Roma Moderna e Contemporanea* 5 (May–December 1997).

Narducci, Enrico. "Di alcuni lavori inediti e sconosciuti di Bartolomeo Marliani." *Atti della R. Accademia dei Lincei, Transunti,* 3rd ser., 8 (1884): 188–190.

Narducci, Pietro. *Sulla fognatura della Città di Roma.* 2 vols. Rome: Forzani, 1889.

Narducci, Roberto. "La riscoperta della rete fognaria romana e la documentazione grafica di fine '800 illustrata dall' ing. cav. Pietro Narducci." In Bianchi, *La Cloaca Maxima,* 42–51.

Navone, Nicola. "Alle origini dell'impresa Fontana." In Curcio, Navone, and Villari, *Studi su Domenico Fontana,* 61–73.

Nenci, Elio. "Camillo Agrippa: Un ingegnere rinascimentale di fronte ai problema della filosofia naturale." *Physis,* n.s., 29 (1992): 71–119.

Nesselrath, Arnold. "Raphael's Archaeological Method." In *Raffaello a Roma,* 357–371.

Ng, Morgan. "New Light on Francesco De Marchi (1504–1576) and His Treatise on Fortification." *Mitteilungen des kunsthistorischen Institutes in Florenz* 58 (2016): 403–410.

Nicolazzo, Vittorio. *Acqua Vergine a Roma: Acquedotti e fontane.* Rome: Colosseo Grafica, 1999.

Nuovo, Angela. *The Book Trade in the Italian Renaissance.* Leiden: Brill, 2013.

———. "Una lettera di Michele Tramezino a Giovanni Bartolomeo Gabiano (1522)." *La Bibliofilia* 115 (2013): 147–156.

Nussdorfer, Laurie. *Civic Politics in the Rome of Urban VIII*. Princeton, NJ: Princeton University Press, 1992.

———. "Men at Home in Baroque Rome," *I Tatti Studies in the Italian Renaissance* 17 (2014): 103–129.

———. "Notaries and the Accademia di San Luca, 1590–1630." In Lukehart, *Accademia Seminars*, 54–67.

———. "The Politics of Space in Early Modern Rome." *Memoirs of the American Academy in Rome* 42 (1997): 161–181.

———. "Il 'popolo romano' e i papi: La vita politica della capitale religiosa." In Fiorani and Prosperi, *Roma, la città del papa*, 239–260.

———. "Priestly Rulers, Male Subjects: Swords and Courts in Papal Rome." In *Violent Masculinities: Male Aggression in Early Modern Texts and Culture*, edited by Jennifer Feather and Catherine E. Thomas, 109–128. New York: Palgrave MacMillan, 2013.

———. "The Vacant See: Ritual and Protest in Early Modern Rome." *Sixteenth Century Journal* 18 (Summer 1987): 173–189.

Nuti, Lucia. "The Perspective Plan in the Sixteenth Century: The Invention of a Representational Language." *Art Bulletin* 76 (March 1994): 105–128.

———. *Ritratti di città: Visione e memoria tra Medioevo e Settecento*. Venice: Marsilio, 1996.

Oberhummer, Eugen. "Der Stadt Plan: Seine Entwickelung und geographische Bedeutung." In Oberhummer, *Verhandlungen des XVI Deutschen Geographentages zu Nürenberg*, 66–101. Berlin: Dietrich Reimer, 1907.

Occhipinti, Carmelo. *Pirro Ligorio e la storia cristiana di Roma da Costantino all'Umanesimo*. Pisa: Scuola Normale Superiore, 2007.

O'Malley, John W. *The First Jesuits*. Cambridge, MA: Harvard University Press, 1993.

———. *Trent and All That: Renaming Catholicism in the Early Modern Era*. Cambridge, MA: Harvard University Press, 2000.

———. *Trent: What Happened at the Council*. Cambridge, MA: Belknap Press of Harvard University Press, 2013.

Orbaan, J. A. F. "Il caso Fontana," *Bollettino d'Arte*, 1st ser., 9 (1915): 165–168.

———. "Die Selbstverteidigung des Domenico Fontana, 1592–1593." *Repertorium für Kunstwissenschaft* 46 (1925): 174–189.

Ostrow, Steven F. "The Discourse of Failure in Seventeenth-Century Rome: Prospero Bresciano's *Moses*." *Art Bulletin* 88 (June 2006): 267–291.

Pace, Pietrantonio. *Gli acquedotti di Roma e il* De Aquaeductu *di Frontino*. Rome: Art Studio S. Eligio, 1983.

Pagani, Valeria. "Documents on Antonio Salamanca." *Print Quarterly* 17 (June 2000): 148–155.

———. "Guido Roberti and Francesco Salamanca." *Print Quarterly* 30 (September 2013): 259–272.

Paglia, Vincenzo. *La morte confortata: Riti della paura e mentalità religiosa a Roma nell'età moderna*. Rome: Edizioni di Storia e Letteratura, 1982.

Pagliara, Pier Nicola. "Le Pincio du XVe au XVIe siècle et la *vigna* de Marcello Crescenzi." In Chastel and Morel, *La Villa Médicis*, 2:92–109.

———. "Vitruvio da testo a canone." In *Memoria dell'antico nell'arte italiana.* Vol. 3, *Dalla tradizione all'archeologia*, edited by Salvatore Settis, 3–85. Turin: Giulio Einaudi, 1986.

Palagiano, Cosimo. "Bufalini, Leonardo." *DBI*, 14:798–799.

Palazzini, Pietro. "Le congregazioni romane da Sisto V a Giovanni Paolo II." In Fagiolo and Madonna, *Sisto V*, 1:19–38.

Palermo, Luciano. "Economia." in Pinelli, *Roma del Rinascimento*, 49–91.

Palma Venetucci, Beatrice. "Pirro Ligorio and the Rediscovery of Antiquity." In *The Rediscovery of Antiquity: The Role of the Artist*, edited by Jane Fejfer, Tobias Fischer-Hansen, and Annette Rathje, 63–68. Copenhagen: Museum Tusculanum Press, University of Copenhagen, 2003.

Palmerio, Giancarlo. "Il reimpiego nella costruzione dell'acquedotto Felice a Roma." In *Il reimpiego in architettura: Recupero, trasformazione, uso*, edited by Jean-François Bernard, Philippe Bernardi, and Daniela Esposito, 373–393. Rome: École française de Rome and "Sapienza," Università di Roma, 2008.

Palumbo, Genoveffa. "'L'assedio delle reliquie' alla città di Roma: Le reliquie oltre la devozione nello sguardo dei pellegrini." In Nanni and Visceglia, "La città del perdono," 377–403.

Papoulia, Evangelia. "Unveiling Gregorian Rome: The Urban and Ecclesiastical Patronage of Pope Gregory XIII, 1572–1585." PhD diss., Courtauld Institute of Art, University of London, 2015.

Parshall, Peter. "Antonio Lafreri's *Speculum Romanae Magnificentiae.*" *Print Quarterly* 23 (March 2006): 3–28.

———. "Prints as Objects of Consumption in Early Modern Europe." *Journal of Medieval and Early Modern Studies* 28 (Winter 1998): 19–36.

Parsons, William Barclay. *Engineers and Engineering in the Renaissance.* Cambridge, MA: MIT Press, 1968.

Partner, Peter. "Papal Financial Policy in the Renaissance and Counter-Reformation." *Past and Present* 88 (August 1980): 17–62.

———. *The Papal State under Martin V: The Administration and Government of the Temporal Power in the Early Fifteenth Century.* London: British School at Rome, 1958.

———. *The Pope's Men: The Papal Civil Service in the Renaissance.* Oxford: Clarendon Press, 1990.

———. *Renaissance Rome: 1500–1559: A Portrait of a Society.* Berkeley: University of California Press, 1976.

Pascucci, Simonetta. "L'acqua corrente nel Cinquecento a Roma: Fontane, fontanelle e abbeveratoi." In Bonaccorso, "Le acque e la città," 43–72.

Passamani, Bruno. "Cavalieri (Cavalleri, Cavallieri, De; Cavalieri, De Cavalleri, De Cavaleriis), Giovanni Battista." *DBI*, 22:673–675.

Pastore, Alessandro. "Il problema dei poveri agli inizi dell'età moderna: Linee generali." In Zamagni, *Povertà e innovazioni istituzionali*, 185–205.

Pastor, Ludwig von. *The History of the Popes from the Close of the Middle Ages.* Translated by F. I. Antrobus, R. F. Kerr, et al. 40 vols. London: Kegan Paul, Trench, Trubner; St. Louis, MO: B. Herder, 1891–1953.

Pastura Ruggiero, Maria Grazia. "La fiscalità pontificia nel Cinquecento: Aspetti e problemi." In Fagiolo and Madonna, *Sisto V*, 1:211–231.

———. *La Reverenda Camera Apostolica e i suoi archivi (secoli XV–XVIII)*. Rome: Archivio di Stato, Scuola di Archivistica Paleografia e Diplomatica, 1984.

Pattenden, Miles. *Pius IV and the Fall of the Carafa: Nepotism and Papal Authority in Counter-Reformation Rome*. Oxford: Oxford University Press, 2013.

Payne, Alina A. *The Architectural Treatise in the Italian Renaissance: Architectural Invention, Ornament, and Literary Culture*. Cambridge: Cambridge University Press, 1999.

Pecchiai, Pio. *Acquedotti e fontane di Roma nel Cinquecento (con documenti inediti)*. Rome: Staderini, 1944.

———. *Roma nel Cinquecento*. Bologna: Licinio Cappelli Editore, 1948.

Pedretti, Carlo. "Leonardo architetto a Imola." *Architectura: Zeitschrift für Geschichte der Architektur* 2 (1972): 92–105.

Pellegrini, Marco. "A Turning-Point in the History of the Factional System in the Sacred College: The Power of Pope and Cardinals in the Age of Alexander VI." In Signorotto and Visceglia, *Court and Politics in Papal Rome*, 8–30.

Pepper, Simon. "Artisans, Architects and Aristocrats: Professionalism and Renaissance Military Engineering." In *The Chivalric Ethos and the Development of Military Professionalism*, edited by D. J. B. Trim, 117–147. Leiden: Brill, 2003.

Peri, Vittorio. "Progetti e rimostranze: Documenti per la storia dell'Archivio Segreto Vaticano dall'erezione alla metà del XVIII secolo." *Archivum Historiae Pontificiae* 19 (1981): 191–237.

Perini, P. Davide Aurelio. *Onofrio Panvinio e le sue opere*. Rome: Tipografia Poliglotta, 1899.

Perocco, Daria. "Pigafetta, Filippo." *DBI*, 83:578–582.

Pessolano, Maria Raffaela. "Idraulica e modernità: Il porto di Napoli nell'immagine della città vicereale." In Curcio, Navone, and Villari, *Studi su Domenico Fontana*, 197–211.

Petraroia, Pietro, and Sergio Lombardi. "Colonna Traiana." In Madonna, *Roma di Sisto V*, 406–407.

Petrucci, Carlo A. *Catalogo generale delle stampe tratte dai rami incisi posseduti dalla Calcografia Nazionale*. Rome: La Libreria dello Stato, 1953.

Petrucci, Giulia. "La città Pia: Un'espansione urbana del Cinquecento." *Storia Urbana* 17 (July–September 1993): 19–48.

Pfister, Max, ed. *Lessico etimologico italiano*. Vol. 3.2. Wiesbaden: Ludwig Reichert, 1991.

Piera Sette, Maria, and Simona Benedetti, eds. *Architetture per la città*. Rome: Multigrafica, 1992.

Pignatti, Franco. "Egio, Benedetto." *DBI*, 42:356–357.

Pinelli, Antonio, "Introduzione." In Pinelli, *Roma del Rinascimento*, v–xl.

———, ed. *Roma del Rinascimento*. Rome: Laterza, 2001.

Pinto, John A. "*Forma Urbis Romae*: Fragment and Fantasy." In Striker, *Architectural Studies*, 143–146.

———. "Origins and Development of the Ichnographic City Plan." *Journal of the Society of Architectural Historians* 35 (March 1976): 35–50.

———. *The Trevi Fountain*. New Haven, CT: Yale University Press, 1986.

Pittoni, Leros, and Gabrielle Lautenberg. *Roma Felix: La città di Sisto V e Domenico Fontana*. Rome: Viviani, 2002.

Pizzamano, Paola, ed. *Giovanni Battista Cavalieri: Un incisore trentino nella Roma dei papi del Cinquecento*. Rovereto, Trent: Nicolodi, 2001.

Popplow, Marcus. "Why Draw Pictures of Machines? The Social Contexts of Early Modern Machine Drawings." In *Picturing Machines, 1400–1700*, edited by Wolfgang Lefèvre, 17–52. Cambridge, MA: MIT Press, 2004.

Porzio, Pier Luigi, ed. *Impronte sistine: Fabbriche civili minori, interventi nel territorio, restauri di monumenti della età di Sisto V*. Rome: Quasar, 1991.

Prodi, Paolo. *Il cardinal Gabriele Paleotti*. 2 vols. Rome: Edizioni di Storia e Letteratura, 1959–1967.

———. *Il sovrano pontefice: Un corpo e due anime; La monarchia papale nella prima età moderna*. Bologna: Il Mulino, 1982. Translated by Susan Haskins as *The Papal Prince: One Body and Two Souls; The Papal Monarchy in Early Modern Europe* (Cambridge: Cambridge University Press, 1987).

Pullan, Brian. "New Approaches to Poverty and New Forms of Institutional Charity in Late Medieval and Renaissance Italy." In Zamagni, *Povertà e innovazioni istituzionali*, 17–43.

———. "Poveri, mendicanti e vagabondi (secoli XIV–XVII)." In *Storia d'Italia: Annali*. Vol. 1, *Dal feudalesimo al capitalismo*, edited by Corrado Vivanti and Ruggiero Romano, 981–1047. Turin: Einaudi, 1978.

Purcell, N. "Forum Romanum (the Republican Period)" and "Forum Romanum (the Imperial Period)." In Steinby, *Lexicon topographicum*, 2:325–336, 336–342.

Quast, Matthias. "Domenico Fontana." *DA*, 11:271–274.

———. *Die Villa Montalto in Rom: Entstehung und Gestalt im Cinquecento*. Munich: Tuduv, 1991.

Quilici, Lorenzo. "Gli acquedotti di Roma: Solidità, utilità, bellezza." *Archeo*. 53 (July 1989): 51–97.

———. "Sull'acquedotto Vergine dal Monte Pincio alle sorgenti." *Studi di Topografia Romano* 5 (1968): 125–160.

Raffaello a Roma: Il convegno del 1983. Rome: Edizioni dell'Elefante, 1986.

Re, Emilio. "Maestri di strada." *ASRSP* 43 (1920): 5–102.

Reinhard, Wolfgang. "Le carriere papali e cardinalizie: Contributo alla storia sociale del papato." In Fiorani and Prosperi, *Roma, la città del papa*, 261–290.

———. *Freunde und Kreaturen: "Verflechtung" als Konzept zur Erforschung historischer Führungsgruppen Römische Oligarchie um 1600*. Munich: Ernst Vögel, 1979.

Rhodes, Dennis E. "Ortensia and Hortolana: With Notes on Angelo degli Oldradi." *Gutenberg-Jahrbuch* 71 (1996): 97–98.

Ribouillault, Denis. *Rome en ses jardins: Paysage et pouvoir au XVIe siècle*. Paris: CTHS / INHA, 2013.

———. "La Villa Giulia et l'âge d'or augustéen." In *Le miroir et l'espace du prince dans l'art italien de la Renaissance*, edited by Philippe Morel, 339–388. Paris: Presses Universitaires François Rebelais and Presses Universitaires de Rennes, 2012.

Riccardi, Maria Luisa, et al. "Gli obelischi sistini." In Piera Sette and Benedetti, *Architetture per la città*, 13–38.

Richardson, Carol. M. *Reclaiming Rome: Cardinals in the Fifteenth Century*. Leiden: Brill, 2009.

Richardson, Lawrence, Jr. *A New Topographical Dictionary of Ancient Rome*. Baltimore: Johns Hopkins University Press, 1992.

Riera, Italo. "Classificazione e funzionamento delle fognature romane." In Bianchi, *La Cloaca Maxima*, 4–21.

Rinne, Katherine Wentworth. Aquae Urbis Romae: *The Waters of the City of Rome*. http://www3.iath.virginia.edu/waters/.

———. "Between Precedent and Experiment: Restoring the Acqua Vergine in Rome (1560–70)." In Roberts, Schaffer, and Dear, *Mindful Hand*, 94–115.

———. "Fluid Precision: Giacomo della Porta and the Acqua Vergine Fountains of Rome." In *Landscapes of Memory and Experience*, edited by Jan Birksted, 183–201. London: Spon, 2000.

———. "Hydraulic Infrastructure and Urbanism in Early Modern Rome." *Papers of the British School at Rome* 73 (2005): 191–222.

———. "Urban Ablutions: Cleansing Counter-Reformation Rome." In Bradley and Stow, *Rome, Pollution and Propriety*, 182–202.

———. *The Waters of Rome: Aqueducts, Fountains, and the Birth of the Baroque City*. New Haven, CT: Yale University Press, 2010.

Robbins, Deborah. "Via della Lungaretta: The Making of a Medieval Street." In Çelik, Favro, and Ingersoll, *Streets: Critical Perspectives*, 165–175.

Roberts, Lissa, Simon Schaffer, and Peter Dear, eds. *The Mindful Hand: Inquiry and Invention from the Late Renaissance to Early Industrialization*. Amsterdam: Koninklijke Nederlandse Akademie van Wetenschappen, 2007.

Robertson, Clare. *"Il Gran Cardinale": Alessandro Farnese, Patron of the Arts*. New Haven, CT: Yale University Press, 1992.

———. *Rome 1600: The City and the Visual Arts under Clement VIII*. New Haven, CT: Yale University Press, 2015.

Rocchi, E. *Le piante icnografiche e prospettiche di Roma del secolo XVI*. 2 vols. Turin: Roux e Viarengo, 1902.

Rodocanachi, Emmanuel. *Les corporations ouvrières à Rome depuis la chute de l'empire romain*. 2 vols. Paris: Picard, 1894.

———. *Les institutions communales de Rome sous la papauté*. Paris: Picard, 1901.

Roland, F. "Un Franc-Comptois éditeur et marchand d'estampes a Rome au XVIe siècle: Antoine Lafrery (1512–1577)." *Mémoires de la Société d'Emulation du Doubs*, 8th ser. 8, 5/6 (1910–1911): 320–370.

Romani, Mario. *Pellegrini e viaggiatori nell'economia di Roma dal XIV al XVII secolo*. Milan: Società Editrice "Vita e Pensiero," 1948.

Romano, Antonella, ed. *Rome et la science moderne entre Renaissance et Lumières*. Rome: École française de Rome, 2008.

Rombai, Leonardo. "Cartography in the Central Italian States from 1480 to 1680." In Woodward, *History of Cartography*, pt. 1, 909–939.

Rosselli, Donatella. "Firmano, Giovanni Francesco." *DBI*, 48:223.

Rowland, Ingrid D. *The Culture of the High Renaissance: Ancients and Moderns in Sixteenth-Century Rome*. Cambridge: Cambridge University Press, 1998.

———. "Raphael, Angelo Colocci, and the Genesis of the Architectural Orders." *Art Bulletin* 76 (March 1994): 81–104.

Rubach, Birte. "Three Prints of Inscriptions: Antonio Lafreri and His Contact with Jean Matal." In Zorach, *Virtual Tourist*, 25–35.

Rurale, Flavio. "Pio IV, papa." *DBI*, 83:808–814.

Russell, Andrew L. and Lee Vinsel. "After Innovation, Turn to Maintenance." *Technology and Culture* 59 (January 2018): 1–25.

Russo, Maria T. B. "Problemi e istituti dell'assistenza romana nel Cinque e Seicento." *Studi Romani* 34 (1986): 230–252.

Salerno, Luigi, Luigi Spezzaferro, and Manfredo Tafuri. *Via Giulia: Una utopia urbanistica del 500*. Rome: Stabilimento Aristide Staderini, 1973.

San Juan, Rose Marie. *Rome: A City Out of Print*. Minneapolis: University of Minnesota Press, 2001.

Sansa, Renato. "Istituzioni e politica dell'ambiente a Roma: Dalle magistrature capitoline alla presidenza pontificia." In *La legislazione medicea sull'ambiente*, vol. 4, *Scritti per un commento*, edited by Giovanni Cascio Pratilli and Luigi Zangheri, 209–224. Florence: Leo S. Olschki, 1998.

———. "Le norme decorose e il lavoro sporco: L'igiene urbana in tre capitali europee; Londra, Parigi, Roma tra XVI e XVIII secolo." *Storia Urbana* 112 (2006): 85–112.

———. "L'odore del contagio: Ambiente urbano e prevenzione delle epidemie nella prima età moderna." *Medicina e Storia* 2 (2002): 83–108.

———. "La pulizia delle strade a Roma nel XVII secolo: Un problema di storia ambientale." *ASRSP* 114 (1991): 127–160.

Sarton, George. "Agrippa, Fontana and Pigafetta: The Erection of the Vatican Obelisk in 1586." *Archives Internationales d'Histoire des Sciences* 8 (July 1949): 827–854.

Scaccia Scarafoni, Camillo. "L'antico statuto dei 'magistri stratarum' e altri documenti relativi a quella magistratura." *ASRSP* 50 (1927): 239–308.

———. *Le piante di Roma possedute dalla biblioteca dell'Istituto e dalle altre biblioteche governative della città*. Rome: La Libreria dello Stato, 1939.

Scaglia, Gustina. "The 'Sepolcro Dorico' and Bartolommeo de Rocchi da Brianza's Drawing of It in the Aurelian Wall between Porta Flaminia and the Tiber River." *Arte Lombarda*, n.s., 96 (1991): 107–116.

Scavizzi, C. Paola. "Fonti per uno studio sulla regolazione del Tevere dal Cinquecento al Settecento fra teoria e pratica." *ASRSP* 102 (1979): 237–313.

———. "Il Ponte Felice al Borghetto nel quadro della viabilità territoriale." In Fagiolo and Madonna, *Sisto V*, 1:621–638.

———. "La rete idrica urbana in età moderna." *Storia della Città*, 29 (1984): 77–94.

Schelbert, George. "Gli acquedotti urbani nelle piante e vedute quattrocentesche e cinquecentesche di Roma." In Bonaccorso, "Le acque e la città," 7–30.

Schiaparelli, L. "Alcuni documenti dei *Magistri Aedificiorum Urbis* (secoli XIII e XIV)." *ASRSP* 25 (1902): 5–60.

Schiffmann, René. Roma felix: *Aspekte der städtebaulichen Gestaltung Roms unter Papst Sixtus V*. Bern: Peter Lang, 1985.

Schnapp, Alain. *The Discovery of the Past*. Translated by Ian Kinnes and Gillian Varndell. New York: Harry N. Abrams, 1997.

Schraven, Minou. "Founding Rome Anew: Pope Sixtus IV and the Foundation of Ponte Sisto, 1473." In *Foundation, Dedication, and Consecration in Early Modern Europe*, edited by Maarten Delbeke and Minou Schraven, 129–151. London: Brill, 2012.

Schreurs, Anna. *Antikenbild und Kunstanschauungen des neapolitanischen Malers, Architekten und Antiquars Pirro Ligorio (1513–1583)*. Cologne: Walther König, 2000.

Schulz, Jürgen. *La cartografia tra scienza e arte: Carte e cartografi nel Rinascimento italiano*. Translated by Tessie Doria De Zuliani. Rev. ed. Modena: Franco Cosimo Panini, 2006.

Scobie, Alex. "Slums, Sanitation, and Mortality in the Roman World." *Klio* 68 (1986): 399–433.

Segarra Lagunes, Maria Margarita. "Le attività produttive del Tevere nelle dinamiche di trasformazione urbana: I mulini fluviali." *Mélanges de l'École française de Rome, Italie et Méditerranée* 118 (2006): 45–52.

———. *Il Tevere e Roma: Storia di una simbiosi.* Rome: Gangemi Editore, 2004.

Settis, Salvatore. "La colonne Trajane: Invention, composition, disposition." *Annales Histoires, Science Sociales* 40 (September–October 1985): 1151–1194.

Setton, Kenneth M. *The Papacy and the Levant (1204–1571).* Vol. 4, *The Sixteenth Century from Julius III to Pius V.* Philadelphia: American Philosophical Society, 1984.

Shearman, John K. G. *Raphael in Early Modern Sources (1483–1602).* 2 vols. New Haven, CT: Yale University Press, 2003.

Sickel, Lothar. "Un progetto per il trasporto dell'Obelisco Vaticano descritto da Camillo Paleotti nel febbraio 1581." *Strenna dei Romanisti* 68 (April 2007): 653–662.

Signorotto, Gianvittorio, and Maria Antonietta Visceglia, eds. *Court and Politics in Papal Rome, 1492–1700.* Cambridge: Cambridge University Press, 2002.

———. "Introduction." In Signorotto and Visceglia, *Court and Politics in Papal Rome,* 1–7.

Simoncelli, Paolo. "Note sul sistema assistenziale a Roma nel XVI secolo." In *Timore e carità: I poveri nell'Italia moderna,* edited by Giorgio Politi, Mario Rosa, and Franco della Peruta, 137–156. Atti del Convegno, Cremona, March 28–30, 1980. Cremona: Biblioteca Statale, 1982.

Simoncini, Giorgio. *Roma: Le trasformazioni urbane nel Cinquecento.* Vol. 1, *Topografia e urbanistica da Giulio II a Clemente VIII.* Florence: Leo S. Olschki, 2008.

———, ed. *Roma: Le trasformazioni urbane nel Cinquecento.* Vol. 2, *Dalla città al territorio.* Florence: Leo S. Olschki, 2011.

———. *Roma: Le trasformazioni urbane nel Quattrocento.* Vol. 1, *Topografia e urbanistica da Bonifacio IX ad Alessandro VI.* Florence: Leo S. Olschki, 2004.

———, ed. *Roma: Le trasformazioni urbane nel Quattrocento.* Vol. 2, *Funzioni urbane e tipologie edilizie.* Florence: Leo S. Olschki, 2004.

———. *"Roma restaurata": Rinnovamento urbano al tempo di Sisto V.* Florence: Leo S. Olschki, 1990.

Sinisi, Daniela. "Lavori pubblici di acque e strade e congregazioni cardinalizie in epoca sistina e presistina." In Spezzaferro and Tittoni, *Il Campidoglio e Sisto V,* 50–52.

———. "I *notarii magistrorum stratarum* nel '500: Nascita di un ufficio notarile privativo per le magistrature di acque e strade." *Roma Moderna e Contemporanea* 4 (May–August 1996): 363–378.

———. "Presidenza Strade." In Pastura Ruggiero, *La Reverenda Camera Apostolica,* 100–118.

Smith, Pamela H. *The Body of the Artisan: Art and Experience in the Scientific Revolution.* Chicago: University of Chicago Press, 2004.

Smith, Strother A. *The Tiber and Its Tributaries: Their Natural History and Classical Associations.* London: Longmans, Green, 1877.

Sonnino, Eugenio. "Le anime dei romani: Fonti religiose e demografia storica." In Fiorani and Prosperi, *Roma, la città del papa,* 327–364.

———, ed. *Popolazione e società a Roma dal Medioevo all'età contemporanea.* Rome: Editrice "Il Calamo," 1998.

———. "The Population in Baroque Rome." In *Rome / Amsterdam: Two Growing Cities in Seventeenth-Century Europe,* edited by Peter van Kessel and Elisja Schulte, 50–70. Amsterdam: Amsterdam University Press, 1997.

Sori, Ercole. *La città e i rifiuti: Ecologia urbana dal Medioevo al primo Novecento*. Bologna: Il Mulino, 2001.

Souffrin, Pierre. "La geometria practica dans le *Ludi Rerum Mathematicarum.*" *Albertiana* 1 (1998): 87–104.

Spagnesi, Gianfranco. *Roma: La basilica di San Pietro, il Borgo e la città*. Milan: Palombi / Jaca Book, 2003.

Spagnesi, Piero. *Castel Sant'Angelo: La fortezza di Roma; Momenti della vicenda architettonica da Alessandro VI a Vittorio Emanuele III (1494–1911)*. Rome: Palombi, 1995.

Spezzaferro, Luigi, and Maria Elisa Tittoni, eds. *Il Campidoglio e Sisto V*. Rome: Edizioni Carte Segrete, 1991.

Stefani, Grete, and Giovanni Di Maio. "Idraulica e modernità: Il canale Conte di Sarno." In Curcio, Navone, and Villari, *Studi su Domenico Fontana*, 213–227.

Steinby, Eva Margareta, ed. *Lexicon topographicum urbis Romae*. 6 vols. Rome: Edizioni Quasar di Severino Tognon, 1993–2000.

Stenhouse, William. "The *Forma Urbis Romae* before Nolli: Antiquarian Scholarship in the Sixteenth and Seventeenth Centuries." In *Giambattista Nolli and Rome: Mapping the City before and after the Pianta Grande*, edited by Ian Verstegen and Allan Ceen, 15–26. Rome: Studium Urbis Rome Center, 2013.

———. "Georg Fabricius and Inscriptions as a Source of Law." *Renaissance Studies* 17 (March 2003): 96–107.

———. "Panvinio and *Descriptio*: Renditions of History and Antiquity in the Late Renaissance." *Papers of the British School at Rome* 80 (2012): 233–256.

———. *Reading Inscriptions and Writing Ancient History: Historical Scholarship in the Late Renaissance*. Bulletin of the Institute of Classical Studies, supp. 86. London: Institute of Classical Studies, School of Advanced Study, University of London, 2005.

Stinger, Charles L. *The Renaissance in Rome*. 2nd ed. Bloomington: Indiana University Press, 1998.

———. "*Roma Triumphans*: Triumphs in the Thought and Ceremonies of Renaissance Rome." *Medievalia et Humanistica*, n.s., 10 (1981): 189–201.

Storey, Tessa. *Carnal Commerce in Counter-Reformation Rome*. Cambridge: Cambridge University Press, 2008.

Stow, Kenneth R. *Catholic Thought and Papal Jewry Policy, 1555–1593*. New York: Jewish Theological Seminary of America, 1977.

———. *Theater of Acculturation: The Roman Ghetto in the Sixteenth Century*. Seattle: University of Washington Press, 2001.

Striker, Cecil L. Ed. *Architectural Studies in Memory of Richard Krautheimer*. Mainz: Philipp von Zabern, 1996.

Stroffolino, Daniela. *La città misurata: Tecniche e strumenti di rilevamento nei trattati a stampa del Cinquecento*. Rome: Salerno, 1999.

Sturm, Saverio. "Carlo Fontana." In Fagiolo and Bonaccorso, *Studi sui Fontana*, 432–438.

Tafuri, Manfredo. "'Cives esse non licere': The Rome of Nicholas V and Leon Battista Alberti; Elements toward a Historical Revision." *Harvard Architecture Review* 6 (1987): 60–75.

Tagliamonte, G. "Forum Romanum (fino alla prima età repubblicana)." In Steinby, *Lexicon topographicum*, 2:313–325.

Tallini, Gennaro. "Giovanni Tarcagnota." *Bibliografia* (January 30, 2012). http://www.nuovorinascimento.org/cinquecento/tarcagnota.pdf.

———. "Tra studio e bottega: Coordinate bio-bibliografiche per Giovanni Tarcagnota da Gaeta (1499–1566)." *Bibliologia* 6 (2011): 15–42.

Tarr, Joel A. "Sewerage and the Development of the Networked City in the United States, 1850–1930." In Tarr and Dupuy, *Technology and the Rise of the Networked City*, 159–185.

Tarr, Joel A., and Gabriel Dupuy, eds. *Technology and the Rise of the Networked City in Europe and America*. Philadelphia: Temple University Press, 1988.

Taylor, Lily Ross. "Degrassi's Edition of the Consular and Triumphal Fasti." *Classical Philology* 45 (April 1950): 84–95.

Taylor, Rabun. *Public Needs and Private Pleasures: Water Distribution, the Tiber River, and the Urban Development of Ancient Rome*. Rome: "L'Erma" di Bretschneider, 2000.

———. "Tiber River Bridges and the Development of the Ancient City of Rome." *Waters of Rome* 2 (June 2002). http://www3.iath.virginia.edu/waters/Journal2TaylorNew.pdf.

Temple, Nicholas. Renovatio Urbis*: Architecture, Urbanism and Ceremony in the Rome of Julius II*. Abingdon, Oxon: Routledge, 2011.

Terpstra, Nicholas. *Cultures of Charity: Women, Politics, and the Reform of Poor Relief in Renaissance Italy*. Cambridge, MA: Harvard University Press, 2013.

———. *Religious Refugees in the Early Modern World: An Alternative History of the Reformation*. New York: Cambridge University Press, 2015.

Testa, Simone. "Peretti Damasceni, Alessandro." *DBI*, 82:340–342.

Thoenes, Christof. "La 'Lettera' a Leone X." In *Raffaello a Roma*, 373–381.

———. "Perché studiare Domenico Fontana." In Curcio, Navone, and Villari, *Studi su Domenico Fontana*, 9–19.

Tiberia, Vitaliano. *Giacomo della Porta: Un architetto tra Manierismo e Barocco*. [Rome]: Bulzoni, 1974.

Tinto, Alberto. *Annali tipografici dei Tramezzino*. Venice: Istituto per la Collaborazione Culturale, 1966.

Tolomeo, Maria Grazia. "L'Acquedotto Vergine (sec. XVI–XVIII)." In *Il trionfo dell'acqua*, 205–208.

Tomasi Velli, Silvia. "Gli antiquari intorno al circo romano: Riscoperta di una tipologia monumentale antica." *Annali della Scuola Normale Superiore di Pisa: Classe di Lettere e Filosofia*, 3rd ser., 20 (1990): 61–168.

Tomory, Leslie. *The History of the London Water Industry, 1580–1820*. Baltimore: Johns Hopkins University Press, 2017.

———. "London's Water Supply before 1800 and the Roots of the Networked City." *Technology and Culture* 56 (July 2015): 704–737.

Tooley, R. V. "Maps in Italian Atlases of the Sixteenth Century, Being a Comparative List of the Italian Maps Issued by Lafreri, Forlani, Duchetti, Bertelli and Others, Found in Atlases." *Imago Mundi* 3 (1939): 12–47.

Torre, Camilla. *Ponti in muratura: Dizionario storico-tecnologico*. Florence: Alinea, 2003.

Travaglini, Carlo M., ed. *La città e il fiume: Secoli XIII–XIX*. Rome: École française de Rome, 2008.

Trevor Hodge, A. "Aqueducts." In *Handbook of Ancient Water Technology*, edited by Orjan Wikander, 39–65. Leiden: Brill, 2000.

———. *Roman Aqueducts and Water Supply*. 2nd ed. London: Duckworth, 2002.

Il trionfo dell'acqua: Acque e acquedotti a Roma, IV sec. a.C.—XX sec. Rome: Paleani Editrice, 1986.

Turci, Marcello. "Gli acquedotti." In Carandini and Carafa, *Atlante di Roma antica*, 1:92–100.

Vagenheim, Ginette. "Appunti per una prosopografia dell'Accademia dello Sdegno a Roma: Pirro Ligorio, Latino Latini, Ottavio Pantagato e altri." *Sassoferrato: Studi umanistici piceni* 26 (2006): 211–226.

———. "Bartolomeo Borghesis, Theodor Mommsen et l'édition des inscriptions de Pirro Ligorio dans le *Corpus Inscriptionum Latinarum* (CIL)." *Journal of the History of Collections* 26 (2014): 363–371.

———. "La falsification chez Pirro Ligorio à la lumière des Fasti Capitolini e des inscription de Préneste." *Eutopia* 3 (1994): 67–113.

———. "La falsificazione epigrafica nell'Italia della seconda metà del Cinquecento: *Renovatio* ed *inventio* nelle *Antichità Romane* attribuite a Pirro Ligorio." In *El monument epigráfico en contextos secundrios: Procesos de reutilización, interpretación y falsificación*, edited by Joan Carbonell Manils, Helena Gimeno Pascual, and José Luis Moralejo Álvarez, 217–226. Belaterra: Universitat Autònoma de Barcelona, 2011.

Valerio, Vladimiro. "Cartography in the Kingdom of Naples during the Early Modern Period." In Woodward, *History of Cartography*, pt. 1, 940–973.

———. "The Neapolitan Saxton and His Survey of the Kingdom of Naples." *Map Collector* 18 (March 1982): 14–17.

Valone, Carolyn. "Giovanni Antonio Dosio: The Roman Years." *Art Bulletin* 58 (December 1976): 528–541.

———. "Matrons and Motives: Why Women Built in Early Modern Rome." In *Beyond Isabella: Secular Women Patrons of Art in Renaissance Italy*, edited by Sheryl E. Reiss and David G. Wilkins, 317–335. Kirksville, MO: Truman State University Press, 2001.

———. "Women on the Quirinal Hill: Patronage in Rome, 1560–1630." *Art Bulletin* 76 (March 1994): 129–146.

Van Deman, Esther Boise, *The Building of the Roman Aqueducts*. Washington, DC: Carnegie Institution of Washington, 1934.

Van der Vleuten, Erik, and Arne Kaijser, eds. *Networking Europe: Transnational Infrastructures and the Shaping of Europe, 1850–2000*. Sagamore Beach, MA: Science History Publications, 2006.

Verde, Paola Carla. "Domenico Fontana." in Fagiolo and Bonaccorso, *Studi sui Fontana*, 421–427.

Verdi, Orietta. "Da ufficiali capitolini a commissari apostolici: I maestri delle strade e degli edifici di Roma tra XIII e XVI secolo." In Spezzaferro and Tittoni, *Il Campidoglio e Sisto V*, 54–62.

———. *Maestri di edifici e di strade a Roma nel secolo XV: Fonti e problemi*. Rome: Roma nel Rinascimento, 1997.

Verner, Coolie. "Copperplate Printing." In *Five Centuries of Map Printing*, edited by David Woodward, 51–75. Chicago: University of Chicago Press, 1975.

Vetrugno, Paolo Agostino. *Antonio Trevisi: Architetto pugliese del Rinascimento*. Fasano: Schena, 1985.

Vian, Nello. "Bordini, Giovanni Francesco." *DBI*, 12:507–508.

Visceglia, Maria Antonietta. "Burocrazia, mobilità sociale e *patronage* alla corte di Roma tra Cinque e Seicento: Alcuni aspetti del recente dibattito storiografico e prospettive di ricerca." *Roma Moderna e Contemporanea* 3 (January–April 1995): 11–55.

———. "Il cerimoniale come linguaggio politico: Su alcuni conflitti di precedenza alla corte di Roma tra Cinquecento e Seicento." In Visceglia and Brice, *Cérémonial et ritual à Rome*, 117–176.

———. *La città rituale: Roma e le sue cerimonie in età moderna*. Rome: Viella, 2002.

———. *Morte e elezione del papa: Norme, riti e conflitti, l'età moderna*. Rome: Viella, 2013.

———, ed. *La nobiltà romana in età moderna: Profili istituzionali e pratiche sociali*. Rome: Carocci, 2001.

———. "Politica internazionale, fazioni e partiti nella Curia Romana del tardo Cinquecento." *Rivista Storica Italiana* 127 (2015): 721–769.

Visceglia, Maria Antonietta, and Catherine Brice, eds. *Cérémonial et ritual à Rome (XVIe–XIXe siècle)*. Rome: École française de Rome, 1997.

Vita Spagnuolo, Vera. "Fonti per lo studio dell'assistenza e della beneficenza a Roma durante il secolo XVI: Gli archivi di alcune confraternite." In Fagiolo and Madonna, *Sisto V*, 1:243–260.

Volpi, Caterina, ed. *Il libro dei disegni di Pirro Ligorio all'Archivio di Stato di Torino*. Rome: Edizioni dell'Elefante, 1994.

———. "Il teatro del mondo: Pirro Ligorio 'architetto' papale nelle Logge Vaticane." *Bollettino d'Arte*, 6th ser., 107 (January–March 1999): 83–102.

Waddy, Patricia. "Barberini Cardinals Need Places to Live." In *I Barberini e la cultura europea del Seicento*, edited by Lorenza Mochi Onori, Sebastian Schütze, and Francesco Solinas, 487–500. Atti del convegno internazionale, Palazzo Barberini alle Quattro Fontane, December 7–11, 2004. Rome: De Luca, 2007.

———. *Seventeenth-Century Roman Palaces: Use and the Art of the Plan*. New York: Architectural History Foundation; Cambridge, MA: MIT Press, 1990.

Wallace, William E. "Michelangelo Engineer." In Conforti and Hopkins, *Architettura e tecnologia*, 96–107.

Wardrop, James. "*Civis Romanus Sum*: Giovanbattista Palatino and His Circle," *Signature*, n.s., 14 (1952): 3–39.

Watkin, David. *The Roman Forum*. Cambridge, MA: Harvard University Press, 2009.

Weiss, Roberto. "Andrea Fulvio antiquario romano (c. 1470–1527)." *Annali della Scuola Normale Superiore di Pisa: Lettere, Storia e Filosofia*, 2nd ser., 28 (1959): 1–44.

———. *The Renaissance Discovery of Classical Antiquity*. 2nd ed. Oxford: Basil Blackwell, 1988.

Westfall, Carroll William. *In This Most Perfect Paradise: Alberti, Nicholas V, and the Invention of Conscious Urban Planning in Rome, 1447–55*. University Park: Pennsylvania State University Press, 1974.

Wilde, Deborah Nelson. "Housing and Urban Development in Sixteenth Century Rome: The Properties of the Arciconfraternita della SS.ma Annunziata." PhD diss., New York University, 1989.

Wilson Jones, Mark. *Principles of Roman Architecture*. New Haven, CT: Yale University Press, 2000.

Wisch, Barbara. "The Matrix: *Le Sette Chiese di Roma* of 1575 and the Image of Pilgrimage." *Memoirs of the American Academy in Rome* 56/57 (2011–2012): 271–303.

———. "The Roman Church Triumphant: Pilgrimage, Penance and Processions Celebrating the Holy Year of 1575." In *Art and Pageantry in the Renaissance and Baroque*, edited by Barbara Wisch and Susan Scott Munshower, 87–117. University Park: Pennsylvania State University, 1990.

Witcombe, Christopher L. C. E. *Copyright in the Renaissance: Prints and the* Privilegio *in Sixteenth-Century Venice and Rome.* Leiden: Brill, 2004.

———. *Print Publishing in Sixteenth-Century Rome: Growth and Expansion, Rivalry and Murder.* London: Harvey Miller, 2008.

Witte, Charles-Martial de. *La correspondance des premiers nonces permanents au Portugal, 1532–1533.* 2 vols. Lisbon: Academia Portuguesa da História, 1986.

Woodward, David, ed. *The History of Cartography*: Vol. 3, *Cartography in the European Renaissance.* 2 pts. Chicago: University of Chicago Press, 2007.

———. *Maps as Prints in the Italian Renaissance: Makers, Distributors and Consumers.* London: British Library, 1996.

———. "Techniques of Map Engraving, Printing, and Coloring in the European Renaissance." In Woodward, *History of Cartography,* pt. 1, 591–610.

Wright, A. D. *The Early Modern Papacy: From the Council of Trent to the French Revolution, 1564–1789.* Edinburgh Gate, Harlow, Essex: Pearson Education Limited, 2000.

Zamagni, Vera, ed. *Povertà e innovazioni istituzionali in Italia dal Medioevo ad oggi.* Bologna: Il Mulino, 2000.

Zannini, Gian Ludovico Masetti. "Rivalità e lavoro di incisori nelle botteghe Lafréry-Duchet e de la Vacherie." In *Les foundations nationals dans la Rome pontificale,* 547–566. Rome: École française de Rome, 1981.

———. *Stampatori e librai a Roma nella seconda metà del Cinquecento.* Rome: Palombi, 1980.

Zapperi, Roberto. "Borromeo, Federico." *DBI,* 13:31–33.

Zorach, Rebecca. "The Public Utility of Prints." In Zorach, *Virtual Tourist,* 63–83.

———, ed. *The Virtual Tourist in Renaissance Rome: Printing and Collecting the* Speculum Romanae Magnificentiae. Chicago: University of Chicago Press, 2008.

Index

Page numbers in italics indicate illustrations.